AF449131

Cell Biology Monographs

Continuation of Protoplasmatologia

Founded by

L. V. Heilbrunn, Philadelphia, Pa., and F. Weber, Graz

Edited by

M. Alfert, Berkeley, Calif.
W. Beermann, Tübingen
G. Rudkin, Philadelphia, Pa.

W. Sandritter, Freiburg i. Br.
P. Sitte, Freiburg i. Br.

Advisory Board

J. Brachet, Bruxelles
D. Branton, Berkeley, Calif.
H. G. Callan, St. Andrews
E. C. Cocking, Nottingham
W. W. Franke, Heidelberg

N. Kamiya, Osaka
G. F. Springer, Evanston, Ill.
L. Stockinger, Wien
B. F. Trump, Baltimore, Md.

Vol. 3

Springer-Verlag
Wien New York

Lysosomes: A Survey

E. Holtzman

Springer-Verlag
Wien New York

Prof. Eric Holtzman

Department of Biological Sciences
Columbia University
New York, N.Y., U.S.A.

With 56 Figures

Library of Congress Cataloging in Publication Data. Holtzman, Eric, 1939—. Lysosomes. (Cell biology monographs; v. 3). Bibliography: p. 1. Lysosomes. I. Title. II. Series [DNLM: 1. Lysosomes. W1 CE128H v. 3 / QH603.L9 H758L]. QH603.L9H64. 574.8′734. 75-30716.

ISBN 3-211-81316-0 Springer-Verlag Wien-New York
ISBN 0-387-81316-0 Springer-Verlag New York-Wien

About twenty years ago, Prof. Friedl Weber (Graz University) and Prof. L. V. Heilbrunn (University of Pennsylvania) conceived the idea for the handbook "Protoplasmatologia" at a time when the state of knowledge in the field of cell biology still permitted one to think of an all-encompassing handbook in the classical sense. Since 1953 fifty-four volumes with a total of about 9,700 pages have been published. The very rapid developments in this area of science, especially during the last decade, have led to new insights which necessitated some alterations in the original plan of the handbook; also, changes in the board of editors since the death of the founders have brought about a reorientation of viewpoints.

The editors, in agreement with the publisher, have decided to abandon the confining limits of the original disposition of the handbook altogether and to continue this work, in a form more appropriate to current needs, as an open series of monographs dealing with present-day problems and findings in cell biology. This will make it possible to treat the most modern and interesting aspects of the field as they arise in the course of contemporary research. The highest scientific, editorial and publishing standards will continue to be maintained.

Editors and publisher

Preface

This brief monograph is intended chiefly for non-specialists and for others interested in a concise introduction to the field. The literature on lysosomes is growing so rapidly that any effort at exhaustive comprehensiveness would be foredoomed to failure. Fortunately, an extensive series of reviews has been published in the past few years (see especially DINGLE and FELL 1969; DINGLE 1972, 1973 a; HERS and VAN HOOF 1973) and the "history" of the organelles is brief enough that major contributors to all stages of that history are still available to provide first-hand discussions (*e.g.*, DE DUVE and WATTIAUX 1966; DE DUVE 1969; NOVIKOFF 1971, 1973; see also VAN FURTH for work on phagocytes and DE REUCK and CAMERON 1963 for useful reviews of early work).

New York, N.Y., September 1975   E. HOLTZMAN

Contents

I. General Considerations and Background

I.1. Perspectives

Hydrolases and other lytic enzymes play essential roles in virtually all living systems. Even a number of viruses code for "lysozymes" and "neuraminidases" that are suspected to participate in viral entry into or exit from cells or tissues (Shatkin 1971, Inouye *et al.* 1973). Extracellular digestive enzymes involved in nutrition or other functions are released by cells as diverse as bacteria and pancreatic acinar cells. Intracellular turnover, of almost all types of macromolecules, is thought to occur continuously within most eucaryotic cell types, and can be fairly extensive in procaryotes as well, at least under some circumstances. Such turnover is based on enzymatic degradation of macromolecules whose constituents are thereby made available for metabolic reuse. Degradative enzymes also participate in processes leading to fertilization in higher plants and animals, selective exclusion of foreign DNA in bacteria, escape of insects from their coccoons, blood-clotting and clot-dissolution, anti-microbial defense systems of higher organisms and an immense variety of other phenomena.

A beginning has been made in the study of the evolution of this diversity— thus, for example, the probable common ancestry of some proteases with differing specificities has been demonstrated (Dickerson and Geis 1969). But at present the evolution of the varying relevant cellular mechanisms is largely a matter for speculation (see *e.g.,* de Duve and Wattiaux 1966). For our considerations, the essential evolutionary step was the establishment of a membrane-delimited "intracellular digestive system" capable of degrading a variety of macromolecules; it is in this system that the lysosomes play a central role. Structures resembling lysosomes probably arose early in the evolution of eucaryotic cells since virtually all animal cell types have such organelles (as usual, mature mammalian red blood cells are among the few exceptions). Increasing evidence also suggests that many lower and higher plant cell types have structures that are quite comparable to animal cell lysosomes. But, procaryotes lack lysosomes, as they do most other intracellular membrane-delimited organelles.

Protozoa possess lysosomes and some of these organisms rely heavily upon these organelles for nutrition; one could argue that this is a major "primitive" role for the lysosomes. In metazoa, as will be seen, the lysosomes have been "adapted" to a variety of functions other than nutrition, and in fact, the

major digestive system of higher animals relies upon the lysosomal enzymes
to a very small extent, if at all. Much more information than is presently
available is needed to sort out the relations between procaryotic and eucaryotic lytic enzymes, between protozoan and metazoan systems and between
intracellular and extracellular digestion. For our purposes, the points to bear
in mind are the following:

A. Many lytic enzymes of eucaryotic cells are associated with lysosomes,
but some common ones such as the digestive enzymes secreted by the pancreas
are not. Most lysosomal enzymes are hydrolases (these catalyze reactions of
the schematic type $A—B + H_2O \rightarrow A—H + B—OH$).

B. Lysosomes carry out most of their known functions within cells although
there are some normal phenomena and some pathological conditions in which
their hydrolases are released to extracellular spaces.

C. Lysosomal enzymes usually function within discrete membrane-delimited
compartments into which materials to be degraded are incorporated. Among
other things, such sequestration normally protects the cell from autolytic
destruction by its own hydrolases.

D. The population of lysosomes of a given cell can participate in degradation of material originating within the same cell (*endogenous* origin), as
well as material originating outside the cell (*exogenous* origin).

I.2. Definitions

In part, because of the heterogeneity of lysosome form and function, the
terminology used has had a complex history. Various names have been proposed for specific types of lysosomes or of related structures in an effort to
point up functionally or structurally unique properties. We will use a system
derived from those codified by others (DE DUVE and WATTIAUX 1966,
NOVIKOFF 1973, NOVIKOFF and HOLTZMAN 1970, STRAUS 1967):

Lysosome: a membrane-delimited structure containing characteristic hydrolytic enzymes, most of which have acid pH optima (see Table 1, p. 9).

I.2.1. General Functional Categories

A. *Primary lysosomes:* structures containing the lysosomal hydrolases in
"unused" state—the enzymes have been "packaged" for future use but have
not yet encountered material to be degraded. The analogy is often drawn
between primary lysosomes and secretion granules, which also are "packages"
of molecules "stored" for subsequent use.

B. *Secondary lysosomes:* structures in which the hydrolases are present
along with material to be degraded, being degraded or already degraded.

C. *Residual body:* a type of secondary lysosome in which much of the
content is residual material not susceptible to, or only slowly susceptible to
further degradation.

I.2.2. Outline of Lysosome Functioning in Phagocytes

To concretize these definitions consider the following example, (Fig. 1). Mammalian phagocytes such as polymorphonuclear neutrophilic (PMN) leukocytes, or macrophages, synthesize lysosomal hydrolases in the ribosome

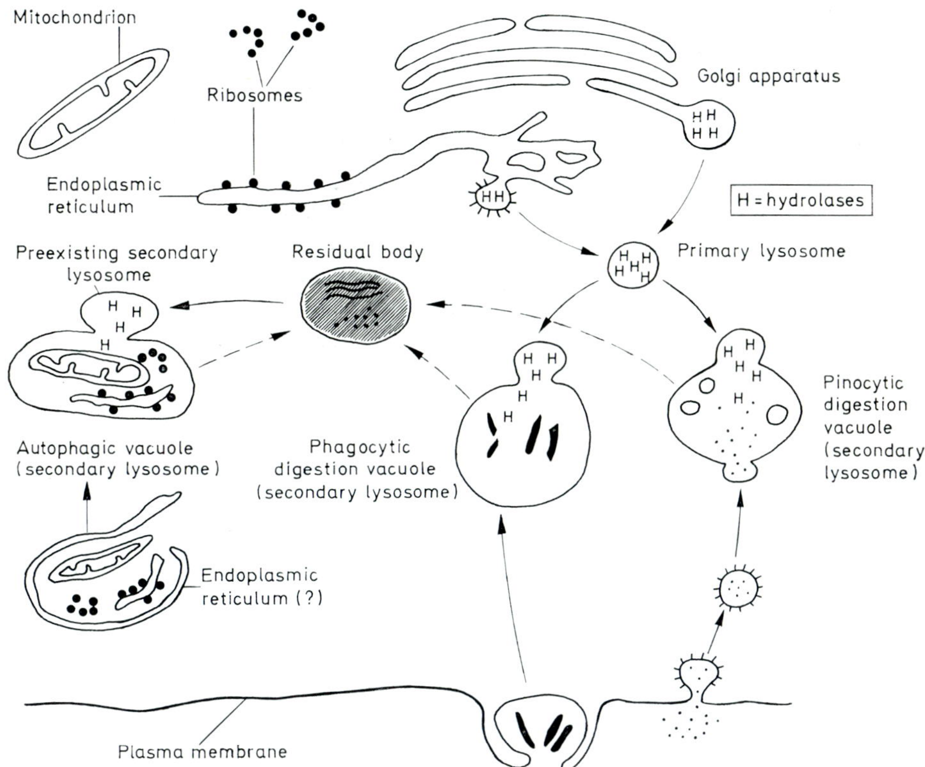

Fig. 1. Some aspects of lysosome formation: (see text, Sections I.2.2 and I.2.3). Newly formed pinocytic, phagocytic and autophagic structures can acquire acid hydrolases through fusion with lysosomes. Such fusions may involve primary lysosomes, formed by budding from the Golgi apparatus or from Golgi-associated endoplasmic reticulum (GERL). In addition, secondary lysosomes can fuse with newly forming endocytic and autophagic vacuoles. As digestion procedes within secondary lysosomes, they transform into residual bodies. Other modes of hydrolase transport and additional interrelations of lysosomes will be discussed and illustrated in Chapter II.

studded (rough) endoplasmic reticulum and package these enzymes within membrane delimited bodies that form by budding from the sacs of the Golgi apparatus or related membrane systems. These bodies are *primary lysosomes*. Upon phagocytosis of foreign matter such as bacteria, the primary lysosomes fuse with the vacuoles containing the foreign material, thereby mixing the hydrolases with material to be degraded and thus producing *secondary lysosomes*. As digestion procedes these transform into *residual bodies* in which what is left of the foreign material is increasingly those components that are resistant to further digestion.

I.2.3. Additional Terms: Heterophagy and Autophagy

The example just given illustrates one class of *digestive vacuole,* the *heterophagic type* in which lysosomal hydrolases are present along with material taken into the cell by *endocytosis.* Endocytosis refers to the very widespread phenomenon of entry of material into the cell by inclusion of the material within vacuoles or vesicles that arise from the cell's surface. *Pinocytosis* refers to such endocytic entry by soluble materials of macromolecular dimensions or smaller, and *phagocytosis,* entry by larger, particulate material.

The exact boundaries between pinocytosis and phagocytosis are not easily drawn and little point would be served in trying to do so in the present work. (Possible metabolic differences will be discussed in section II.1.4.2 and we will take up "micropinocytosis" in sections II.1.1. and II.1.4.2.) Similarly we will deal with *vacuoles* and *vesicles* simply as larger and smaller variants of the same general class of more or less spherical, membrane-delimited structures.

The second category of digestive vacuole is the *autophagic vacuole.* These are common types of lysosomes in which the materials being degraded are of endogenous origin: in other words, they are structures in which the cell digests portions of its own cytoplasm, by incorporating cytoplasm within a membrane delimited vacuole along with lysosomal hydrolases.

Some additional useful definitions will emerge from the discussion below of lysosome morphology and cytochemistry [1].

I.3. Characterization of Lysosomes

The study of lysosomes has represented an interesting symbiotic relationship between microscopists and biochemists since the earliest days of the field when the relevant structures in an acid-hydrolase-rich cell fraction prepared from rat liver were identified by electron microscopy as hepatocyte "pericanalicular dense bodies", a characteristic group of membrane delimited electron dense intracellular inclusions that tend to be found adjacent to bile canaliculi (see NOVIKOFF 1961, DE DUVE 1969 for first-hand accounts). The realization that cytochemical methods permitted microscopic visualization of the intracellular locations of some of the hydrolases cemented the relationship between morphological and chemical approaches and permitted the extension of the lysosome concept to cells and tissues difficult to study by conventional biochemical procedures. While there still are some important unresolved tensions between

---

[1] The following terms are also in fairly widespread use: *Pinosome* and *phagosome* are sometimes used to refer to pinocytosis vesicles and phagocytosis vacuoles that have not yet fused with lysosomes. The generic term proposed for such structures is *heterophagosome* (DE DUVE and WATTIAUX 1966). *Autophagosome* refers to analogous "prelysosomes" thought by some to occur in the autophagic line. *Telolysosome* is a term proposed (GORDON *et al.* 1965) for late stages in the evolution of a lysosome including hypothetical stages in which the hydrolases themselves are permanently inactivated. See also ERICSSON (1969 a) for an alternative system of terminology (*e.g.,* ERICSSON uses *cytosegresome* in place of *autophagosome*).

microscopists' and biochemists' views of lysosomes, these are much less impressive than the numerous concordances.

I.3.1. Basic Biochemical Characteristics of Lysosomes

The vast majority of biochemical studies on the organelles has been done with lysosomes isolated from livers of rats or of a few other mammals. Spleen, kidney and some leukocytes have also received appreciable attention and other tissues have been studied, but to a much lesser extent. Plant cells have been worked on as well but we will defer consideration of these to a later point (see Section V.1).

Lysosomes represent on the order of 1% or less of liver mass and many fall in the same size range as the more numerous mitochondria and peroxisomes (see BEAUFAY 1972). Thus, their isolation and purification poses some formidable problems. Reviews of basic methodology and some perspectives on the problems are presented in TAPPEL (1968 a), DE DUVE (1971), and DINGLE (1972; see especially BEAUFAY's contribution). In general, the procedures used represent modifications of modern methods for homogenization and centrifugation (differential, isopycnic, zonal etc.) supplemented on occasion by special methods such as the employment of Millipore filters to isolate cell structures of particular sizes (BAUDHUIN and BERTHET 1967), or the use of preparative electrophoresis (STAHN et al. 1970). We will not be concerned in this book with details of the methodology. But several points must be borne in mind in considering results of biochemical studies. Most important is the fact that even with the most favorable tissues it is exceedingly difficult with conventional methods to obtain purified lysosome fractions. By sacrificing yield and using extreme care it may sometimes be possible to prepare fairly pure lysosomes from hepatocytes (see e.g., SAWANT et al. 1964, TAPPEL 1968 a, STAHN et al. 1970) and particularly from cells such as polymorphonuclear leukocytes (BAGGIOLINI et al. 1969). But usual lysosome preparations show very substantial contamination by mitochondria, peroxisomes, microsomes or other material; this is true even of those cell fractions with specific activities of hydrolases that are quite high by comparison with the starting homogenate. DE DUVE and his co-workers stress the fact that "lysosome" fractions are usually really only somewhat enriched in lysosomes, through consistent presentation of their results by means of histograms showing the distribution of enzymes of several organelles and of protein in all subcellular fractions (see e.g., Fig. 2 and BEAUFAY 1972).

There is an isolation "trick" that has proved quite useful in improving purity. By virtue of their heterophagic roles, lysosomes can be made to acquire contents that make them differ greatly from normal in buoyant density, or other properties affecting their centrifugation characteristics. For phagocytes, indigestible particles such as latexes have been used (e.g., STOSSEL et al. 1971). For liver, dextran and iron-containing compounds are sometimes employed (e.g., BENGT et al. 1974) but far more widespread is the use of the non-ionic detergent Triton WR1339 (WATTIAUX et al. 1963). Livers from animals injected with this detergent contain lysosomes that are readily

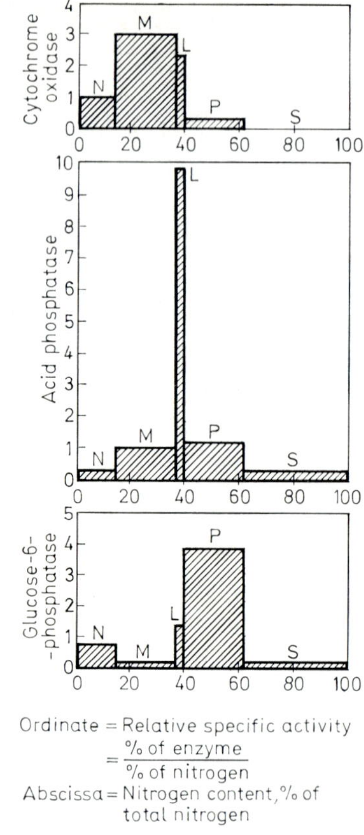

Fig. 2. Distribution of "marker" enzyme activities in the cell fractions obtainable from rat liver by differential centrifugation. Cytochrome oxidase is a characteristic enzyme of mitochondria, acid phosphatase of lysosomes and glucose-6-phosphatase of endoplasmic reticulum. N = nuclear fraction; M = mitochondrial fraction; L = lysosomal fraction; P = microsomal fraction; S = post-microsomal supernatant (soluble fraction). (Figure based on one from POOLE, B., and C. DE DUVE, 1973: Proc. Symp. on Intracellular Protein Turnover, Friedrichroda [May 1973; in press].)

separable on density gradients from other organelles (Fig. 3). Such Triton-loaded lysosomes ("tritosomes") can be used to evaluate questions that require pure preparations, but obviously caution is required in extrapolation of findings on such bodies. They contain large amounts of a quite abnormal indigestible material, (the detergent), and DAVIES (1973) and others have reported preliminary findings suggesting that tritosomes may be relatively inactive in intracellular digestion. The detergent also seems to induce the formation of autophagic vacuoles and other pertinent changes in hepatic cells (see *e.g.*, the comments by NOVIKOFF following WATTIAUX *et al.* 1963).

Problems also are generated from the fact that most sources of subcellular fractions are heterogeneous in cell content. In general, unless tissue culture cells are used, or, as with leukocytes, initial cell separation is possible, the content of lysosome fractions is derived from more then one cell type. For

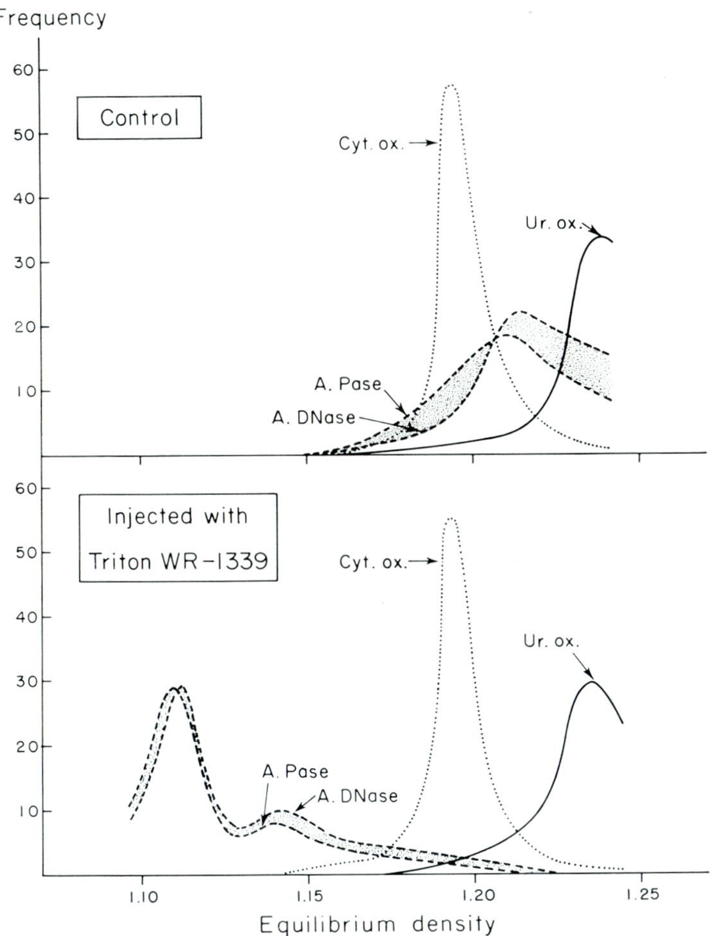

Fig. 3. Distribution of organelles of rat liver homogenates centrifuged on aqueous sucrose density gradients. The organelles are detected by marker enzymes. The "injected" preparation was from a rat injected four days previously with Triton WR-1339. The "control" rat was not specially treated. Triton brings about a selective shift of lysosomes (acid phosphatase, acid DNase) to lower density positions of the gradient. Mitochondria (cytochrome oxidase) and peroxisomes (urate oxidase) are not affected by the detergent. Note the slight differences in sedimentation behavior of the two lysosomal hydrolases; this may reflect some form of heterogeneity in the lysosome population. (From DE DUVE, C., 1965: Harvey Lects. **59**, 49—87.

example, liver is one of the most homogeneous of conventionally used vertebrate tissues but while hepatocytes number a bit more than half of the cell population, there are also substantial numbers of other cells. Of these, the Kupffer cells are especially significant for out considerations since they are quite abundant, (Fig. 36) and possess numerous large lysosomes important for major hepatic functions (see Section III). Both hepatocyte and Kupffer cell lysosomes accumulate Triton WR1339. (Other liver cells such as those of gall bladder and bile duct also contain large lysosomes [SALTHOUSE and PFEFFER 1965] but these can be partially excluded by careful dissection.)

Finally, the lysosomes within a given cell are usually heterogeneous in size, content, role and other properties. Among other implications, this means that isolation procedures adequate for some of the lysosomes may fail to obtain others. For example, it is suspected that primary lysosomes in some tissues may be quite small and relatively scarce (see Section II.2.1) so these may not be included in fractions containing the bulk of the cell's hydrolases or, equally important, if they are included, they may be impossible to separate from their larger or more numerous relatives. The only cell types from which structures thought to be primary lysosomes can be obtained with reasonable facility are some of the leukocytes which specialize in lysosome-dependent phagocytic activities; PMN leukocytes have proved particularly valuable in this regard (Section II.1.3).

I.3.1.1. Key Features of Lysosomes

The essential biochemical features that permit one to identify a population of organelles as lysosomes are the presence of acid hydrolases and properties indicating the sequestration of these enzymes within a discrete "structure". The latter characteristic is most often defined in terms of the cosedimentation of hydrolases, and of "latency" (DE DUVE 1969). Latency refers to the fact that potential substrates added to a population of lysosomes isolated by careful procedures, will not be degraded unless the lysosomes are exposed to disruptive agents or treatments such as freeze-thawing, aging, osmotic stress, ultrasonification or detergents such as Triton X-100. This is most simply explained by the concept that the enzymes are contained within a structure bounded by a semipermeable lipoprotein membrane that is impermeable to large molecules and must be "damaged" if enzymes and substrates are to intermingle *in vitro*. (Other possible contributory factors are discussed in Section I.3.2.3.) Among other things, the latency of hydrolases helps demonstrate that the enzymes are not merely adsorbed to the surface of a structure with which they sediment. (BEAUFAY 1972). Lysosome preparations exposed to Triton X-100 are often used as the fully active "baseline" for comparison with other preparations. Commonly, freshly isolated particles show non-latent activities of as much as 10–20% or even more of this baseline. At least part of this reflects disruption of some lysosomes during preparation (Section IV.4.1).

I.3.1.2. The Lysosomal Enzymes

Table 1 is a reasonably up-to-date list of hydrolase activities thought to be present in lysosomes; other useful published lists are those by TAPPEL 1969, BARRETT 1972, and VAES 1973. BARRETT and DINGLE (1971) have edited a very valuable set of discussions on the proteolytic enzymes. BARRETT's list (1972), among others, includes the International Union of Biochemistry classifications and points out that some of the "hydrolases" are classified as transferases.

Table 1. *Major Enzymatic Activities of Lysosomes.* Based Primarily on BARRETT 1972, BARRETT and DINGLE 1971, DE DUVE and WATTIAUX 1966, TAPPEL 1968 a, 1969, VAES 1973. Macromolecular substrate classes are indicated at the left.

Lipids	Acid lipase Ceramidase Phospholipase A 1 Phospholipase A 2 Sphingomyelinase (phospholipase C)
	Esterases (several types active against various thiol-, indoxyl-, naphthyl-, fatty acid-, and cholesterol-esters
Nucleic acids	Acid deoxyribonuclease (DNase II) Acid ribonuclease
	Acid phosphatase Acid pyrophosphatase Phosphodiesterase
Complex lipids polysaccharides and carbohydrate side chains of glycoproteins	N-acetyl-α-galactosaminidase N-acetyl-α-glucosaminidase N-acetyl-β-hexosaminidases Aspartylglucosylaminidase Fucosidase(s) Galactosidases (β- and probably α-); some can hydrolyze galactocerebrosides Glucosidases (α- and β-); some can hydrolyze glucocerebrosides β-glucuronidase Hyaluronidase Iduronidase(s) Mannosidases (α- and probably β-) Neuraminidase (includes sialidase activity) O-seryl-N-acetyl-α-galactosamidase β-xylosidase
	Aryl sulfatases (A and B) Chondrosulfatase Sulfamidase
Proteins	Acid carboxypeptidase Amino acid naphthylamidase (peptidase?) Cathepsin A (probably carboxypeptidase) Cathepsin B (B 1 and B 2; endopeptidases) Cathepsin C (dipeptidylaminopeptidase) Cathepsin D (acid "proteinase") Dipeptidylaminopeptidase II

The following enzymes are suspected to be present in the lysosomes of some tissues:
Cathepsin E
Dipeptidase (uncertain)
"Elastase"
Enzymes that can activate kininogens, plasminogen, and angiotensinogen
Exonuclease
Lysozyme
Naphthylamidase (*in vivo* activity uncertain)
Neutral proteases (neutral proteinases)
Phosphoprotein phosphatase (lysosomal and non-lysosomal enzymes probably exist)
Phosphatidate phosphatase

It is likely that a number of the hydrolase activities listed in Table 1 will turn out to be limited to a few types of cells or tissues or to special classes of organisms, and it is not impossible that one or two will be "removed" from the lysosomes as a result of work with more highly purified fractions. But, the present consensus is that lysosomes of many tissues and of many organisms do possess the ability to degrade most macromolecules of all major classes (Fig. 4). The actual number of distinct enzymes (as opposed to activ-

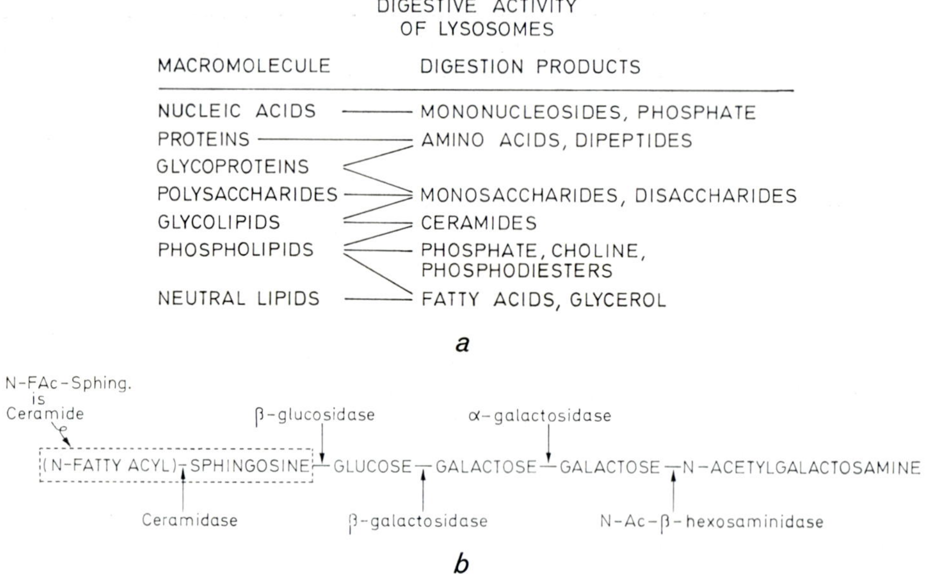

Fig. 4 *a*. Products of lysosomal digestion of macromolecules. (Courtesy of POOLE, B., and C. DE DUVE.) *b* Probable pathway for the lysosomal degradation of globoside, a glycosphingo-lipid of red blood cell membranes. The molecule apparently is degraded by sequential removal of the sugars (right to left in the diagram) followed by splitting of the ceramide. The sites at which the various enzymes act are indicated by the corresponding arrows. (Modified from VAES, G., 1973: In: Lysosomes and storage diseases. [HERS, H. G., and F. VAN HOOF, eds.], pp. 43—77. New York: Academic Press.)

ities) in lysosomes is not known. Particularly with the advent of fluorogenic (methylumbelliferyl) synthetic substrates for some of the enzymes hydrolyzing complex lipids, reports of new enzymatic activities detectable in lysosomes are still coming in at a rapid rate. What is required now is more thorough purification work with techniques such as affinity chromatography and immunological methods, plus further investigations of the range of effects on lysosome capabilities of mutations affecting single enzymes (see Section IV. 1.2); such studies should clarify the substrate specificities of the hydrolases.

It is already clear that the enzymes overlap somewhat in capabilities and that some of them may occur in multiple "isozymic" forms; different forms vary in electrophoretic behavior and other properties, such as the ease with which particular "isozymes" can be solubilized from the particles (see *e.g.*,

Axline 1968, for acid phosphatases). Since many of the hydrolases are known or thought to be glycoproteins, differences among isozymes have been attributed to variations in the saccharide side chains (*e.g.*, Konne and Ellis 1973). In line with this, treatment of the hydrolases with enzymes such as neuraminidase can alter their electrophoretic behavior so that forms of an enzyme that were initially relatively anionic migrate with more cationic forms (see Touster 1973, for a skeptical discussion and Goldstone *et al.* 1971, Sellinger *et al.* 1973, and Srivastava *et al.* 1974 for typical findings). From studies of this type it also has sometimes been argued that enzymes with somewhat similar, but still quite distinct hydrolytic properties, such as aryl sulfatases A and B, or certain of the hexosaminidases might differ primarily in their saccharide chains. But more recent work suggests that the relations among such hydrolases are more complex (*e.g.*, some may share common polypeptide subunits, while differing in other subunits, as discussed in Sections II.2.2 and IV.1.2). The general lack of detailed studies of amino acid sequences and other features of the polypeptide chains has impeded adequate analysis of the relations among the enzymes and the "isozymes".

A technical point meriting emphasis here is that there are non-lysosomal intracellular hydrolases that are active at acid pH. Included among these are some phosphatases (Neil and Horner 1964, 1965), which is particularly unfortunate since acid phosphatase is often used as a diagnostic or "marker" enzyme to follow the sedimentation of lysosomes or their behavior under varying experimental conditions. As Beaufay (1972) stresses, the lysosomal acid phosphatase can be distinguished from some important non-lysosomal phosphatases by its substrate and inhibitor specificities. The lysosomal hydrolase splits both β-glycerophosphate and p-nitrophenyl phosphate (these are widely employed analytical substrates), whereas enzymes such as the glucose-6-phosphatase of liver microsomes (pH optimum of 6) are relatively inactive against glycerophosphate but do split the nitrophenyl substrate. The widespread use of nitrophenyl phosphate substrates for lysosomal acid phosphatase determinations would thus seem unwise unless appropriate controls are utilized. Non-lysosomal esterases (Holt 1972) and β-glucuronidases (Section II.2.2) have also proved troublesome.

Another point to bear in mind in evaluating the data discussed in the following chapters is the obvious one that measured enzyme activity does not always accurately reflect enzyme amount. A few naturally occurring inhibitors of certain acid hydrolases are known [see *e.g.*, Lesca and Paoletti (1969) for DNase, and Lenney *et al.* (1974) for proteases] and information about the specific requirements for optimal activity of the different enzymes under varying conditions *in vivo* and *in vitro* is scanty.

I.3.1.3. Digestion in Lysosomes

Direct evaluation of digestive capacities predicted from the lysosomal enzyme complements (Fig. 4) has been attempted through mixing of crude or somewhat purified lysosome extracts with potential substrates such as macromolecules or isolated organelles (*e.g.*, Aronson and Davidson 1968,

Aronson and de Duve 1968, Coffey and de Duve 1968, Fowler and de Duve 1969, Mahadevan *et al.* 1969, Maunsbach 1969, Sawant *et al.* 1964 b, Tappel *et al.* 1963.) In addition, lysosomes within endocytically active cells have been loaded with radioactively-labelled macromolecules such as albumin iodinated with I^{125} or I^{131}. For protozoa (Brownstone and Chapman-Andresen 1971) or cultivated macrophages (Ehrenreich and Cohn 1967) digestion can then be evaluated by following the release of low molecular weight molecules from the cells (the iodinated tyrosines that carry the label are not reused for new synthesis). For liver (Mego 1973 a) and kidney (Davidson 1973) studies have been done on release of digestion products from lysosomes isolated and incubated in suspension after intravenous administration of radioactive proteins. With *Paramecium*, Berger and Kimball (1964) employed autoradiography to follow the loss of tritium-labelled DNA from phagocytosed bacteria, and the subsequent incorporation of the label in the nuclei of the protozoa.

The results of such investigations confirm the broad-ranging abilities of the hydrolases, although especially with lipid substrates there are still difficulties in duplicating with *in vitro* systems some of the capacities that have been deduced from consideration of likely *in vivo* roles. This is not entirely surprising. Presumably, the lysosomal hydrolases have evolved along lines that permit them to function simultaneously or sequentially within a common organelle (and to survive for appreciable periods, the presence of active proteases in the same organelle; de Duve 1973). Unfortunately little is known of the internal milieu that supports such cooperation within functioning lysosomes. Present knowledge of pH regulation in the organelles (Section II.1.4.5) and of the properties and influences of non-enzymatic components (Section I.3.2.3), is very sketchy. Beyond some confirmation of the suspected acid pH within the organelles little is really known about the intimate conditions (ions, pH, binding to other material etc.) under which enzymes and substrates meet.

Much of what gains access to lysosomes is in the form of large aggregates or microscopically distinctive structures and it is well known that degradation of such materials may be far different from digestion of their separate constituents. Binding of enzymes and substrates to surfaces can promote more rapid interaction by altering substrate conformations or by holding molecules in proximity to one another (the latter can be thought of as equivalent to raising the "effective" concentration of potential interactants). Perhaps more germane are inhibitory effects that are most readily observed with lipids. Even if they do not start out as parts of large structures, lipids often form spontaneously into micelles or other structured arrays, and enzymes capable of hydrolyzing bonds of the individual molecules when they are adequately dispersed may encounter difficulties in penetrating the structures to "reach" the bonds (Gatt *et al.* 1971). For *in vitro* systems this is reflected in the need to include detergents, or specific proteins or lipids in the incubation mixtures—these "additives" are thought to facilitate entry of the enzymes into the lipid structures. (Bangham and Dawson 1962, Ho 1973, Ho and Light 1973, Ho *et al.* 1973, Quarles and Dawson 1969, Teng and Kaplan

1974, VENGER *et al.* 1973.) It is also observed that proteins can be far more resistant to hydrolysis when in native conformation that they are when denatured (see *e.g.*, COFFEY and DE DUVE 1968).

Many of the materials that are degraded by lysosomes under normal or experimental conditions require several enzymes for digestion to small molecules (BARRETT 1972, TAPPEL 1969, VAES 1973). Even ordinary proteins are digested by action of several distinct cathepsins and peptidases (BARRETT and DINGLE 1971; the suspicion is currently growing that Cathepsin B_1, which can degrade a wide assortment of proteins and acts over a fairly broad range of pH's often initiates the attack [WIBO and POOLE 1974]). Glycoproteins, complex lipids or the interlinked proteins and polysaccharides of connective tissue matrixes are digested by action of a spectrum of hydrolases. For example, the "proteoglycan" of cartilage matrix is composed of sulfated polysaccharides linked covalently to protein, and sulfatases, several different saccharidases and several proteases are suspected to participate in its degradation (MUIR 1973, VAES 1973). For glycolipids such as the globoside of red blood cells, sequential action of a number of saccharidases apparently removes sugars one by one and this is followed by degradation of the lipoidal parts of the molecules. (Fig. 4). Enzymes outside the lysosomes are suspected to play some interesting roles related to lysosomal digestion. For example dipeptidases in the "hyaloplasm" (the unstructured "cell sap", "cytosol" or "soluble fraction" of the cell) may be responsible for breaking down some dipeptides produced within lysosomes (COFFEY and DE DUVE 1968). However, relatively few such cases are known and for major aspects of macromolecule breakdown the lysosomes appear to be enzymatically "self-sufficient" (Fig. 4). Such viewpoints are always in danger of being shown up as superficial and we will return to this "self-sufficiency" question later when we take up turnover and possible initial steps in degradation (Section III.4).

I.3.1.4. Lysosomal "Permeability"

These last considerations raise another important point—at what stage of digestion can products escape from the lysosome? Little is known directly about the membranes that bound the organelles (Sections II.1.4.3 and II.1.4.4) and in any event these membranes probably vary a good deal from cell to cell and at different stages of digestion. However, the available data do indicate that, as might be predicted from the enzymatic abilities of the lysosomes, only small molecules normally pass across the membranes. This view is supported by two related lines of investigation. First, as implied above, lysosomes are osmotically sensitive organelles suggesting that their membrane is of the semipermeable type, as are the other lipoprotein membranes of the cell (BERTHET *et al.* 1951). LLOYD (1973), and LEE (1972) among others have used the osmotic properties of isolated lysosomes to provide clues to the permeability properties of the membranes; compounds that prevent osmotic swelling can be presumed not to penetrate rapidly while those that do penetrate, should not offer such osmotic "protection". There are important caveats; for example, given the likely existence within most digestive vacuoles, of substantial amounts of charged macromolecules (Section I.3.2.3), Donnan equilibrium effects may greatly influence the distribution of inorganic ions. And, some compounds may directly alter the membrane (see *e.g.*, LEE 1972, ROMEO *et al.* 1967); since what is presumed to be osmotic rupture is generally measured by a decrease in lysosomal latency, such alterations could produce misleading outcomes. Nonetheless, the results of osmotic studies are quite suggestive. Thus, sucrose offers good osmotic protection to lysosomes isolated from liver or kidney (these are probably primarily heterophagic vacuoles), Glucose, and many amino acids do not. Inorganic ions, such as Na^+ or K^+ apparently can penetrate, although there is still debate especially about possible differences in this regard among lysosomes of different types or from different sources (see *e.g.*, BERTHET *et al.* 1951, BOWERS 1969, LLOYD 1973, REIJNGOUD and TAGER 1973). In designing media for isolation of lysosomes, sucrose or other components are generally added, in part to control osmotic responses to the salts. But in some cases, for reasons that are unclear, this is not necessary (*e.g.*, BOWERS 1969, has used 0.2 M KCl for isolating spleen and thymus lysosomes). DAVIDSON and SONG (1975) assert that the permeability to salts, of isolated mouse kidney lysosomes, is quite low at 37 $^{\circ}$C but rises markedly as the temperature approaches 0 $^{\circ}$C. They detect the change by preloading the lysosomes *in vivo* with radioactive RNAse and then, with varying suspension media and temperatures, evaluating the release of high molecular weight RNAse from a lysosome-enriched fraction or assessing the rates of proteolysis in such fractions (osmotic rupture presumably diminishes intralysosomal degradation). REIJNGOUD and TAGER (1975) report that the Na^+ and K^+ permeability of hepatic tritosomes is much lower at 25 $^{\circ}$C than at 0 $^{\circ}$C; at both temperatures the particles are permeable to protons. Hopefully, more direct studies of the salt permeability of lysosomes will help clarify some of the present ambiguities; it might be a useful simplification to work with phagocytic vacuoles containing inert materials, such as latex beads (*cf.*, Section II.1.4.3).

In general, disaccharides and larger dipeptides fall above the size limit for molecules that can enter mammalian cell lysosomes readily, while many monosaccharides, dipeptides with molecular weights less than 200, amino acids and so forth fall within this limit. Nucleotides such as adenosine monophosphate cannot pass across the membrane but there is suspicion that nucleosides can. Charged molecules encounter more difficulty in entering liver lysosomes than do comparable uncharged ones, and carbohydrates with many hydroxyls (*e.g.*, hexitol) enter poorly into lysosomes isolated from heart (ROMEO *et al.* 1967). Overall, as is true for permeability through the plasma membrane and other semipermeable lipoprotein membranes, lipid soluble molecules penetrate the lysosomal surface more readily than do hydrophilic ones.

Comparable results to those obtained with mammalian preparations have been compiled for *Tetrahymena* lysosomes (LEE 1970, 1971) although there are some interesting differences in detail such as the greater permeability of the protozoan organelles to hexitols. (Lee points out that this may relate to the presence of such compounds in *Tetrahymena*'s food materials.) Systematic extension of such studies to additional cells types would be valuable.

The other major line of investigation of lysosome permeability also relies on osmotic effects. Thus, when cells are permitted to endocytose into their lysosomes high concentrations of molecules such as sucrose, which cannot be readily hydrolyzed, the lysosomes swell (Fig. 5). Presumably the swelling reflects osmotic influx of water. COHN and his co-workers (see COHN 1970 b) have exposed cultured macrophages to a variety of relatively small molecules, some digestible and some not (*e.g.*, peptides made of D-amino acids and of L-amino acids, the former being non-hydrolyzable since the hydrolases as specific for the L-isomers; EHRENREICH and COHN 1969). The state of the lysosomes in the macrophages can be observed in the living cells by phase microscopy. The results agree in general with those described above for work on isolated organelles—*e.g.*, small dipeptides do not produce sustained swelling, whereas dipeptides of D-glutamic acid or tripeptides of D-alanine, and various indigestible disaccharides and oligosaccharides such as sucrose, cellobiose or raffinose do.

Finally, there are some molecules whose behavior is problematic. An example is cystine, an amino acid whose size is comparable to dipeptides that do not pass through the lysosome membrane. There are some disorders in which cystine accumulated in lysosomes (probably in insoluble forms; SCHULMAN and BRADLEY 1970, SEEGMILLER 1973). But cystine does not engender lysosomal swelling in experiments of the types decribed above, suggesting either that normally some special system for transmembrane movement exists or that cystine is reduced in the lysosomes to cysteine, which can cross the membrane [*]. The existence of active or facilitated membrane transport systems at the lysosome surface has been suggested (Section II.1.4.4) as has the presence in lysosomes of possibly relevant non-hydrolytic enzymes such as reductases (see *e.g.*, STRAUSS 1967, STOSSEL *et al.* 1971; discussion following TAPPEL 1963). However, unequivocal evidence along these lines is exceptionally difficult to obtain, largely as a result of the problems in discriminating among metabolically significant lysosomal activities, residual activities of molecules being degraded (*e.g.*, HENNING and STOFFEL 1972), non-enzymatic oxidations (*e.g.*, GOLDFISCHER *et al.* 1966, NOVIKOFF 1971) and activities of contaminant organelles present in lysosome preparations.

I.3.2. Some Cytochemical and Morphological Characteristics

Structures identifiable as lysosomes come in diverse forms and sizes. This is due in large part to the heterogeneity of lysosome roles and the attendant variations in the mechanisms and rates with which hydrolases and their potential substrates are brought together. Morphological and cytochemical

---

[*] The experiments referred to, actually were not done with cystine itself, due to solubility problems with this compound. Rather, SCHULMAN and BRADLEY (1970) exposed human fibroblasts to compounds, such as mixed disulfides of cysteine and penicillamine, which are similar to cystine in size and other properties. Normal fibroblasts did not show extensive vacuolation with such compounds, whereas fibroblasts from cystinotic patients did. (Vacuolation presumably is due to lysosome swelling, and can be induced in normal fibroblasts by appropriate non-degradable molecules.)

criteria for identification of bodies as probable lysosomes have been essential in sorting out the variety of structures and in overcoming the difficulties derived from tissue or cell heterogeneity. Microscope-based approaches have also been crucial for the study of many aspects of lysosome dynamics—modes of formation, interrelations among different structures and so forth.

I.3.2.1. Cytochemical Methods

A reasonable basis for *tentative* identification of a structure seen in the microscope as a lysosome is the presence of cytochemically demonstrable acid hydrolase activity plus the presence of a delimiting membrane; these criteria are articulated and evaluated by NOVIKOFF (1963). When possible, many microscopists also use microscopically evident degradation of material in a putative lysosome to help sustain their identification.

In most cases surprisingly few difficulties or ambiguities arise from careful application of such criteria. But care is needed since there are important restrictions on confident interpretation of cytochemical findings in more conventional biochemical terms. The most significant of these restrictions derive from the limits of present cytochemical methodology. Thus a brief outline of relevant techniques seems warranted.

Methods are now available for light microscopic demonstration of a number of acid hydrolases retained *in situ* in cells and tissues. The procedures are based on use of incubation mixtures containing appropriate substrates and other components such that the enzymes produce reaction products which are visible in the microscope and which precipitate rapidly and bind to the tissue, thus marking the sites of enzyme action. Usually the enzymes are kept in place by fixation of the cells with aldehyde fixatives (formaldehyde or glutaraldehyde) and tissue sections are produced through frozen-sectioning or tissue chopping (*cf.*, SMITH and FARQUHAR 1965). Techniques such as frozen-sectioning may disrupt permeability barriers and thus facilitate passage of the incubation mixture into the cells. For light microscopy, the incubation mixtures commonly include substrates that are altered by the corresponding enzymes to forms that can then bind covalently to dyes. For example, the enzyme *acid phosphatase* splits naphthylphosphates to produce naphthols which react readily with dyes of the diazonium-salt type to form colored precipitates (see *e.g.*, BARKA and ANDERSON 1962, BECK *et al.* 1972, HAYASHI 1965, HOLT and HICKS 1966, SMITH and FISHMAN 1969 for discussion of such "azo-dye" methods and of some other comparable procedures). A crucial,

---

Fig. 5. Phase-contrast photomicrographs showing a mouse peritoneal macrophage maintained *in vitro*. The cell was grown for 24 hours in a medium containing sucrose and then it was placed in a medium containing invertase. During the exposure to the sugar, numerous sucrose-containing vacuoles accumulate. The exogenous enzyme leads to a disappearance of these bodies, probably by virtue of its entry into the vacuoles through pinocytosis and its hydrolysis of the sucrose. The four micrographs show successive stages in the disappearance of the vacuoles. ×2,500. (From COHN, Z. A., and B. A. EHRENREICH, 1969: J. exp. Med. **129,** 201—226.)

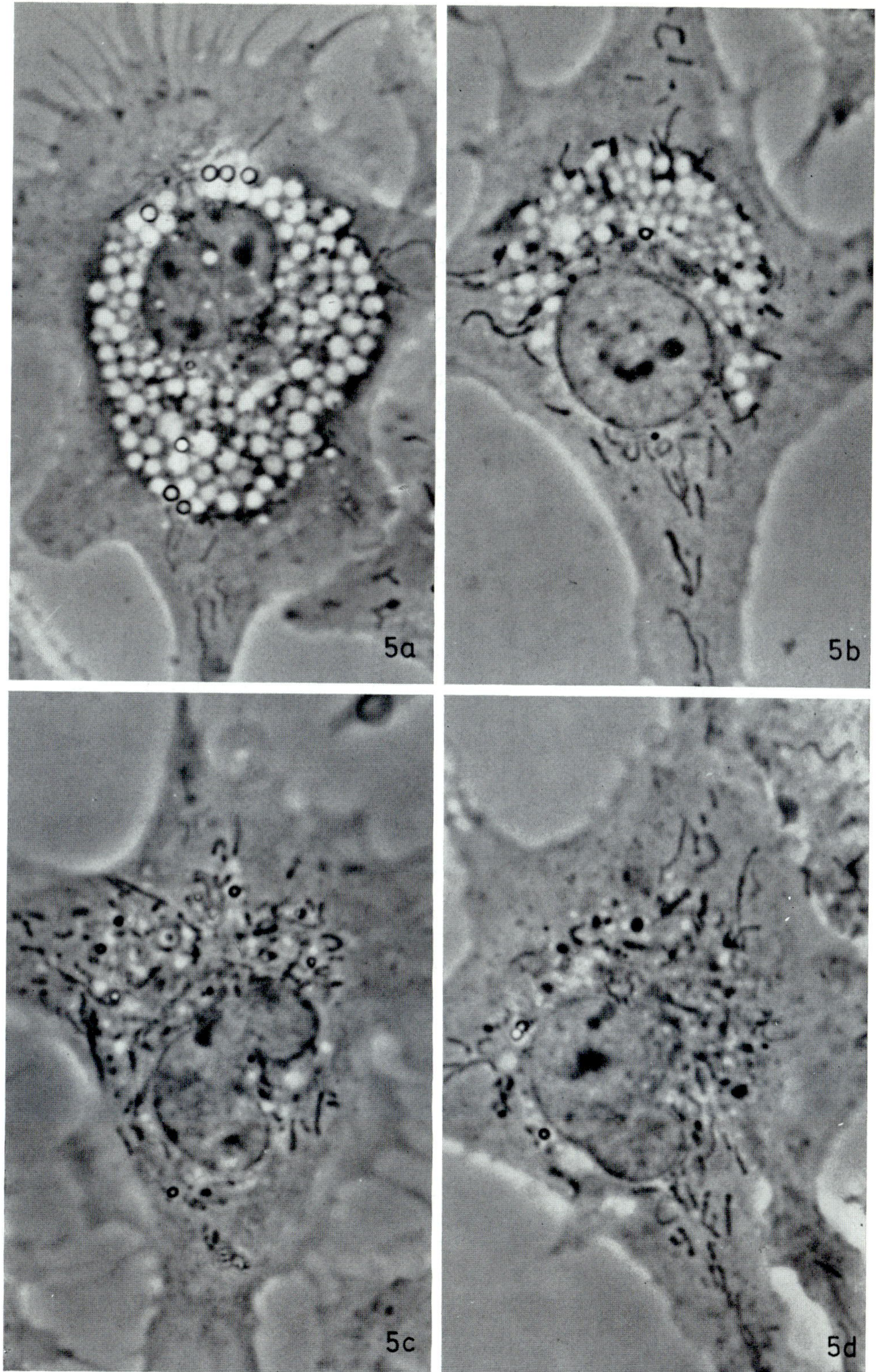

Figs. 5 *a–d*.

different cytochemical approach explored vigorously by GOMORI (1952) and his successors is exemplified by another set of acid phosphatase methods. When fixed tissue is incubated at acid pH in a mixture containing lead ions and β-glycerophosphate, the glycerophosphate is hydrolyzed and the liberated phosphates precipitate rapidly in the form of lead phosphates; lead phosphates are rendered visible for light microscopy by exposing them to sulfide ions which converts the precipitate to an insoluble, dark brown lead sulfide. "Lead-salt" methods of this type are widely used for light microscopy but their greatest advantage lies in the fact that lead phosphates survive preparation of tissues for electron microscopy and are sufficiently "electron dense" (electron opaque) to be readily visible. Efforts have been made to adapt some of the dye procedures for the electron microscope but while there have been some advances along these lines that give hope for the future (BOWEN 1973, HOLT and HICKS 1966, SMITH and FISHMAN 1969) the vast majority of electron microscope studies of hydrolase localization have utilized the lead salt techniques, or close relatives.

For light microscopy, reasonably reliable techniques are presently available for localization of several enzymes at the subcellular level. Most frequently used are those for demonstration of acid phosphatase, aryl sulfatase (Fig. 6) β-glucuronidase, N-acetyl-β-glucosaminidase and "non-specific" esterase (Fig. 56) (this last enzyme activity is detected with substrates such as indoxyl acetates and thiolacetates). Procedures for a few other enzymes are at varying states of development [see *e.g.*, VORBRODT 1961 for discussion of nucleases and BECK *et al.* (1972) for discussion of aminopeptidases]. Probably the most important set of enzymes for which adequate methods are not widely available are the cathepsins. Immunohistochemical approaches for localizing cathepsin D seem very promising (POOLE *et al.* 1972, 1974, ROJAS-ESPINOSA *et al.* 1974) but have not yet been broadly applied. More "conventional" cytochemical procedures may also be on the horizon (see the work by SMITH with naphthylamide substrates reported in McDONALD *et al.* 1971).

For electron microscopy, there are useful procedures available for demonstration of "esterase", (HOLT and BARROW 1972) and of aryl sulfatase (Fig. 7; GOLDFISCHER 1965, HOPSU-HAVU *et al.* 1967) and some others are being explored (BOWEN 1973, HAYASHI *et al.* 1968, McDONALD *et al.* 1971). But by comparison with acid phosphatase techniques, these procedures have been employed very little.

The ability to demonstrate acid phosphatase by electron microscopy in structures such as those of the Golgi region, which are difficult to isolate in purified form, accounts for some of the differences in emphasis that occasionally crop up between the cytochemists' lysosomes and the biochemists'. But along with its benefits, the traditional very heavy reliance by cytochemists on acid phosphatase as the lysosome "marker" enzyme also leads to some potential problems. Often it is reasonable to extrapolate from the demonstrated presence of this enzyme to the presumed presence of the other lysosomal hydrolases, and such extrapolation actually is quite common among biochemists as well as among cytochemists. Sometimes the presumptions can be checked through use of cytochemical procedures for other hydrolases

(Section II.6) or by judicious comparison of microscopic and biochemical information for the same cell type. However, in the absence of such supporting evidence, identification of a structure as a lysosome based on the presence of acid phosphatase must *always* be regarded as somewhat tentative. Several other limitations deserve strong emphasis. Most cytochemistry is done with fixed tissue, and both fixation and the presence of some common components of incubation media (notably heavy metal ions such as lead) can

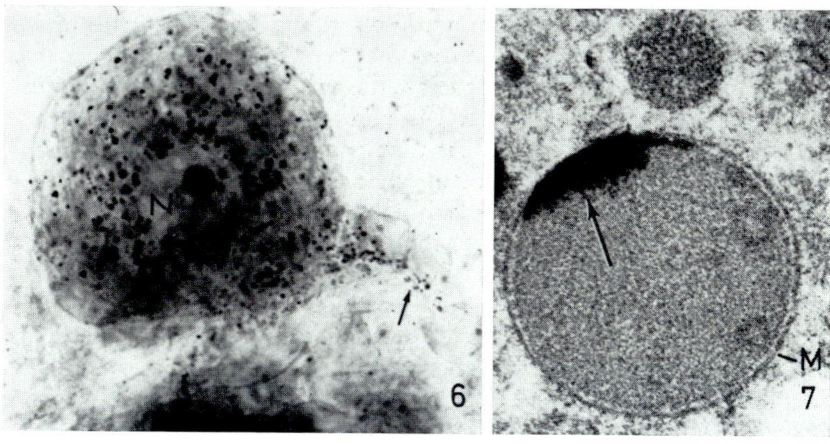

Figs. 6 and 7.

Fig. 6. Light micrograph of a neuron from a cultured mouse dorsal root ganglion incubated to demonstrate aryl sulfatase activity. The nucleus is indicated by *N* and the beginning of the axon by the arrow. Numerous lysosomes containing reaction product are seen as dark granules. ×1,000. (From Holtzman, E., 1969: In: Lysosomes in biology and pathology. [Dingle, J. T., and H. B. Fell, eds.], Vol. 1, pp. 192—216. Amsterdam: North-Holland Co.)

Fig. 7. Lysosome from an epinephrine (adrenaline) cell of rat adrenal medulla incubated to demonstrate aryl sulfatase activity. Reaction product is seen at the arrow. Note also the zone of low electron density ("halo") that separates the delimiting membrane (*M*) from the moderately electron-dense content of the body. ×50,000. (From Holtzman, E., and R. Dominitz, 1968: J. Histochem. Cytochem. **16**, 320—336.)

markedly inhibit enzyme activities (*cf.*, Barka and Anderson 1962, Brunk and Ericsson 1972, Essner 1973, Goldfischer *et al.* 1964, Gomori 1952) or affect the permeability of structures to components of the incubation medium (Bainton and Farquhar 1968). Much remains unclear about the intimate details of formation of cytochemical reaction products and their subsequent precipitation in tissues. For instance, the fact that a membrane-delimited granule shows reaction product chiefly near its surface need not imply that the corresponding enzyme is bound to the membrane. It might also be that the levels of enzyme activity are so high within the granule, that substrate penetrating from the medium is completely hydrolysed before it can reach the deeper portions of the structure (see *e.g.*, de Duve 1965), or that reaction product produced in the structure has an affinity for sites near the surface and diffuses to these sites, or even that the trapping agent (*e.g.*, lead) cannot penetrate deep into the granule, so that precipitation of reaction product takes place only near the surface (FARQUHAR *et al.* 1972).

That negative cytochemical findings (the absence of reaction product) must be interpreted with special caution is given added weight by observations such as those made by BAINTON and FARQUHAR on PMN leukocyte granules, which are known to contain acid phosphatase but which seem to react in the lead salt procedures only during or shortly after their formation (BAINTON and FARQUHAR 1968, FARQUHAR *et al.* 1972). Since the enzyme again becomes detectable when the granules fuse with phagocytic vacuoles the unreactive state may reflect some special feature of enzyme storage related perhaps to formation of multimolecular aggregates within the granules or to permeability properties of the structures (BAINTON and FARQUHAR 1968, SEEMAN and PALADE 1967.) It is also conceivable that enzyme inhibitors are present (see *e.g.*, SEEMAN and PALADE 1967) although direct evidence for this is lacking and the granules that do not stain with the lead salt methods do, with azo-dye precedures (BAINTON and FARQUAR 1968), an observation that has also been made for some other types of lysosomes (KATAYAMA and YAM 1972, NOVIKOFF 1973).

Another important point is that only a few types of compounds are suitable for use as cytochemical substrates. Thus, it sometimes is impossible to define in adequate fashion the relations between an enzyme activity demonstrable cytochemically and its suspected biochemical counterpart. This is especially true if several biochemical candidates exist. For some enzymes, inhibitors have proved useful in this regard—for example, fluoride ions inhibit both the biochemically defined lysosomal acid phosphatase and the cytochemically demonstrable enzyme. Features such as pH optima can also be valuable—thus, it has been argued that the two lysosomal aryl sulfatases may be distinguished to some extent, by use of the same cytochemical incubation mixture buffered at different pH's (see GOLDFISCHER 1965). However, there are some vexing cases for which no such simple aids exist. Esterase activities have been particularly difficult to interpret, which is unfortunate since electron microscope methods for these activities are available. It is not clear what the cytochemically demonstrable esterases hydrolyze *in vivo*. At one point it was thought that cathepsin C contributed to the cytochemical activity but this now seems quite unlikely (BARRETT 1969, HOLT 1963, VANHA-PERTTULIA *et al.* 1965). Part of the problem derives from present uncertainties as to the roles of some of the biochemically demonstrable esterases (see *e.g.*, BARRETT 1969, 1972, TAPPEL 1969, VLADUTIN and ROSE 1974). Obviously, one would guess that some of the lysosomal esterases attack

---

Fig. 8. Cells from rat liver exposed to intravenously administered horse-radish peroxidase for 10 minutes then incubated to demonstrate peroxidase activity. The cell in the center of the field is one of those that border hepatic sinusoids. Peroxidase is present in extra-cellular spaces (*E*) and in endocytic structures within the cells (*V* indicates a vesicle and the arrows, multivesicular bodies, a class of lysosome that very often is involved in handling exogenous macromolecules). ×14,000. (From FAHIMI, H. D., 1970: J. Cell Biol. **47**, 247—262.)

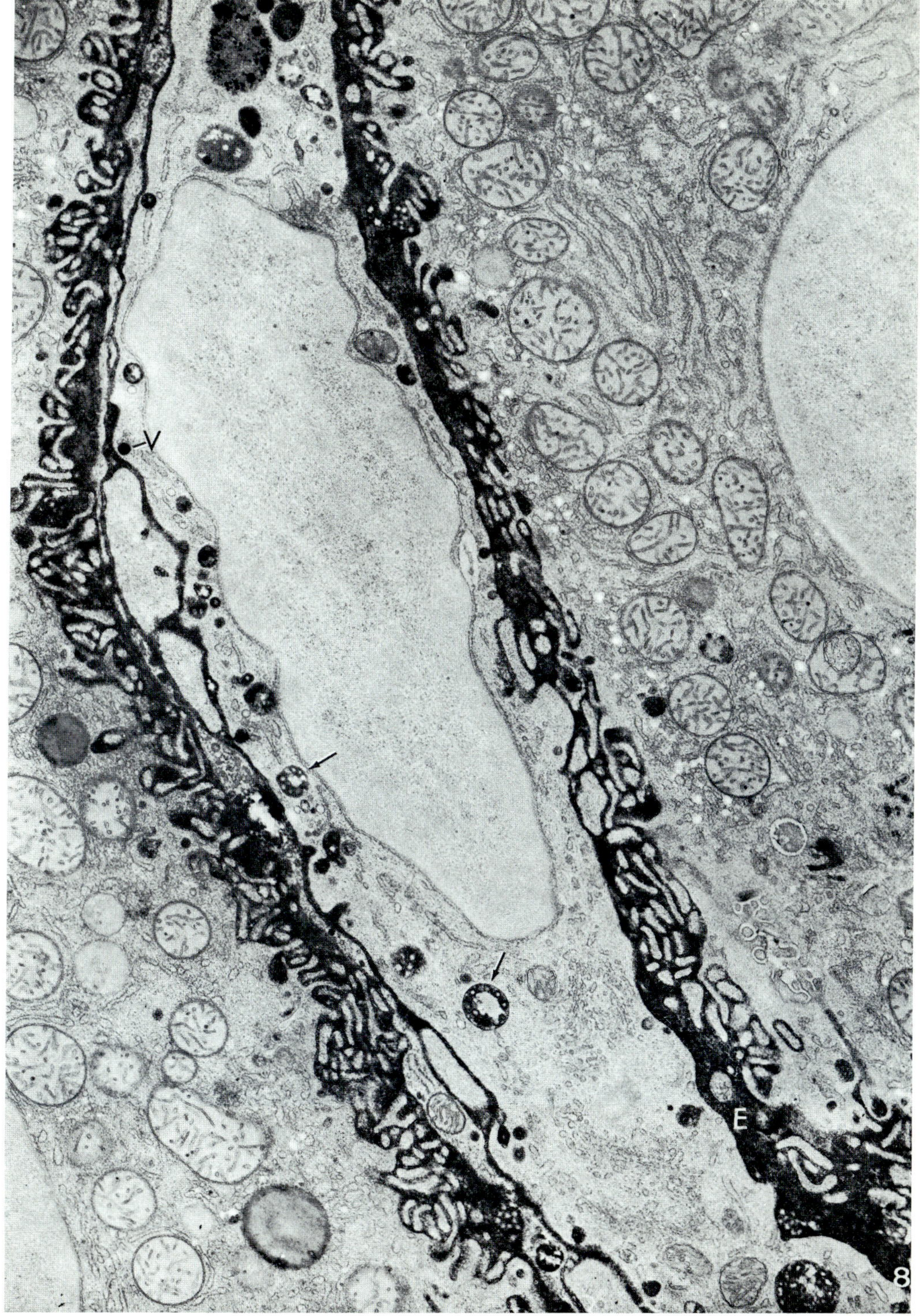

Fig. 8.

lipids, *in vivo*, but this has not been established for the cytochemically demonstrable enzyme.

Some efforts have been made to quantitate enzyme activities *in situ* by microspectrophotometric or other determination of the rates of deposition of reaction product (BITENSKY *et al.* 1973, HIRSCH 1968). Factors such as those discussed in this section (see also Section IV.4.1) can greatly complicate interpretation of results; on the whole, quantitative relations between enzyme activity and cytochemical reaction product formation are not well understood. But, for example, in evaluations of possible changes in intensity of lysosomal staining in cells responding to altered conditions, such objective measurements can help overcome serious problems that affect subjective judgements (*e.g.*, alterations in the size or distribution of stained structures often produce false impressions of changed staining intensities).

I.3.2.2. Cytochemical Studies of Exogenous Tracers

The use of enzymatically active proteins to follow the uptake and processing of exogenous materials by cells has proved extremely valuable for study of lysosomes. The usual non-enzymatic macromolecular endocytosis tracers such as ferritin, or colloidal metals (gold, iron, thorium dioxide and so forth; Fig. 14) produce at best a single electron dense "dot" per molecule when viewed in the electron microscope and they are not detectable by light microscopy. In contrast, suitable enzymatically active proteins can give rise to heavy deposits of cytochemical reaction product, visible by light or electron microscopy. Some effort has been made to use phosphatases as tracers but peroxidatic activities of heme-containing proteins have been best exploited for this purpose. A number of such proteins can simultaneously reduce H_2O_2 and oxidize benzidine compounds via enzymatic reactions or sometimes by non-enzymatic ones; when suitable benzidine derivatives are used, the reaction product is colored and insoluble and it can react with osmium tetroxide giving rise to an electron dense deposit. Most work has been done with horseradish peroxidase as the tracer protein (see STRAUS 1967 for reviews of pioneering light microscope work). This enzyme is soluble, readily available, fairly small (mol. wt.: 40,000; radius about 25 Å) and even after exposure to aldehyde fixatives, it is strongly active with cytochemical substrates. At present the most widely uses benzidine is the 3,3'-diamino derivative included in an incubation medium devised by GRAHAM and KARNOVSKY (1966).

With peroxidase methods and with other tracers it was early demonstrated that exogenous proteins taken up by cells remain within membrane-delimited structures (Fig. 8); usually tracers are soon sequestered largely (perhaps almost exclusively) within lysosomes. For example, when horseradish peroxidase is introduced into the lumen of a renal tubule, it soon accumulates primarily in large vacuole-like "droplets" readily seen within the tubule cells by light microscopy (Fig. 48 shows an extreme case of this). These vacuoles can be shown also to contain acid hydrolases by sequentially incubating the tissue in cytochemical media for demonstration of peroxidase and for demonstration of hydrolase; when the media are chosen to give different colors, the admixing of the tracer and the endogenous enzyme can be detected (STRAUS 1967).

Peroxidatically active proteins (*e.g.,* catalase, lactoperoxidase, hemoglobin) and smaller molecules (*e.g.,* "microperoxidase", a peptide derived from cytochrome C) of differing sizes and properties can be used to supplement horseradish peroxidase in study of endocytosis and of penetration of exogenous molecules into tissues, between cells and so forth.

Needless to say, the peroxidase methods are not without their problems. For our concerns the most important difficulty to bear in mind is that one can only assume that the routes and mechanisms by which horseradish peroxidase or any other foreign tracer is processed by a given cell or tissue correspond to those that would apply to less alien molecules, such as homologous proteins. To some extent this can be checked—thus mammalian kidney tubule cells have been studied with a number of macromolecular tracers including radioactively labelled mammalian proteins that are detectable both autoradiographically (MAUNSBACH 1969) and by biochemical study of isolated lysosomes (DAVIDSON 1973). The same basic intracellular routes for processing seem to apply in all cases. However, it is already known that in some mammals, including rats, horseradish peroxidase can produce changes in blood-vessel permeability via induction of histamine release and perhaps through other effects (*e.g.,* COTRAN and KARNOVSKY 1966) and it is possible that peroxidases or other tracers can have other subtle effects on cells. For instance, investigations on cultured cells and protozoa have shown that some proteins can stimulate endocytosis (see Section II.1.2).

I.3.2.3. Additional Cytochemical Features of Lysosomes: "Matrix" Materials and Some Other Non-Enzymatic Components

Lysosomes have frequently been studied with cytochemical methods other than those used to demonstrate enzymatic activities but unambiguous interpretation of the results has usually been difficult. Lysosomes of several cell types stain with the periodic-acid-Schiff reaction (PAS) and with other light and electron microscope methods thought to demonstrate polysaccharides or other forms of carbohydrate (Fig. 9); (see DAEMS *et al.* 1969, HARDIN and SPICER 1971, KOENIG 1969, NOVIKOFF 1961, STRAUS 1967). Some of this "staining" probably reflects the presence of saccharide side chains of the hydrolases. In addition, in the case of secondary lysosomes, carbohydrate components may derive from material taken up for degradation (*e.g.,* Fig. 9). However, it it widely agreed that some primary lysosomes contain non-enzymatic components, including polysaccharides or other macromolecules to which saccharide residues are linked. Spicer and his co-workers have developed cytochemical evidence for the presence of sulfated "muco-substances" (presumably polysaccharides) in the lysosomes of PMN leukocytes (*e.g.,* HARDIN and SPICER 1971) and studies on appropriate granules isolated from the leukocytes also indicate a content of acidic polysaccharides (FEDORKO and MORSE 1965, FARQUHAR *et al.* 1972).

Lysosomes also often stain with several types of vital dyes (*e.g.,* neutral red) and particularly with fluorescent dyes, such as acridine orange, striking metachromatic effects can be observed. High concentrations of small basic dyes like neutral red or acridine orange can accumulate in lysosomes within a few minutes. This supports the view that such dyes enter the cell by passage through the plasma membrane and then penetrate the lysosome mem-

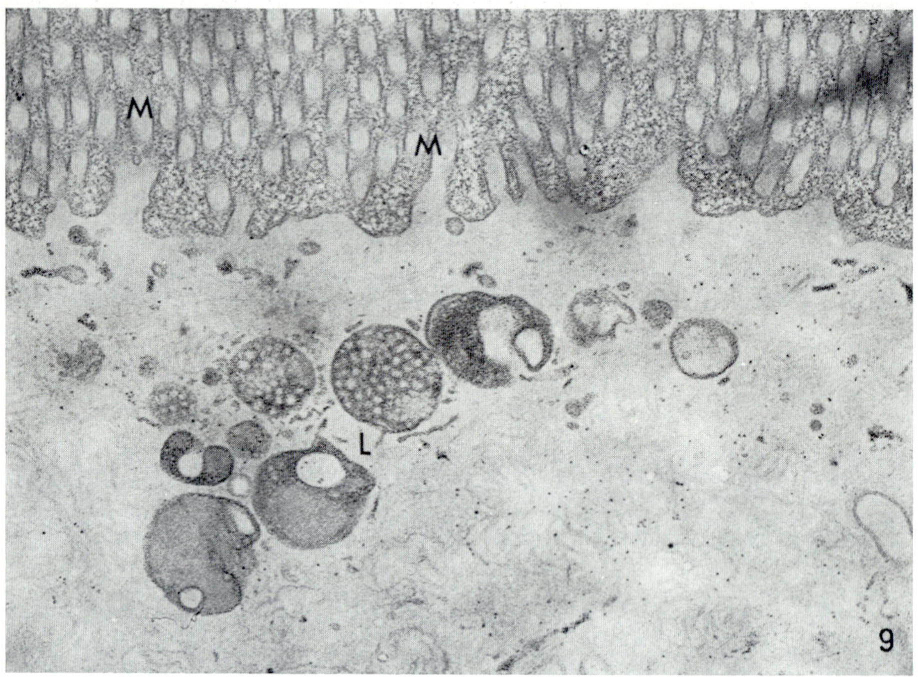

Fig. 9. Apical portion of a human intestinal absorptive cell stained with a method thought to demonstrate carbohydrates (Thiery's silver proteinate technique). The moderately electron dense stain is seen on the extracellular coat surrounding the microvilli (*M*) and within several lysosomes (*L*) clustered in the cytoplasm. ×20,000. (From DAEMS, W. TH., J. J. VAN GEMUND, P. M. A. VIO, R. G. J. WILLINHAGEN, and W. R. DEN TANDT, 1973: In: Lysosomes and storage diseases. [HERS, H. G., and F. VAN HOOF, eds.], pp. 575—598. New York: Academic Press.)

brane; but especially since the molecules are not visible in the electron microscope, contributions of endocytosis are difficult to evaluate (ALLISON and YOUNG 1969, DE DUVE and WATTIAUX 1966). Endocytosis probably is the primary mode of entry into lysosomes for some other vital dyes such as Trypan Blue which binds tightly to protein in the circulation or incubation media (see BECK *et al.* 1972).

The accumulation of basic dyes in lysosomes may partly reflect the acid pH in the organelles (Section II.1.4.5). But it also has been suggested that the dyes bind to acidic lipoproteins or saccharide-rich molecules, concentrated in the lysosomes (BARRETT and DINGLE 1967, and BARRETT 1969, KOENIG 1969, WITTEKIND *et al.* 1974). According to some investigators (DAVIES *et al.*

1971, Goldstone *et al.* 1970, Koenig 1969, Murata *et al.* 1973, Wolman 1965) such acidic macromolecules, especially polysaccharides and lipoproteins are the key constituents of a lysosome "matrix" with which the hydrolases are presumed to associate via electrostatic bonds (see also Section II.2.2). The interaction of hydrolases and matrix is thought to contribute to lysosome latency, the enzymes being immobilized or otherwise inactivated. This theory is very difficult to evaluate adequately especially since the acidic molecules in question have yet to be thoroughly characterized. In some cases the key biochemical or cytochemical analyses adduced as evidence were done on secondary lysosomes with contents of complex origin. The presence of appreciable amounts of lipids within secondary lysosomes and the possible binding of lipids with hydrolases, if such binding actually is extensive, might simply reflect slow digestion and some of the other features of enzyme-lipid interactions alluded to earlier (Section I.3.1.3). De Duve has pointed out (1963) that hydrolases can diffuse out of isolated lysosomes relatively easily once the structures are disrupted by osmotic or other means, and it would seem likely that the digestion within lysosomes depends upon diffusion that is sufficiently unrestricted to permit enzymes and substrates to readily inter-mix. Thus if there is general intralysosomal binding of hydrolases to some sorts of non-enzymatic "matrices" the bonds involved apparently are weak and might actually contribute little to latency.

Lysosomes of a number of tissues show autofluorescence with spectral characteristics suggesting the presence of flavin nucleotides among other components (*e.g.*, Tappel 1969). In addition, lysosomal staining reactions often indicate a content of metals such as iron or copper (Brunk *et al.* 1968, Goldfischer *et al.* 1970). Most of the metals and nucleotides almost certainly represent residual materials; for example, lysosomal heme compounds include what seem to be fragments of mitochondrial cytochromes formed through proteolytic cleavage (Tappel 1969).

From this brief summary it should be apparent that little is really known about the non-enzymatic components that might accompany the acid hydrolases. Future work on lysosomes that accumulate large quantities of particular materials, as occurs in some disorders (Section IV.1) should help clarify the organization of digestion vacuoles and residual bodies. But at present we have very limited insight into central questions, such as the state of activity of hydrolases in primary lysosomes.

I.4. Morphological Categories of Lysosomes

The only structural feature shared by all known types of lysosomes is the presence of a surface membrane with dimensions and appearance expected for a cellular lipoprotein membrane. Structures believed to be lysosomes vary in diameter from a few tens of nanometers (*e.g.*, the vesicles of the Golgi region discussed below) to several microns or more (*e.g.*, secondary lysosomes such as protozoan food vacuoles [Fig. 11]) or lysosomes of kidney tubule cells and those found in some pathological conditions (Section IV.1); the rat hepatocyte lysosomes that have been so extensively studied by biochemists measure roughly 0.5 microns in diameter. The variations in appearance are correspondingly great. However the acid hydrolase-containing organelles do fall into morphological categories related to their roles and it is useful to be able to recognize the different populations. Some caution is required in identification particularly since there may be few, if any differences in appearance between a "precursor structure" such as a newly formed endocytic vacuole that has not yet acquired hydrolases and a similar structure in which digestion has just begun.

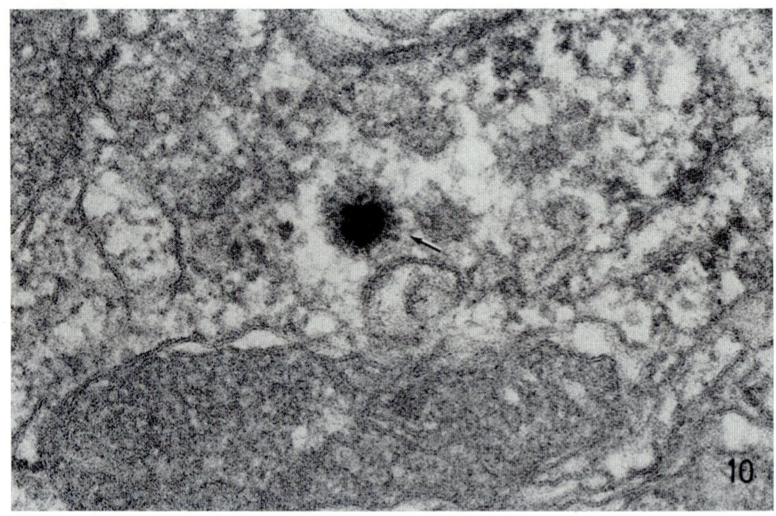

Fig. 10. From an epithelial cell of rat vas deferens incubated to demonstrate acid phosphatase activity. Reaction product is seen in a vesicle in the center of the field. The arrow indicates a portion of the vesicle surface along which "coating" (radiating projections) is visible. ×100,000. (From FRIEND, D. S., and M. G. FARQUHAR, 1967: J. Cell Biol. **35**, 357—376.)

For most cell types the *primary lysosomes* are the most difficult to recognize. From cytochemical evidence it is suspected that primary lysosomes in many tissues are small vesicles formed from the Golgi apparatus or from Golgi associated endoplasmic reticulum. Without cytochemical hydrolase demonstration these are often impossible to distinguish from other vesicles in the same region. Sometimes the lysosomal vesicles are of the "coated" type (HOLTZMAN *et al.* 1967). FRIEND and FARQUHAR (1967) stress that acid phosphatase-containing coated vesicles of cells in vas deferens (Fig. 10) and probably of other tissues as well, are smaller (50–75 nm diameter) than the similar-looking structures that participate in endocytosis (100 nm diameter).

Coating, however, is not a good "membrane marker" since it is present on membranes of diverse types (FRIEND and FARQUHAR 1967, HOLTZMAN *et al.* 1967, KANESAKI and KADOTA 1969, ROTH and PORTER 1964). In addition coated vesicles are often observed to transform into membrane delimited vesicles of ordinary appearance; apparently the coat can be lost or modified into a form no longer readily visible in the microscope (BUNT 1969, BREARLEY 1973, JONES and ROCKSEL 1973, KANESAKI and KADOTA 1969, OGAWA 1972, ROTH and PORTER 1964).

The primary lysosomes of some leukocytes are large enough, abundant enough and distinctive enough in appearance to be readily recognized. As shown in Fig. 12 the ones in rabbit PMN leukocytes contain an electron dense material, giving them an overall appearance comparable to that of secretion granules of many gland cells. Lysosomes with electron dense content are often referred to by the generic term *dense body* and it is believed that primary lysosomes of the dense body type may occur in some cells other than leukocytes (Sections II.1.3.2 and II.2.1). But as will be seen below, secondary lysosomes can also appear as dense bodies and the term is not adequately discriminatory as an indicator of function.

Phagocytic digestive vacuoles can often be identified by their content of recognizable exogenous structures such as microorganisms (Fig. 50) or cellular debris (Fig. 52). *Pinocytic digestive vacuoles* can similarly be identified if the exogenous macromolecules or particles they contain are electron dense (Fig. 14) or otherwise distinguishable. It is found that pinocytosed materials very often accumulate in a morphologically distinctive type of lysosome known as a *multivesicular body* (Figs. 26–28); these structures contain several or many small vesicles enclosed within a delimiting membrane.

Autophagic vacuoles (Figs. 18–22) are identifiable by their content of cytoplasmic organelles; occasionally one must worry about endocytosis as a possible mode of origin of such material, but this can often be ruled out on the basis of knowledge of the cell type, its surroundings and the circumstances. In addition, the structures that give rise to autophagic vacuoles may differ markedly from those involved in phagocytosis (*e.g.*, in their initial stages of formation autophagic vacuoles may be delimited by a sac, rather than a single membrane; *cf.*, Figs. 19 and 21) which facilitates recognition by "context".

Residual bodies (Fig. 34) can also frequently be identified by their morphology and by the circumstances in which they are observed. Such bodies often have an electron dense content and thus are sometimes referred to as a type of dense body; they may contain lamellar material, lipid droplets, electron dense grains and other structured material embedded in an amorphous "matrix". Lysosomes are commonly encountered with appearances expected for intermediate stages in the development of autophagic or heterophagic structures into residual bodies (*e.g.*, Fig. 34).

II. Lysosome Formation, Functioning and Fate

Our principal concerns in this chapter will be with general features of the mechanisms by which lysosomes participate in normal processes of intracellular digestion. We will consider the routes and mechanisms by which substrates and hydrolases are brought together and will also have occasion to take up some features of the lysosome surface, aspects of pH regulation within the organelles and the ways in which cells deal with digestive residues. Subsequent sections of the monograph will discuss extracellular events in which lysosomal hydrolases participate, and pathological processes.

II.1. Heterophagy

II.1.1. Background

The fact that protozoa utilize lysosomal hydrolases to degrade food materials taken into the cell through endocytosis has been demonstrated for a number of classes, including several familar types of amebae and ciliates

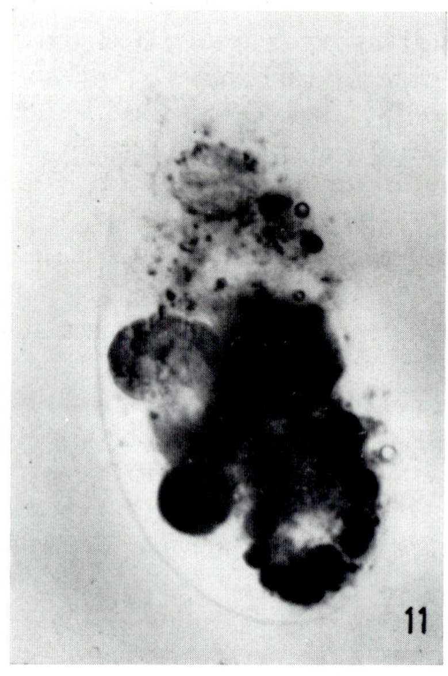

Fig. 11. Light micrograph of a *Tetrahymena* incubated to demonstrate acid phosphatase activity. Reaction product is seen in numerous vacuoles. ×1,000 (approx.). (Courtesy of MÜLLER, M.)

(for reviews see AARONSON 1973, DUTTA 1973, ECKHOUT 1973). The prominent food vacuoles in which endocytosed substances accumulate, contain cytochemically demonstratable acid hydrolases (Fig. 11; see *e.g.*, CHAPMAN-ANDRESEN and LAGUNOFF 1966, HAUSMANN and STOCKEM, ROSENBAUM and WITTNER 1962). Lysosomes have also been isolated from a few protozoa, notably *Tetrahymena* (*e.g.*, LEE 1970, LLOYD *et al.* 1971, MÜLLER *et al.* 1966, 1971, ROTHSCHILD 1966). The isolated organelles possess the essential characteristics described in the previous chapter although some interesting variants may exist (see for example MÜLLER's [1973] description of the parasitical trichomonad flagellates, many of whose "lysosomal" hydrolases have pH optima in the range 6–6.5 and whose other organelles also have quite atypical properties).

Few details are well established for unicellular organisms about the mechanisms through which hydrolases are packaged and those by which enzymes and endocytosed nutrients are brought together. By both light and electron microscopy it is common to find numerous smaller bodies in the vicinity of newly formed food vacuoles—swarms of small structures stainable with neutral red are sometimes observed (KITCHING 1956) and microscopic evidence suggests that these bodies can fuse with the food vacuoles. Electron microscope histochemical methods have shown the presence of acid phosphatase in small membrane delimited vesicles near newly formed food vacuoles (*e.g.*, ELLIOTT and CLEMMONS 1966, RICKETTS 1970, HAUSMANN and STOCKEM 1973) and it is reasonable to believe that these vesicles fuse with and thus contribute hydrolases to the vacuoles. However, interpretation is complicated by observations that small membrane-delimited structures also commonly bud from the food vacuoles, as can be demonstrated with endocytosed tracers. RUDZINSKA has also pointed out (1969) that in protozoa that feed on other protozoa (*e.g.*, the suctorian *Tokophyra*, which can capture *Tetrahymena*) the prey's own enzymes may be released into the digestive vacuole when the cell dies, and these might supplement the predators' hydrolases.

In some unicellular organisms such as *Paramecium* (ESTEVE 1970) *Euglena* (BRANDES 1965), amebae (HAUSMANN and STOCKEM 1973) and others (*e.g.*, *Trichomonas;* NIELSON 1974) cytochemically demonstrable acid phosphatase is present in sacs of the Golgi apparatus, which could reflect participation of the apparatus in lysosome formation, perhaps via the budding off of primary lysosomes. In the ciliate *Campanella* (GOLDFISCHER *et al.* 1963) there is little reason to suspect substantial involvement of the Golgi apparatus; acid phosphatase is demonstrable in the endoplasmic reticulum and hydrolases might move directly from the reticulum to nutrient-containing vacuoles, perhaps by transport in small vesicles. For *Campanella* (CARASSO *et al.* 1964) and many other organisms (*e.g.*, FAVARD and CARASSO 1963, HAUSMANN *et al.* 1972) it is believed that digestion takes place not only in the large food vacuoles but also in the smaller vesicles that separate from them; the latter may obtain their hydrolases, along with the rest of their contents, from the parent vacuole but the possibility that they also receive hydrolases subsequent to separation warrants consideration.

Metazoan intestinal cells do show extensive endocytosis, and some investigators believe that their lysosomes participate to an extent in completing digestion initiated in the intestinal lumen (HSU and TAPPEL 1964). But most higher metazoans probably do not rely extensively on heterophagy for primary degradation of foodstuffs (see KITCHING 1956, PARSONS and BOYD 1972, TIFFON 1971, TIFFON *et al.* 1973 for studies on lower metazoa which may represent evolutionary intermediates in these regards and see also JENNINGS 1957, and MCLEAR and HOLLAND 1973 for cases in which somewhat higher metazoa, such as gastropods and flatworms, seemingly utilize heterophagy in their digestive systems). Despite this seeming "downgrading" of nutritional roles, virtually all nucleated animal cells are capable of pinocytosis, and some are quite active phagocytes. The materials internalized by these endocytic processes usually wind up in lysosomes although there are a few exceptions such as the vesicle-mediated transport of macromolecules across capillary endothelia (Section IV.5.3). Endocytosis in higher organisms has been made use of in a number of ways. Phagocytosis and pinocytosis are central to defense, "scavenging" and turnover mechanisms in which extracellular materials come to be degraded within the lysosomes of cells such as macrophages, PMN leukocytes and hepatocytes (Sections III.1, III.2, IV.2). Some specialized cell types, such as oocytes, or the cells of thyroid follicles, employ specific endocytic steps in the accumulation or processing of characteristic products such as thyroid hormone (Section V.2.1) and yolk (Section V.3). But, the ubiquity of heterophagic capacities is as yet unexplained; some hypotheses suggest a link to turnover of surface membranes (Sections II.4.2.2, III.4.3.3).

The type of endocytosis that is virtually ubiquitous is the formation of small vesicles (diameter on the order of 100 nm) by budding from the cell surface. This sometimes is referred to as "micropinocytosis". Its widespread occurrence was established through electron microscopy especially when coupled with the use of macromolecular tracers such as ferritin. The vesicles involved often are coated (ROTH and PORTER 1964). Pinocytosis as originally described by LEWIS is restricted to a relatively few cell types, notably amebae, macrophages and tissue culture cells; these are found to form fluid-filled vacuoles, large enough to be readily observed by light microscopy. The word "pinocytosis" was intended initially to convey the notion of "cell drinking" as distinct from phagocytosis, in which uptake of discrete particulate material is characteristic. But the force of this distinction was diminished by the recognition that pinocytosis can accomplish uptake of large macromolecules. There is merit for some considerations in distinguishing the processes of "micropinocytosis" and "pinocytosis" and when relevant to the subsequent text we will do so. However, in general we will use the term pinocytosis in the inclusive sense to cover both processes and to imply bulk uptake of material of molecular dimensions.

II.1.2. Specificity

It is important for the roles of lysosomes in the economies of cells and organisms that endocytosis is a somewhat selective process (see *e.g.*, CHAPMAN-ANDRESEN 1964, GORDON 1973, JACQUES 1969, RYSER *et al.* 1971). Phagocytes for example, do not merely move about at random and take up everything present in their vicinity. Rather, chemotactic and other "stimuli" (as well as barrier systems such as blood vessel walls in higher animals) control the migration of phagocytes, and batteries of cell-surface receptors strongly influence the specificities and the rates of uptake. In protozoa, the size, nature and motility of potential prey are among the factors that affect uptake (see KITCHING 1956 for some of the complexities). In higher vertebrates an elaborate immunological apparatus participates in determining what is to be phagocytosed—bacteria or other foreign particles react with antibodies which in turn are recognized by receptors at the surfaces of macrophages or PMN leukocytes. Such stimulation of phagocytosis by antibodies or other "opsonins" (roughly defined as molecules that promote phagocytosis upon combining with material to be taken up) exemplify chemical recognition mechanisms that are found in animals from protozoa on up, although these mechanisms vary greatly in detail (RABINOVITCH 1970, STOSSEL 1973, STUART 1973).

Pincytosis also can be partly selective. Molecules that adsorb to the cell surface can be incorporated in pinocytic vesicles at concentrations that are substantially elevated above their concentration in the bulk medium immediately adjacent to the cell (see *e.g.*, JACQUES 1969 for theoretical comparisons and *e.g.*, BOWERS and OLZEWSKI 1972, RYSER 1971, and STEINMAN and COHN 1972 for experimental data on uptake when such adsorption does or does not occur). The binding sites responsible for adsorption can be highly specific, as is true for receptors on leukocytes that bind antibodies or antigens. The chief candidates for highly selective sites are cell surface proteins (immunoglobulins for lymphocytes, and perhaps other glycoproteins for other cell types, such as hepatocytes). Or, binding can be relatively non-selective, such as the electrostatic attachment of charged macromolecules observed with protozoa and some other cells; ionizable groups in the polysaccharides of surface coats and perhaps in sialic acid and other components of surface glycolipids and glycoproteins are held responsible for this type of association. In higher organisms, specific circulating factors in addition to the immunoglobulins can bind to other macromolecules producing complexes that are rapidly endocytosed presumably due to "recognition" of the complexes by cell surfaces; this is true, for example of haptoglobin which binds with hemoglobin and of the α-macroglobulin which complex with proteases.

Furthermore, *rates* of pinocytosis are increased by the presence of "inducing" components in the medium. These vary for different cell types and range from salts and small molecules such as nucleotides, to polyelectrolytes and proteins [see reviews by CHAPMAN-ANDRESEN (1964) on amebae, by GORDON and COHN (1973) on macrophages and by RYSER *et al.* (1971) on some tissue culture cells]. Thus, for example pinocytosis by amebae is strongly influenced by the concentrations of Na^+ and Ca^{++} (BRANDT and FREEMAN 1967) and also is stimulated by cationic proteins (CHAPMAN-ANDRESEN 1964). Macrophages show increased pinocytosis in the presence of a number of anionic proteins and also when exposed to an immunoglobulin, occurring naturally in some sera, that binds to their plasma membranes (COHN and PARK 1967, COHN 1970 b). And, thyroid stimulating hormone engenders markedly enhanced endocytosis by thyroid epithelial cells (Section V.2.1).

II.1.3. The Transport of Acid Hydrolases

The mechanisms by which endocytosed materials are handled once inside the cell have been most thoroughly studied in the specialized vertebrate phagocytes, the macrophages and PMN leukocytes (see VAN FURTH 1970, GORDON and COHN 1973 and VERNON-ROBERTS 1972 for reviews). The latter cell type has proved especially favorable for study of hydrolase packaging.

II.1.3.1. The Formation of Primary Lysosomes in PMN Leukocytes

Since the early days of hematology it has been evident that polymorpho-nuclear neutrophilic leukocytes contain at least two distinct prominent classes of cytoplasmic granules (other than mitochondria) which differ from one another in staining characteristics. Further, at least in some species, such as the rabbit, two granule types can be differentiated on morphological grounds by electron microscopy (Figs. 12 *a* and *b*). From work with classical blood stains (*e.g.*, WRIGHT's stain) the names *azurophilic* granules and *specific* granules have come into wide use. The azurophilic granules clearly are lysosomes; they contain acid hydrolases demonstrable cytochemically (BAIN-TON and FARQUHAR 1968, DUNN *et al.* 1968, FARQUHAR *et al.* 1972, WETZEL *et al.* 1967) and biochemically (BAGGIOLINI *et al.* 1969). This type of granule also contains *myeloperoxidase*, an enzyme that may be quite important in bactericidal activities (Section IV.2). The specific granules are not lysosomes; they contain cytochemically demonstrable alkaline phosphatase (BAINTON and FARQUHAR 1968, WETZEL *et al.* 1957) and much of the cells' lysozyme (BAG-GIOLINI *et al.* 1969); some other proteins are also present but most of these are not yet well understood (they include cationic proteins that may have anti-bacterial effects [Section IV.2] and an iron-binding protein, lactoferrin; BAGGIOLINI *et al.* 1970). Some specific granules stain with cytochemical methods for acid phosphatase (BAINTON *et al.* 1971) but cytochemical and biochemical studies have made it clear that they lack most of the acid hydro-lases. Some lysozyme also is present in the azurophilic granules (BAGGIOLINI *et al.* 1969). The significance of this, and of the acid phosphatase in specific granules, is obscure; perhaps there is some "accidental" intermingling of granule components during packaging of the enzymes (see also Section V.2 where we discuss the presence of acid phosphatase in secretion granules).

During proper stages in the maturation of PMN leukocytes, one can demon-strate, by cytochemistry, the presence of characteristic granule enzymes in the endoplasmic reticulum (ER) (BAINTON 1972, BAINTON *et al.* 1971, DUNN *et al.* 1968). This strongly suggests that the granule contents are synthesized on ribosomes of the rough endoplasmic reticulum and then passed into the ER cisternae; apparently the route followed is similar to that of proteins destined

---

Fig. 12. Portions of developing polymorphonuclear leukocytes from rabbit bone marrow. *G* = designates Golgi sacs, *L* = mature azurophilic granules, *S* = mature specific granules, *I* = immature granules and *C* = centrioles. Note that the azurophilic granules are larger and more electron dense than are the specific granules. Fig. 12 *a* shows a progranulocyte at the stage when azurophilic granules are forming. The granules seem to form at the inner (concave) surface of the Golgi apparatus. At the arrow, a dilated region of a Golgi sac, resembling an immature granule, seems to be budding off. ×35,000. Fig. 12 *b* shows a myelocyte at the stage when specific granules are forming. The Golgi sacs in which dense material accumulates at this stage are at the outer (convex) surface of the apparatus (compare with 12 *a* noting the curvature of the Golgi elements and the locations of the centrioles). Thus, specific granules appear to form at the surface of the Golgi apparatus opposite to the one involved in formation of azurophilic granules. (From BAINTON, D. F., and M. G. FARQUHAR, 1966: J. Cell Biol. **28**, 277—301.)

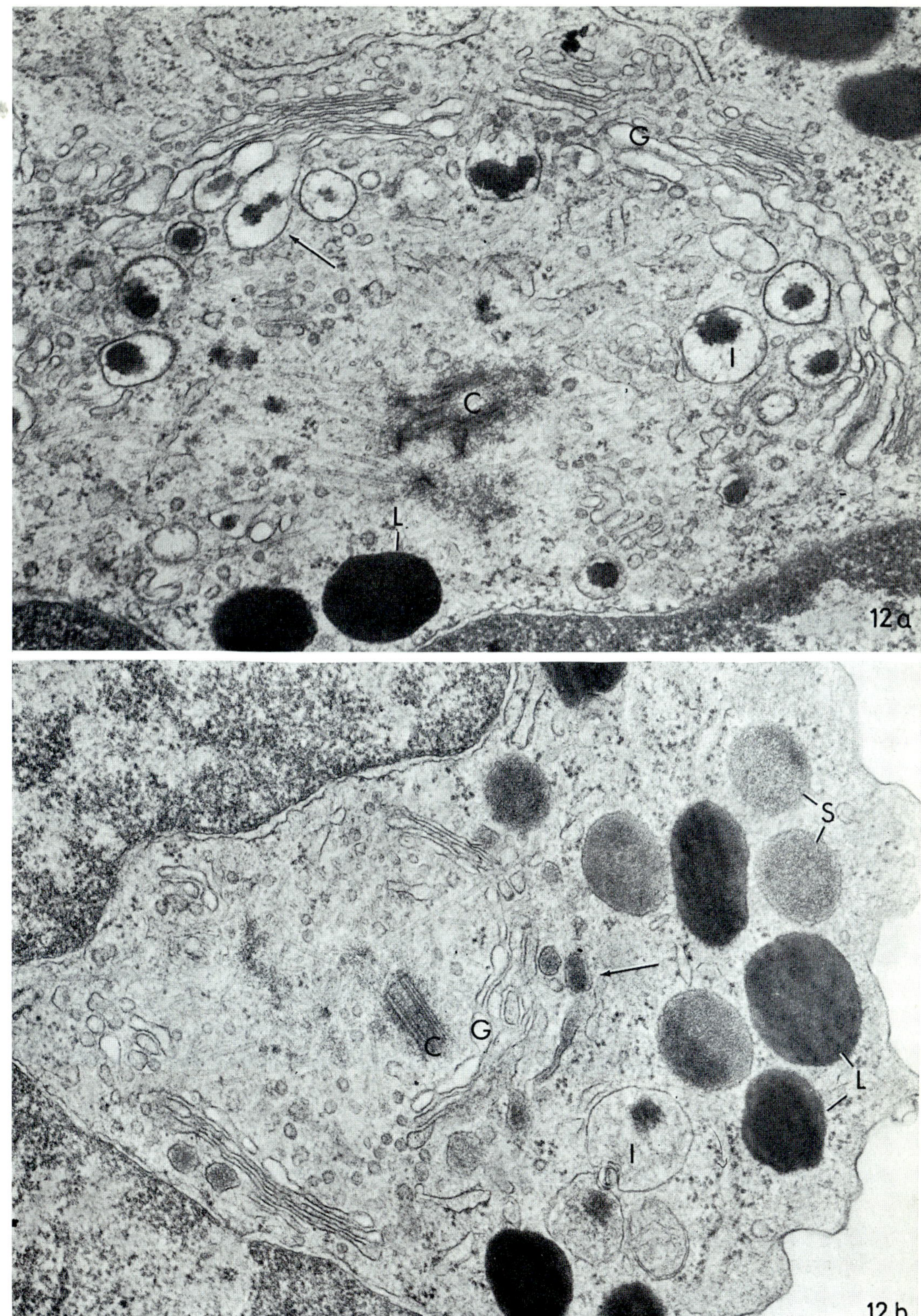

Figs. 12 *a* and *b*.

for secretion to extracellular spaces. Granule enzymes also are demonstrable cytochemically in sacs of the Golgi apparatus and one can readily observe configurations indicating that the membrane-delimited granules arise by separation of dilated "buds" from these sacs. Again, as is true for secretory proteins, concentration or "condensation" of granule contents appears to occur during the packaging process. Autoradiographic evidence confirming that proteins do move from the ER to lysosomes via the Golgi apparatus has been presented by FEDORKO and HIRSCH (1966).

Azurophilic granules form at an earlier stage in granulocyte maturation than do the specific granules (ACKERMAN 1968, BAINTON et al. 1971). This temporal separation apparently is accompanied by a difference in spatial aspects of packaging; the observations of BAINTON and FARQUHAR lead them to conclude that the sacs giving rise to azurophilic granules are located at the opposite surface of the Golgi apparatus from those that give rise to specific granules. As further developed in section II.2.1, both of these observations are of general interest to cell biologists, since they may shed light on the mechanisms by which cells in general are able to utilize the same basic machinery (endoplasmic reticulum and Golgi apparatus) to handle a variety of proteins destined to be included in distinct packages (lysosomes, secretion granules, peroxisomes and so forth).

We have already mentioned the proposal that some important non-enzymatic macromolecules such as polysaccharides may accompany the enzymes in PMN granules (Section I.3.2.3). The significance of these components is not established. But, they might bind to the hydrolases (e.g., MURATA et al. 1973) to produce multimolecular aggregates and this could contribute to maintaining the enzymes in inactive states, as well as to the formation of a highly concentrated storage structure (cf., the work by JAMIESON and PALADE 1971 a, on pancreatic secretion granules which seem to become concentrated through osmotic efflux of water attendant upon the aggregation of the proteins).

From electron microscopic studies of the cells or in work on sub-cellular fractions, additional granule types have sometimes been described as minor components of PMN leukocytes (WETZEL et al. 1967, MURATA and SPICER 1973). The status of such granules is quite uncertain. In one series of cell fractionation investigations it developed that a possible third PMN granule type actually was a contaminant derived from the small number of monocytes present in the cell population used as starting material for the fraction-

---

Fig. 13. *a* Rabbit peritoneal macrophage incubated to demonstrate aryl sulfatase activity. Reaction product is present in the endoplasmic reticulum (*er*) including portions of the nuclear envelope (*pc*). Product is also seen in small vesicles in the Golgi region (*G*). The insert at the lower left is from another cell and shows sulfatase activity in Golgi sacs (*GC*) and vesicles (*v*) as well as in a digestion vacuole (*dv*). *M* = indicates mitochondria and *nu* = part of a nucleus. Main figure ×20,000; insert ×37,000. *b* Promonocyte from human bone marrow incubated to demonstrate endogenous peroxidase activity. Reaction product is seen in azurophil granules (*a*), in endoplasmic reticulum (*er*) and in a sac of the Golgi apparatus (*G*; arrow). ×15,000. (From NICHOLS, B. A., D. F. BAINTON, and M. G. FARQUHAR, 1971: J. Cell Biol. **50**, 498—515.)

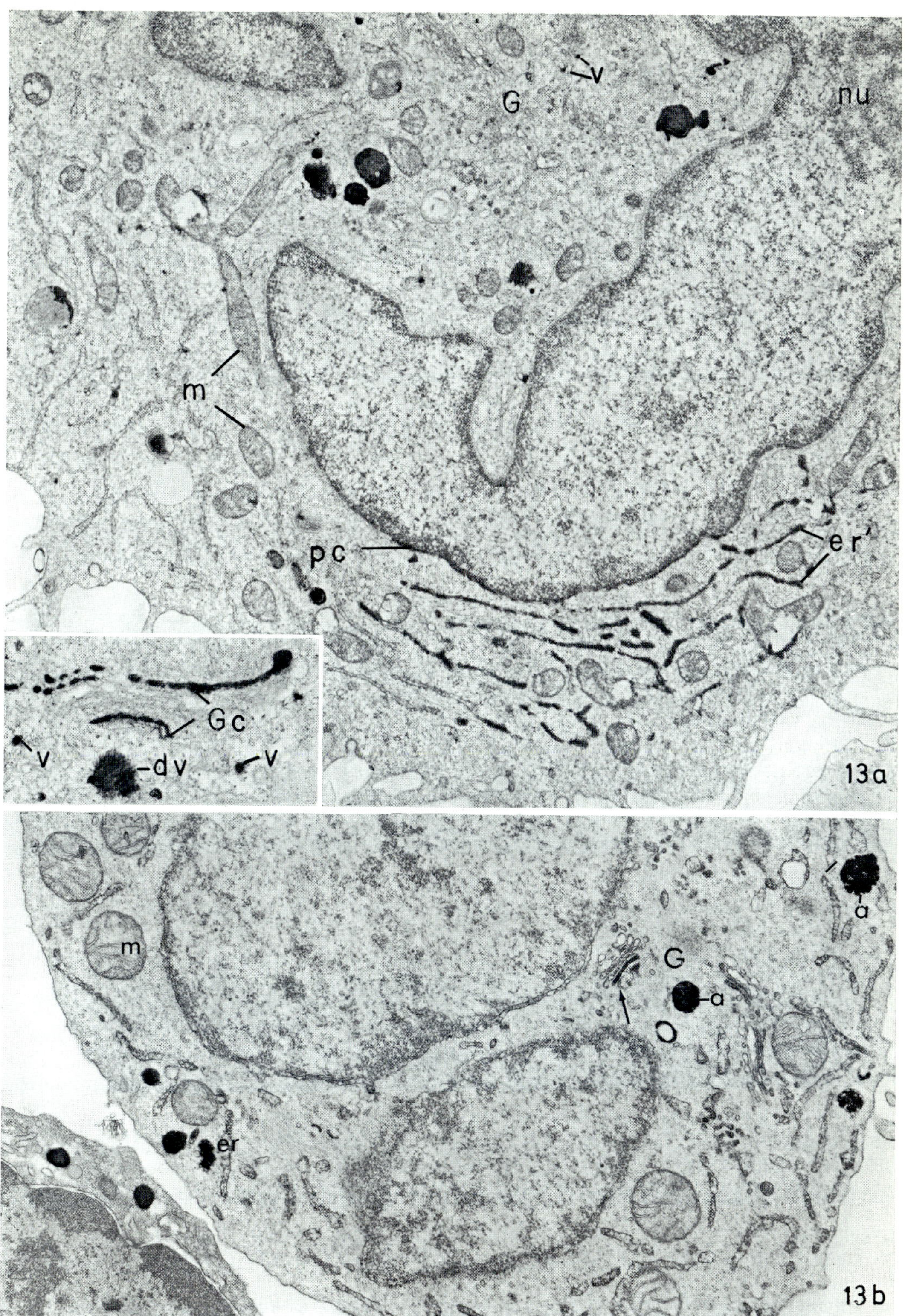

Figs. 13 *a* and *b*.

ation (Farquhar *et al.* 1972) Perhaps some of the suspected third granule types seen in the microscope represent residual bodies or variants of the major granule classes rather than special, distinctive entities. However, the possibility still is open, at least for some species, that there is more heterogeneity in granule types than is implied in the present discussion, which is based chiefly on rabbit material. For example Bretz and Baggiolini (1974) and Spitznagel *et al.* (1974) have recently shown that alkaline phophatase-containing structures of human PMN leukocytes are separable on density gradients from the specific granules rich in lysozyme (see also Section II.6). And a preliminary report of an autoradiographic study by Payne and Ackerman (1974) indicates that the "tertiary" granules distinguishable in human neutrophils on morphological grounds can incorporate radioactive sulfate as they form; thus the granules may contain newly synthesized mucopolysaccharides.

The distinctive granules of eosinophilic leukocytes contain acid hydrolases and a peroxidase that differs from the PMN myeloperoxidase (Archer and Hirsch 1963, Cotran and Litt 1969, Desser *et al.* 1972, Gessner *et al.* 1973, Parmley and Spicer 1974, Seeman and Palade 1967). The hydrolases appear to be packaged via ER and Golgi apparatus (Bainton and Farquhar 1970, Miller and Herzog 1969, Wetzel *et al.* 1967) and thus it is proper to regard them as a form of primary lysosome, quite comparable to the azurophilic granules of leukocytes.

II.1.3.2. Primary Lysosomes in Macrophages

Much remains uncertain about the primary lysosomes of macrophages, especially since purified fractions have not yet been obtained. However, recent cytochemical findings do suggest that the situation may be similar in its essentials to that just described for PMN leukocytes. Monocytic precursors of macrophages show acid hydrolases in their ER and Golgi apparatus (Fig. 13), and as monocytes differentiate they form a set of azurophilic granules (lysosomes) with moderately dense content (Cohn 1970 a, Hirsch and Fedorko 1970, Nichols *et al.* 1971). Once these bodies are used for digestion they are not replaced as such. Rather, the mature macrophage seems to utilize other small vesicles formed from the Golgi apparatus as primary lysosomes (Fig. 13): some of these are coated (Nichols *et al.* 1971). (In contrast to macrophages, PMN leukocytes, which are much shorter lived, do not replenish their store of cytoplasmic granules and generally do not survive very long after the granules have been depleted.) Actively endocytosing macrophages have been used to demonstrate through autoradiography, that some newly synthesized proteins follow the expected route from ribosomes to lysosomes—that is labelled amino acids move from ER to Golgi region and thence to the (secondary) lysosomes (Cohn and Fedorko 1969).

Cohn and his co-workers have determined that the rate of synthesis of acid hydrolases by macrophages increases when the cells are actively endocytosing (Axline and Cohn 1970, Cohn 1970 a). Interestingly, endocytosis of non-digestible materials such as latex is not effective in promoting hydrolase synthesis, suggesting that digestion products may mediate the effect.

Ricketts (1971) has made similar observations on *Tetrahymena* but Lagunoff (1964) failed to find much stimulation of hydrolase synthesis by endocytosis of proteins in amebae. Beyond such findings and some general information (*e.g.*, about altered hydrolase levels in tissues undergoing pathological
developmental or other dramatic changes) there is little known of the controls
of lysosomal enzyme synthesis in any cell type. Pesanti and Axline (1975)
report that the endocytosis-induced hydrolase increase in macrophages is
inhibited by colchicine, but why this is so is not clear.

In some species (*e.g.*, mice, humans) monocytes in marrow are reported to
include a peroxidase in the primary lysosomes they form (Fig. 13), but as
the cells mature into macrophages, they apparently cease elaborating the
enzyme (Dunn *et al.* 1968 b, Hirsch and Fedorko 1970, Nichols and Bainton 1973, Nichols *et al.* 1971, van Furth *et al.* 1970). However peroxidase activity is detectable cytochemically in the ER and Golgi apparatus of
resident peritoneal macrophages in guinea pig (Cotran and Litt 1970,
Daems and Brederoo 1973) and of rat liver Kupffer cells (Fahimi 1970,
Widmann *et al.* 1972) although normally there is not much activity found
in the lysosomes of these cells. (It should be borne in mind that unlike the
granulocytes of blood, most cell types seem not to have peroxidases in their
lysosomes though non-enzymatic peroxidatic activities are often seen by
cytochemistry. Peroxidases are found in the ER in several cell types.)

Neither for macrophages, nor for PMN leukocytes, is much known of the
details of hydrolase transport from the ER to the Golgi region; see Section
II.2.1 for general discussion, and the footnote on p. 56 for some pertinent
preliminary findings.

II.1.4. Fusion Phenomena, Lysosome "Reuse" and Lysosome Membranes

In macrophages and in PMN and eosinophilic leukocytes, lysosomes fuse
rapidly with incoming endocytic vacuoles. When studied in living granulocytes during phagocytosis this process can produce an impression of rapid
degranulation of the cells. Fusion may actually commence before phagocytic
vacuoles have completed their separation from the cell surface; as will be
seen, the consequent leakage of lysosome contents to extracellular spaces may
have profound implications (Section IV.3).

When viewed by phase-contrast microscopy of living cultures, macrophages
illustrate nicely the phenomenon of movement of pinocytosed materials from
the cell surface toward the region of the Golgi apparatus. During this movement fusion of pinocytic vacuoles with lysosomes takes place. That comparable migration to the Golgi region occurs in pinocytosis by many cell
types, can be shown through use of macromolecular tracers, which permit
study at the level of "micropinocytosis".

II.1.4.1. Lysosome "Recycling"

There is increasingly good reason to believe that not only primary lysosomes, but also some secondary lysosomes can fuse with structures formed

through endocytosis. A demonstration of this is illustrated in Fig. 14. GORDON
and his co-workers (1965) permitted cultured mouse fibroblasts to take up
a colloidal iron suspension which became sequestered in secondary lysosomes;
after several hours these evolved into fairly compact residual structures containing quantities of iron in a form recognizable by electron microscopy.

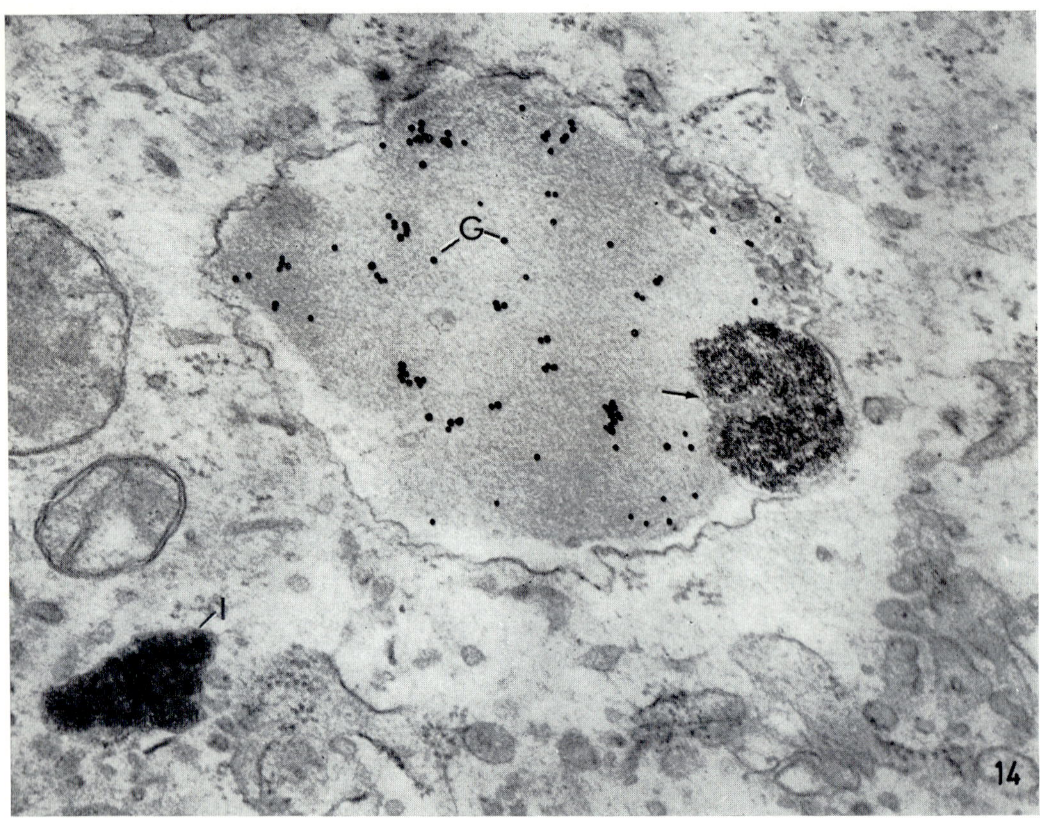

Fig. 14. From an L-strain cultured fibroblast exposed to saccharated iron for two hours,
then incubated without tracer for two hours and then exposed to a suspension of DNA-
colloidal gold particles. *I* indicates an iron-containing body, probably a secondary lysosome.
At the arrow a similar body seems to have fused with an endocytic vacuole that contains
gold particles (*G*). ×50,000. (From GORDON, G. B., L. M. MILLER, and K. G. BENSCH,
1965: J. Cell Biol. **25**, 41—55.)

After removing the iron-containing medium, the cells were exposed to another
distinctive-looking tracer, colloidal gold, which also is taken up via endocytosis. Both gold and iron were soon present within the same intracellular
vacuoles, implying that some "old" iron-containing secondary lysosomes had
fused with the "new" gold-containing vacuoles.

It is not known whether enzymes that remain active in residual bodies are
quantitatively important in subsequent rounds of degradation in which they
might participate as a result of such recycling. It is also unclear whether in

usual circumstances lysosomes eventually reach a point at which they no longer can fuse with newer structures (Sections II.5.1, II.5.2).

II.1.4.2. Energetics and Control of Movements in Heterophagy

A. Energy use in heterophagy: Heterophagic processes seem to require the expenditure of metabolic energy. The lysosomal hydrolases themselves are not dependent upon ATP or other sources of chemical energy. But low temperature, or inhibitors of glycolysis (*e.g.*, iodoacetic acid, fluoride) or of respiration (dinitrophenol, azide, cyanide, N_2-atmosphere) can markedly inhibit endocytosis. The inhibition occurs at some unknown step or steps subsequent to adsorption of materials to cell surface receptors since such adsorption is insensitive to the inhibitors and is minimally affected by low temperature. Interestingly, the inhibitor studies suggest that glycolysis can supply the energy for endocytosis by some cells, such as phagocytosis by PMN leucocytes or by macrophages isolated from the peritoneal cavity (COHN 1970 b, KARNOVSKY *et al.* 1970) while in other cell types such as alveolar macrophages (KARNOVSKY *et al.* 1970) and perhaps kidney tubule cells (MILLER *et al.* 1965) endocytosis seems to depend more heavily on oxidative metabolism. STEINMAN *et al.* (1974) found both glycolytic and respiratory inhibitors to have some effect on pinocytosis by cultured fibroblasts, but it was necessary to use both types of inhibitors to depress pinocytosis (and ATP stores) to very low levels.

With peritoneal macrophages, respiratory inhibitors that do not greatly affect phagocytosis can substantially inhibit pinocytosis of the type visible by light microscopy (COHN 1970 b, GORDON and COHN 1973, KARNOVSKY *et al.* 1970). However, even when uptake of particles larger than 0.1 micron is suppressed (by use of low temperature or very high concentrations of glycolytic or respiratory inhibitors) micropinocytosis reportedly continues (CARR 1973, CASLEY-SMITH 1964, 1969) although quantitative measurements of the uptake of a soluble protein (horseradish peroxidase) indicate that the amounts of material endocytosed by macrophages at low temperature or in the presence of fluoride are quite small (STEINMAN and COHN 1972).

This sort of information cannot yet be adequately interpreted. In very few cases has there been thorough direct determination of the actual effects of inhibitors on ATP levels or on other aspects of energy metabolism. The inhibitors may have side effects, and in any event, interruption of energy metabolism would be expected to affect a broad range of cellular processes, which greatly complicates distinguishing primary causes from secondary phenomena. However, the findings do hint at possible cellular abilities to compartmentalize or channel use of energy when ATP levels are low or when "competing demands" exist. Eventually they may provide clues to the still mysterious sequence of events that lead to the internalization of a portion of the cell surface in the form of an endocytic structure. As summarized in the next few paragraphs, one can think of plausible steps in the alteration of surfaces, or of events in the subjacent cytoplasm, that might depend upon energy and that might vary with the size or nature of the material being taken up. Experimental approaches are being developed that may soon provide crucial leads. For example, when divalent antibodies bind to the membranes of lymphocytes, or plants lectins such as concanavalin A bind to PMN leukocytes, the antibodies or lectins eventually are endocytosed. But this occurs subsequent to an extensive rearrangement of the cell surface that collects the bound molecules at one cell pole. The latter movements do not occur at low temperature or in the presence of inhibitors such as iodoacetic acid (*e.g.*, DE PETIS and RAFF 1973, RYAN *et al.* 1974; see also Section III.4.3.3 for some related behavior of plasma membrane molecules that may turn out to be germane).

JAMIESON and PALADE (1968) have shown for pancreatic acinar cells, that inhibitors of oxidative metabolism interupt the vesicle-mediated transport of proteins from the ER to the secretion granules forming in the Golgi region, and also the exocytic release of secretion (*exocytosis* is the fusion of the membranes surrounding secretion granules with the plasma membrane). By analogy, it would not be surprising if future studies showed that aspects of hydrolase packaging or processes of membrane fusion or budding during heterophagy require an energy input (see *e.g.*, BERTINI 1969). Another important focus for further investigation is on the replacement or recycling of cell surface membrane or receptors during endocytosis (Section III.4.3.3). WERB and COHN (1972 b) note that when macrophages phagocytose latex particles, the cells show a marked decline in particle uptake after a few hours, but they can subsequently resume rapid phagocytosis apparently after replenishing their surface membrane (see also GOLDMAN 1974 b).

B. Movement of Heterophagic Structures: Movement of endocytic structures and of lysosomes within cells might be expected to involve some input of metabolic energy, but there is only a little pertinent information (*e.g.*, in macrophages, movement of pinocytic vacuoles toward the Golgi apparatus is much less sensitive to inhibitors of energy metabolism than is endocytosis; GORDON and COHN 1973). As is usual for intracellular motion, microtubules and microfilaments are the prime suspects as potential participants in generating or orienting the motion (FREED and LEBOWITZ 1970, REBHUN 1972, SILVERBLATT *et al.* 1974). Oriented arrays of tubules and filaments are present near the surfaces of cells such as macrophages, (see *e.g.*, REAVEN and AXLINE 1973). STOSSEL and POLLARD (1973) have found a myosin-like protein in leukocytes, and it has been known for some time that microfilaments in a wide variety of cell types can bind heavy meromysin, presumably because they are composed of proteins resembling actin (see *e.g.*, LAZARIDES and WEBER 1974 for recent immunohistochemical conformation). Given the roles of actin and myosin in muscle fibers, the presumption is widespread that these proteins somehow participate in generating cellular and intracellular movements for many cell types. It may soon be possible to interpret in such terms, the close associations that are sometimes seen between, for example, endocytic structures, and microtubules or arrays of microfilaments (*e.g.*, ALLEN 1975).

One set of still unexplained observations that might reward more intensive follow-up are the occasional demonstrations within lysosomes of impressive oriented groups of structures with the morphology of microtubules (BOLEY and ARHEIGER 1966, JOURNEY 1964). Some of these arrays lie just below the lysosome surface.

The customary agents employed to affect microtubules (colchicine, vinblastine) or microfilaments (cytochalasin B) have at least some of the effects upon heterophagic processes that might be anticipated. For example colchicine inhibits movement of endocytic vesicles toward the Golgi apparatus of macrophages (BHISEY and FREED 1971) and colchicine and cytochalasin B can alter the movement of ligands on lymphocyte and leukocyte cell surfaces referred to above (the influences are complex and are still being debated; see RYAN et al. 1974 for references and discussion). ROBBINS and GONATAS (1964 a, b) report that in dividing HeLa cells, lysosomes are found in clusters around the cell periphery, whereas in interphase they are grouped near the nucleus; exposure of interphase cells to colchicine leads to a lysosomal distribution similar to that in dividing cells. (In dividing hepatocytes of regenerating liver, lysosomes leave their pericanalicular distribution and cluster near the spindle poles with roughly equal numbers entering the two daughter cells [KENT et al. 1965]. The spindle of course, is largely of microtubules.) Colchicine may inhibit PMN granule fusions (Ann. N.Y. Acad. Sci. **253**, 738).

As always in studies with such agents one must be cautious, for example about direct action of cytochalasin B or colchicine on membrane transport systems unrelated to microfilaments or tubules (AXLINE and REAVAN 1974, CZECH et al. 1973). Also, when microtubule disruption alters an aspect of cellular motion it is often impossible to decide whether this means that the tubules themselves generate the motion, whether other structures bring about the motion but microtubules orient it or whether the overall disarrangement of cell organization resulting from interference with the architectural or cytoskeletal roles of the tubules is exerting an indirect influence.

According to descriptions from several groups, cytochalasin B prevents cell movement and endocytosis by PMN leukocytes and some other cell types (ALLISON 1974, AXLINE and REAVAN 1974, DAVIS et al. 1971, MALAWISTA et al. 1971, WEISSMANN et al. 1973, 1974). Interestingly, phagocytosis of red blood cells or bacteria by macrophages can be substantially inhibited by levels of cytochalasin B that apparently have much less effect on pinocytic uptake of proteins or other macromolecules by the same cells (KLAUS et al. 1973, WILLS et al. 1972). ALLISON (1974) points out that such effects may actually reflect dissociation of "micropinocytosis" from larger scale processes. If this is the case, perhaps the dissociation will provide a means for distinguishing localized membrane events of endocytosis from more elaborate processes responsible for reorganization of large regions of cell surface or translocation of vesicles and vacuoles. The analysis of phagocytosis by SILVERSTEIN and co-workers (1975) suggests that these distinctions may be experimentally approachable from other angles; they propose that the process by which a macrophage engulfs a particle depends on the formation of bonds between the macrophage surface and the particle's surface with such bond-formation proceeding progressively around the particle in a manner reminiscent of a zipper. Phenomena such as antigen-antibody interactions could account for the surface binding. But, at least at first glance, it seems unlikely that the movement of cytoplasm around the forming vacuole would follow simply as a passive concomitant of such binding. Perhaps it will prove possible to perturb, in selective fashion, different aspects of vacuole formation.

When endocytosis is inhibited through use of cytochalasin B, one can induce lysosomes to fuse with the plasma membrane by exposing the cells to materials that normally would be taken up. Thus, cytochalasin-treated PMN leukocytes were found by ZURIER *et al.* (1973 a, b) to release hydrolases to the extracellular medium when the cells are exposed to zymosan, a preparation of yeast cell walls (see also DAVIES *et al.* 1973 b and SKOSEY *et al.* 1974 for similar studies). The release of granule contents is sensitive to colchicine. In addition, it requires Ca^{++}, is accompanied by a rise in intracellular concentrations of cyclic GMP and is inhibited by cyclic AMP, and by agents such as prostaglandin E_1 or epinephrine that are thought to operate via cyclic AMP (IGNARRO and GEORGE 1974, ZURIER *et al.* 1973 a, b). Exposure to cyclic AMP also reduces rates of phagocytosis by PMN leukocytes (COX and KARNOVSKY 1973) although SEYBARTH *et al.* (1974) report that little change in cAMP levels is observable in active phagocytes. [ZABUCCHI *et al.* (1975) report that Ca^{++} plus a divalent cation ionophore induce discharge of enzymes from human PMN cells not treated with cytochalasin B.]

From such findings, WEISSMANN and his co-workers have developed the hypothesis that microtubules are the agents responsible for bringing together the lysosomes and the structures with which they will fuse (internalized plasma membrane in normally phagocytosing cells, surface membrane in cytochalasin-treated cells; ZURIER. For contrary facts: J. Exp. Med. **142**, 903). Protein kinases responding to cyclic nucleotides might control the assembly of the microtubules. ALLISON (1974) has presented the complementary theory that microfilaments provide a physical barrier to fusion of lysosomes and other membrane systems; disruption of the filament barrier would permit membranes to approach one another and such disruption might occur experimentally via cytochalsin B effects or normally, through Ca^{++}-induced contraction of filaments. One can also speculate about more direct roles of actin-microfilaments in moving other structures.

II.1.4.3. Lysosome Membranes; Fusion

Fusions involving lysosomes must be selective in some sense. For example, lysosomes merge readily with structures of endocytosis and autophagy but they seem rarely, if ever, to fuse with nuclei, mitochondria or peroxisomes. Correspondingly, endocytosis vesicles or vacuoles fuse only with a restricted set of intra-cellular organelles, chiefly lysosomes. The basis for these specificities is not known. Nor is it understood, for example why actively endocytosing cells may accumulate very large numbers of vacuoles, which lie side by side without fusing (Fig. 5), as though constrained by some controls related to size (BUCKLEY 1972). Significant clues as to mechanisms that might eventually prove relevant come from studies on exocytotic release of secretions; both for gland cells and for neurons, Ca^{++} ions seem to play a key role in exocytosis, cyclic nucleotides may also be involved and in morphological terms, membrane fusion seems to occur by straight-forward processes of approximation of surfaces and sequential alterations in the layers of which membranes are composed (*e.g.*, DOUGLAS *et al.* 1971, PALADE 1972). Special

arrangements of the particles visible in membranes with freeze-etching techniques have also been reported to characterize initial stages of exocytosis
(SATIR *et al.* 1973).

At present, however we lack even an adequate static picture of the membranes delimiting lysosomes. There have been studies of sublysosomal fractions referred to as "lysosome membranes", but it is not at all clear how most
such studies are to be interpreted. Usually they have been done with impure
fractions. Even when this is not the case, the structures studied generally are
secondary lysosomes, and many of these (*e.g.*, multivesicular and residual
bodies, and autophagic vacuoles) have membranes within them derived from
material being digested. When one removes "soluble" components from such
lysosomes what is left is a mixture of internal and external membranes (see
e.g., SHIBKO *et al.* 1965). One preparation in which this problem is minimized
are the phagocytic vacuoles of macrophages or leukocytes that have taken up
non-digestible particles such as latex (NACHMAN *et al.* 1971, STOSSEL *et al.*
1971, WERB and COHN 1972 b); these persist in the cells for prolonged periods,
and as indicated earlier the content of latex facilitates isolation.

Some progress is being made. For example, through use of antisera
obtained from animals injected with the appropriate cell fractions TROUET
(1969) has shown that membranes from lysosomes isolated by the Triton
WR 1339 method possess some "antigens" in common with the plasma membrane and some in common with other cellular membranes. But there appear
also to be a few "antigens" peculiar to the lysosomes. Triton-isolated hepatic
lysosomes also resemble the plasma membrane in their relatively high content
of sphingomyelin, cholesterol and saturated fatty acids and in the presence
of much sialic acid (GERSTEN *et al.* 1974, HENNING *et al.* 1973, MILSOM and
WYNN 1973, THINES-SEMPOUX 1973). Some data pointing in the same direction is available for rat liver lysosomes isolated without Triton (HENNING
and HEIDRICH 1974, THINES-SEMPOUX 1973). Presumably these similarities
reflect the incorporation of plasma membrane into the organelles during processes such as heterophagy. This certainly also is the most obvious explanation of the findings that phagocytic digestion vacuoles of macrophages are
bounded by membranes resembling the plasma membrane in composition.
(NACHMAN *et al.* 1971, WERB and COHN 1971, 1972 a, b.) On the other
hand, HENNING *et al.* (1973) maintain that some important details of membrane composition (such as the relative prominence of various gangliosides)
are sufficiently different in plasma membranes and hepatic "tritosomes" as
to argue for caution. Conceivably the membrane delimiting forming endocytic vesicles differs in its composition from other plasma-membrane regions
(Section III.4.3.3); analyses of plasma membrane fractions provides only
averages that might mask such heterogeneity. Or, perhaps extensive compositional modifications occur once an endocytic structure has fused with a
lysosome (Section II.1.4.4).

Work on primary lysosomes has been initiated with PMN leukocytes. One
recent finding that may illuminate interesting features of leukocyte functioning is that the membranes bounding specific granules differ from those
of azurophilic granules in the proteins that are present and in the relative

proportions of cholesterol and phospholipid (*e.g.,* azurophils have a higher proportion of cholesterol; NACHMAN *et al.* 1972). During phagocytosis the specific granules seem to fuse with phagocytic vacuoles at a slightly earlier time than do the azurophilic granules (BAINTON 1972, 1973). This might be due in part to differences in granule frequency resulting in different probabilities of encounters with incoming vacuoles (in mature rabbit PMN cells, there are roughly three times as many specific granules as azurophilic granules although the total *volumes* of the populations are similar). But BAINTON suggests that the predominance in frequency of specific granules still may not account for the thirty second to three minute lag before peroxidase (a marker enzyme for the azurophil granules) is demonstrable in phagocytic vacuoles (see the discussion following BAINTON 1972). One can construct speculative theories to explain the timing of fusions on the basis of specificities residing in the granule surfaces. And it may soon become possible to mount a direct experimental attack since WHITE and ESTENSEN (1974 a) and ESTENSEN *et al.* (1974) claim that when human PMN leukocytes are exposed to phorbol myristate acetate their specific granules, but not the azurophilic ones, release their contents from the cell. A preliminary communication by GOLDSTEIN *et al.* (1974) reports that Ca^{++} ions can induce a similar selective specific-granule discharge. (See also J. Cell Biol. **66**, 647 [1975].)

Several general hypotheses relating to lysosome fusions are under current consideration. For example, DINGLE (1968, 1969) has argued that differences in surface tension forces among different membranes or changes in local surface tension in membrane regions might help control fusions and budding. POSTE and ALLISON (1971) suggest that local changes in surface potential related to curvature and to interactions of inorganic ions and ATP with the membrane, affect fusions by influencing the closeness with which membranes can approach one another and through inducing molecular reorientation within membranes. Such "physical-chemical" hypotheses are sustained largely on the basis of theoretical considerations and of experiments with model systems, such as emulsions. Only very indirect evidence is presently available that might permit extrapolation to lysosomes and endocytic vesicles. POSTE and ALLISON (1971, 1973) cite the observations that the fusions involving PMN leukocyte granules require cellular energy and divalent cations (see above and WOODIN 1973, WOODIN and WIENEKE 1964) and they speculate that this is indicative of a cycle of association and dissociation of ATP and Ca^{++} with membranes. Agents such as vitamin A or cholesterol that can affect surface tensions, also have interesting influences on lysosome behavior but as we will see (Section IV.4.3) the analysis of their effects is still quite controversial.

DE DUVE and his co-workers (DE DUVE and WATTIAUX 1966) propose that membranes fuse on the basis of chemical "similarities"—these are not yet readily definable in detail but one can think in terms of preferential fusion of membranes of similar composition or of those bearing certain components in common. Membrane thickness might serve as a crude guide to similarities (DE DUVE 1969), since thickness of membranes as viewed in the electron microscope does reflect such features as the relative proportions of cholesterol (for examples and references see *e.g.*, Colbeau *et al.* 1971, Poste and Allison 1973, Werb and Cohn 1971). These features can be quite important for membrane properties (*e.g.*, artificial membranes rich in cholesterol show lower permeability to water than do those poor in cholesterol, due probably to differences in the internal viscosity of membrane lipid layers). It is commonly found that plasma membranes, the membranes bounding secondary lysosomes and some Golgi membranes are thicker (75–100 Å) and probably richer in cholesterol and sphingomyelins that are, for example, ER membranes of the same cells (50 Å). Such similarities make sense in terms of lysosome formation and functioning. On the other hand the plasma membranes of secretory cells and synaptic nerve terminals differ considerably in protein composition and other features from the membranes delimiting secretory structures (zymogen granules, synaptic vesicles etc.) that fuse with these plasma membranes during exocytosis. (Adequate data is not available on the specific portions of the plasma membranes where such fusions occur, so this is not a decisive test of the importance of membrane similarities for fusion, although it does suggest that the relations, if they exist, may be complex.)

Lucy (1969) has developed a more detailed set of hypotheses. He asserts that lysosomal surface membranes, and some other membranes that undergo fusion, are relatively rich in lysolecithin. This component could alter the bilayer structure of membranes to promote micelle formation which, in turn, might facilitate fusion of one membrane with another. This is an interesting idea with some support for model systems but its applicability to lysosomes is impossible to determine at present. Lysosomes do contain lysolecithin, but whether this is significant for their fusion, or merely reflects the presence of phospholipids and phospholipases within the organelles is still unresolved. Poste and Allison (1973) suggest that proteins or polysaccharides are better candidites for promoters of specificity and localization of fusion than are components such as lysolecithin. (For this, see also J. Cell Biol. **66**, 183.)

It is probably premature to expect success in induction of specific lysosomal fusions in preparations of isolated organelles although initial efforts to study fusions *in vitro* have been made. For example, when a crude preparation of lysosomes was incubated at 37 °C for several hours by Raz and Goldman (1974), the average diameter of the membrane-delimited structures in the suspension increased markedly. Apparently this increase reflects fusions among organelles, but in view of the release of enzymes into the medium (hence the possibility of artifactual changes in organelle surfaces) and other uncertain factors in the experiment, the results require a good deal more analysis before they can be interpreted in terms of *in vivo* events. Similar strictures apply to the report by Poznansky and Weglieki (1974) that lysophospholipids induce lysosome fusions *in vitro*.

Situations in which fusion of endocytosis structures with lysosomes is absent or atypical might be exploited to advantage for future studies of the problems under discussion in this section. Fusion is minimal or absent in transendothelial movement of macromolecules in capillaries and transepithelial movement of antibodies in some newborn animals (Section IV.5.3) and it is absent, reduced or modified in aspects of yolk formation (Section V.3), in infection of cells with certain microorganisms (Section IV.2.2) and in some cells exposed to lectins (Section IV.2.2).

II.1.4.4. Other Features of the Lysosome Surface: Enzymes, Changes and Stability

It is often reported that some hydrolases are bound to lysosome membranes in a manner stronger than might be expected for simple adsorption *in vivo* or for comparable "spurious" effects of isolation (see *e.g.*, AXLINE 1968, and SLOAT and ALLEN 1969 for acid phosphatase and LUCY 1969, TAPPEL 1969, TOUSTER 1973; for various other enzymes). Association of enzymes with surfaces can greatly influence their functioning; for example an enhancement of activity is seen when pancreatic lipase is bound to glass beads, perhaps due to substrate concentrating at the bead surfaces, or to conformational effects (BROCKMAN *et al.* 1973). Unfortunately, once again there is usually no strong evidence as to whether apparent hydrolase binding involves the lysosome surface as opposed to internal structures. Observations that cyto-chemical reaction product for an acid hydrolase may sometimes be restricted to a rim near the lysosome surface have occasionally been adduced as evidence for the concentration of the hydrolase at this location, but we have already mentioned the difficulties with such interpretations (Section I.3.2.1). Binding of hydrolases to intralysosomal membranes might simply reflect interactions of enzymes and substrates; particularly with lipid degradation, one might anticipate complexities in such interactions. But it is not certain whether this sort of effect can explain all of the differences that are observed between hydrolases associated with lysosome membranes and corresponding "free" intralysosomal forms of enzymes with similar specificities identifiable in the same preparations (see *e.g.*, SLOAT and ALLEN 1969 for differences in heat stabilities of acid phosphatases, and DE DUVE'S comments which follow this paper and in which he suggests that some such differences might be accounted for by changes in given enzymes engendered by the other hydrolases present in the same organelle).

Some authors (*e.g.*, LLOYD 1969) have speculated that the nonlatent enzyme activities demonstrable to some extent in even the most carefully isolated lysosome fractions might correspond to hydrolases built into the organelle surface and having direct access to substrates outside the particles. Since doing some damage is unavoidable during isolation it is difficult to see how one might obtain critical evidence to sustain, or rule out such a concept.

Enzymes other than acid hydrolases are sometimes reported to be components of the lysosome surface membrane. For example, 5'-nucleotidase is active in phagocytic digestion vacuoles of macrophages and in secondary lysosomes of hepatocytes (COFFEY and PLETSON 1971, WERB and COHN 1972 b, WIDNELL 1972). The situation is complicated by the existence of possible intracellular sources of 5'-nucleotidase (FARQUHAR *et al.* 1974, among other, have identified an enzyme of this type in the Golgi apparatus). And, KAULAN *et al.* (1970) claim that the enzyme in liver lysosomes differs in its responses to inhibitors from that at the plasma membrane. But at least in macrophages, the likelihood is strong that plasma membrane enzymes can remain active for some time subsequent to fusion of endocytic structures with lysosomes (WERB and COHN 1972; analogous considerations also may account for the presence of alkaline phosphatase in the lysosomes of intestinal epithelial cells whose brush border is rich in this enzyme; DAEMS *et al.* 1969).

The possibility that plasma membrane-derived enzymes might influence the character of heterophagic lysosomes is an intriguing one. There is tentative evidence for selectivity in endocytic enzyme-internalization; that is, the cell surface membrane regions that participate in endocytosis may be specialized, or at least lack some of the enzymes or other components present at the surface (see Section III.4.3.3). Thus lysosomes might acquire a selected set of surface macromolecules. Particular interest attaches to transport enzymes; these could be important for the passage of material into or out of the lysosomes. (Note that the geometry of endocytosis and fusion with lysosomes is such that a membrane surface originally facing the exterior of the cell, comes to face the interior of a lysosome, which could affect transport directions.) To account for some effects of ATP on biochemical events in lysosome fractions, it has been suggested that ATPases in the lysosome membrane pump ions into or out of the organelles; this might, for example, affect the intralysosomal pH (see Section II.1.4.5). Perhaps the Na^+—K^+ ATPase of cell surfaces sometimes persists in lysosomes (DAEMS *et al.* 1972); if the enzyme retained the same orientation it has in the plasma membrane, it could function to move Na into and K out of the organelles. Some ATPase activities are demonstrable in lysosome-enriched cell fractions (see *e.g.*, ALLISON 1968) but tritosomes contain very little or no detectable Na^+—K^+ ATPase (KAULEN *et al.* 1970, THINES-SEMPOUX 1973) and there is no present evidence definitively locating this or other possible transport enzymes at the lysosome surface. In a brief paper, SCHNEIDER (1974) reports that ATPase activities, demonstrable in hepatic tritosomes, are markedly enhanced by disruption (sonication) of the particles; he questions whether this is the behavior expected for a transport ATPase.

The only enzyme currently known to be associated with membranes of PMN granules is alkaline phosphatase, which in rabbit seems to be tightly bound in the surface membrane of the specific granule (BRETZ and BAGGIOLINI 1973, see also Section II.1.3.1 and BRETZ and BAGGIOLINI 1974, and ZEYA and SPITZNAGL 1974 for complications in other species).

We turn next to the somewhat puzzling fact that the lysosome surface remains intact despite the presence within the organelles of degradative enzymes that can digest membranes. Under very few, if any circumstances, do the hydrolases chew their way out into the cytoplasm. One can only conjecture as to the basis of this apparent resistance of the surface. Perhaps limited degradation does occur but it is not extensive enough to open a "hole"; this might result from the restriction that the hydrolases can attack only from one surface, coupled with the difficulties encountered by enzymes in entering organized lipid domains. Or, there might be some kind of repair or restitution process such that the dynamics of the lysosome surface prevent enzymatic attack from seriouly disrupting the structure. Alternatively barriers may be interposed between the hydrolases and the membrane. Notions of special matrices (Section I.3.2.3) are relevant here. Further, it is commonly observed, particularly for lysosomes of the "dense-body" type that a narrow band of relatively low electron density (a "halo"; *cf.*, DAEMS *et al.* 1969) separates the surface membrane from the interior (Fig. 7). It may eventuate that this is due to some special component or organization that protects the surface from the enzymes. TAPPEL (1969) suggests that some of the hydrolases themselves might be bound near the surface as a protective layer. Or perhaps the halo is rich in carbohydrates that can bind and "inactivate" hydrolases. [According to electron microscope studies by HENNING *et al.* (1973) colloidal iron "stains" chiefly the interior surface of the membranes delimiting isolated tritosomes; the sites to which the iron binds are neuraminidase sensitive and thus probably involve sialic acid residues.]

Finally, there are a few observations indicating that important characteristics of the surface membranes of endocytic vesicles and heterophagic lysosomes may change during the functioning of the organelles. To some extent this is obvious. Heterophagic digestion vacuoles undergo extensive modifications in volume, presumably reflecting, in part, alterations in osmotic properties associated with the progress of digestion and the departure of digestive products. Accompanying the volume changes are changes in surface area—in general, the vacuoles shrink and membrane is lost by budding of vesicles into the cytoplasm, as in protozoa (Section II.1.1) or by internalization of lysosome surface membrane within the lysosome itself, as is apparently the case with multivesicular bodies (Section II.4.1). In some protozoa, the reduction in area is preceded by a stage in which the vacuole membrane shows elaborate folding and thus presents an extensive surface area for exchanges with the cytoplasm (*e.g.*, ULLIG *et al.* 1965).

It is not known whether there are specificities in the loss of lysosome membrane. Secondary lysosomes probably often start out bounded by a mosaic of membrane regions; this would be the result, for example, of the merger of a newly formed endocytic vesicle with a primary or secondary lysosome. Perhaps as the lysosome evolves, differences in surface regions are obliterated through interdiffusion of membrane macromolecules (SINGER and ROTHFIELD 1973). Or perhaps the processes just described selectively remove one or another type of membrane. The fragmentary information available does not permit a decision between such alternatives. Nevertheless, it is interesting that proteins such as 5′-nucleotidase presumed to derive from the plasma membrane, persist in recognizable form at the surfaces of latex-loaded macrophage phagocytic vacuoles for only a few hours (NACHMAN *et al.* 1971, HUBBARD and COHN 1975, WERB and COHN 1972 b). It is thought that cell surface coats may also be modified or degraded fairly rapidly subsequent to their entry into lysosomes (EHRENREICH and COHN 1969, JACQUES 1969, STRAUS 1967).

From their autoradiographic studies, CHAPMAN-ANDRESEN and HOLTER (1964) have reported that exogenous radioactive glucose diffuses rapidly throughout the cytoplasm of amebae that are active in endocytosis, whereas ordinarily the ameba plasma membrane is impermeable to glucose. RASMUSSEN

(1973) and Hoffman *et al.* (1974) have shown that when *Tetrahymena* is induced to endocytose non-digestible particles, the organism requires lower concentrations of external sugars and nucleosides to sustain its growth. One plausible interpretation of such results is that phagocytic vacuoles carry soluble molecules into the cell and, probably by virtue of permissive permeability, release them into the cytoplasm. One might even imagine that the addition of lysosomes' membranes to an incoming phagocytic vacuole makes the resulting digestion vacuole more freely permeable, perhaps simply because the lysosomes' membranes are less restrictive than the plasma membrane. However, the findings require further analysis since, for example, Brandt and Freeman (1967) and Brandt and Hendil (1972) have demonstrated that when pinocytosis is induced in amebae there is an enhancement of the permeability of the plasma membrane itself to ions and small non-electrolytes.

Fromme (1968) among others has made the point that the membranes of phagocytic vacuoles in phagocytes that specialize in sequestering and degrading potentially injurious material (*e.g.*, macrophages and leukocytes) might differ from the vacuole membranes in cells such as protozoa that employ phagocytosis for nutrition—in the one case evolution may have operated to minimize permeability and maximize protection, and in the other, to maximize extraction of useable nutrients. In line with this, Smolen and Shohet (1974) assert that granulocyte phagocytic vacuole membranes are especially rich in saturated fatty acids which might promote a relatively low permeability. There may be the elements of a quite interesting story here since Elsbach *et al.* (1972) are convinced that PMN leukocyte vacuole membranes contain acylating enzymes which could participate in remodelling their phospholipids.

Another kind of membrane modification has been proposed by Zahlten *et al.* (1972) who found that treatment of rats with glucagon (a drug that engenders autophagy in hepatocytes) leads to P^{32} uptake into membranes that is particularly evident in the mitochondrial, lysosomal and microsomal fractions of liver. Both lipid phosphates and protein phosphates show the label, and since glucagon effects are mediated by cyclic AMP one might speculate that the nucleotide influences phosphorylation reactions that alter the membranes. How seriously such ideas should be taken must emerge from future work.

Modern tools of membrane research, such as fluorescent probes of molecular conformation have not yet been systematically applied to lysosomes (see *e.g.*, Howinger and Timmons 1973 for some first steps); eventually these may help answer some of the questions left unresolved in this and the previous section.

II.1.4.5. Acidification

The pH of intracellular structures is notoriously difficult to analyze (see *e.g.*, the reviews by Waddell and Bates 1969 and Wiggins 1969). This is true in terms both of techniques and of interpretations. There are few methods for accurately measuring hydrogen ion concentrations in structures as small as many organelles; and there may be so few free ions within objects of the sizes and compositions of mitochondria, lysosomes and so forth as to create difficulties in applying the usual definitions of pH that relate to concentration of free hydrogen ions per unit volume. Further, the concentration of diffusible ions is often quite different at surfaces than in the bulk medium bathing the surfaces: Thus when enzymes are bound to membranes or other structures, measuring the pH of the soluble phases of the system may give only an indirect indication of the state of affairs in the local environment of the enzymes (*e.g.*, LUCY 1969).

The only class of lysosomes whose internal pH has been studied to a considerable extent, are heterophagic secondary lysosomes. These are accessible to exogenous indicator dyes whose color chages can be monitored by light microscopy; most studies have been based on phagocytosis of particles such as yeast prestained with the dyes. Since adsorption of dyes to structures can affect their properties and the dyes themselves might also exert some buffering effects within the lysosomes, the results are of severely limited precision (see *e.g.*, the comments following the paper by MÜLLER *et al.* 1963).

Through use of dye methods it has long been known that the interior of phagocytic vacoles in metazoa and of protozoan food vacuoles becomes acid soon after the vacuoles form [see KITCHING (1956), DOGIEL (1965) for early work]. Subsequently, the pH may return to neutrality or even to slightly alkaline levels [this is reported for protozoa by several workers; see MÜLLER *et al.* (1963) and KITCHING (1956) for discussion and references]. But it seems clear that the interior of lysosomes does become acid enough to sustain high levels of activity by the lysosomal hydrolases, and that such acidity is maintained at least for many minutes. It has been pointed out for some protozoa that much of the microscopically obvious intralysosomal digestion (and much of the accumulation of cytochemically demonstrable hydrolase activity) occurs when food vacuoles apparently have returned or are returning to neutral or alkaline pH's. If the dye methods that suggest this prove reliable, it raises questions about the course of degradation of phagocytosed material. It is hard to believe that most of the digestion depends on operation of hydrolases at very non-optimal pH's (*cf.*, MÜLLER *et al.* 1966) although it should be noted, for example that once cathepsin D has acted on proteins, further degradation by other acid proteases is facilitated even at pH's fairly far from their optima (GOETTLICH-REIMANN *et al.* 1971).

For PMN leukocytes, indicator dye studies, using stained yeast, suggest that there is a delay of a few minutes between phagocytosis and strong acidification of the interior of the phagocytic vacuole (JENSEN and BAINTON 1973). This correlates nicely with the fact mentioned earlier that components of specific granules seem to accumulate in phagocytic vacuoles earlier than do the enzymes of azurophilic granules. Presumably the initial period of pH's near neutrality promotes activity of the specific granule enzymes, and then as the pH drops the acid hydrolases come into play.

How low does the pH in lysosomes become? From studies with indicator dyes on various cell types some report figures as low as 1–3 (KITCHING 1956) and others, lower limits as high as 6–6.5 (MANDELL 1970). Figures in the range of 3–5 (*e.g.*, HIRSCH 1972, JENSEN 1973, KITCHING 1956, SPRICK 1956) are fairly common. To a degree these variations probably reflect differences among cells and in technique or timing of measurements but future studies might profitably be concerned with determinations of the extent to which the intralysosomal pH in a given cell type varies with different phagocytosed materials and under different conditions of cellular metabolism. In any event, if the pH does become extremely low, this might directly facilitate digestion by helping kill phagocytosed organisms and by promoting denaturation or other alterations in potential substrates (see *e.g.*, VAES 1973).

An interesting technique for measuring the pH within isolated lysosomes has been used by REIJNGOUD and TAGER (1973; see also GOLDMAN and ROTTENBERG 1973). They studied the penetration of the weak base, methylamine into triton-isolated liver lysosomes. The uncharged form of this compound should penetrate membranes far more readily than the charged form. Since the equilibrium between charged and uncharged species depends on pH, methylamine will distribute between compartments at concentrations that reflect the relative pH's. The extent to which it accumulates in tritosomes suspended in a medium of pH 7.5 suggests that the interior of the organelles is slightly acid (pH 6.5). A similar line of reasoning can be applied to the accumulation in lysosomes within living cells of vital dyes such as neutral red or acridine orange, and of other weak bases such as the drug, chloroquine (DE DUVE 1969). These compounds also may cross the lysosome membranes in uncharged form and then be "trapped" inside in the relatively acid environment; weak acids will assume the opposite distribution. (There is no necessary contradiction between this picture of the accumulation of vital dyes and the notion that the dyes bind to intralysosomal macromolecules; both could take place.)

How is a low pH achieved within lysosomes and how are possible osmotic effects of the ionic imbalances that may be implied by the pH differences across lysosome membranes dealt with? Several hypotheses have been put forth but none is universally accepted. Broadly speaking they fall into two categories—those that propose special metabolic inputs to drive acidification, and those that do not.

Donnan equilibrium effects occur whenever compartments separated by semi-permeable membranes contain non-diffusible charged groups and diffusible ions; these effects can generate ionic gradients, including pH gradients, across membranes (see *e.g.*, COFFEY and DE DUVE 1968, DE DUVE and WATTIAUX 1966, GOLDMAN and ROTTENBERG 1973, HENNING *et al.* 1973). Since charged macromolecules are present within lysosomes (Section I.3.2.3) one would expect Donnan effects to influence the distribution of diffusible ions between the organelle interior and the surrounding cytoplasm or suspending medium. The work on distribution of methylamine just cited is evidence that this is the case, since it indicates that metabolically quiescent isolated lysosomes contain diffusible ions (charged forms of methylamine molecules, and presumably protons) at concentrations different from the surrounding medium.

HENNING *et al.* (1973) stress that many sialic acid residues originally present at the cell surface may be added to the interior of heterophagic vacuoles as

4*

a consequence of endocytosis, and components such as the "acid mucosubstances" of PMN leukocyte granules (Section I.3.2.3) also enter newly formed digestion vacuoles. With suitable subsidiary proposals to explain the timing, and some thinking about the changes in concentrations that would accompany the alterations in vacuole volume during digestion (KITCHING 1956) one can construct schemes by which influxes of charged macromolecules could control (or at least strongly influence) the vacuole acidity.

If the pH inside lysosomes is maintained for prolonged periods at levels as low as 3 or 5 even in the face of substantial entry of potentially neutralizing components such as the weak bases, it appears unlikely that Donnan effects could be exclusively responsible—the gradients seem too steep, although we do not have really adequate information about the concentrations of relevant macromolecules and about the "buffering" characteristics of the lysosome interior to be certain about this (DE DUVE et al. 1974). Among other factors that might contribute to acidification, one interesting possibility is the release of acidic groups (e.g., phosphates, carboxyls) through the enzymatic activities of the hydrolases (see e.g., LUCY 1969). Studies on model systems have demonstrated that hydrolytic enzymes acting upon their substrates can affect the pH of compartments in which they are located. For example, when trypsin plus suitable substrates are incorporated in polyacrylamide spheres along with glucose oxidase, the "apparent pH optimum" of the oxidase (the pH of the external medium needed to sustain maximal activity of the enzyme in the sphere) rises, as expected if protons "liberated" by trypsin tend to drive down the pH in the immediate vicinity of the oxidase (GASTRELIUS et al. 1973). The effects are on the order of tenths of pH units, but perhaps the simultaneous operation of many hydrolases could be more dramatic.

According to ROBBINS et al. (1964) the accumulation of acridine orange in HeLa cell lysosomes requires cellular energy. And, HAWKINS et al. (1972) have found that acridine orange, after having concentrated in the lysosomes of tissue culture cells, diffuses out very rapidly when the cells are exposed to iodoacetic acid or to cyanide. Such observations might constitute indirect evidence that the maintenance of low intralysosomal pH is based on active participation by the cell. (On the other hand, once macrophage lysosomes have acquired endocytosed molecules, the degradation of these molecules, which one would expect to vary with pH, is not very sensitive to inhibitors of energy metabolism; PARKS and COHN 1973.) Studies on the effects of ATP on isolated lysosomes have led MEGO and his co-workers to propose that the lysosome membrane contains an energy dependent proton pump involved in acidification of the interior. As indicated in section I.3.1.3, lysosomes isolated after intravenous administration of proteins, can continue degradation if suspended in a suitable medium. MEGO (1973 a) has shown that such degradation is responsive to the pH of the suspending medium, as expected since the lysosome surface is permeable to ions to at least some extent. Mildly alkaline pH's inhibit degradation. The finding that this inhibition is reversed if ATP is added to the medium is central to Mego's analysis; the ATP supposedly permits the lysosome to create pH gradients.

The interpretation of the results just summarized is clouded. For example, REIJNGOUD and TAGER (personal communication) have found that the addition of ATP to a suspension of Triton-isolated hepatic lysosomes does not alter the distribution of methylamine across the lysosomal membrane, suggesting that the internal pH is not affected. Mego's studies were done on hepatic or renal lysosomes not subjected to Triton-WR 1339 and his fractions are not highly purified; neither the fate of ATP nor the effects on intralysosomal pH were determined directly. There are also some disputes about the stability of isolated lysosomes in the presence of ATP. Mego's data on enzyme latency convince him that at alkaline pH, ATP does not increase, and may actually diminish the disruption of lysosomes and leakage of hydrolases that invariably occurs to some extent during the incubations (MEGO *et al.* 1972). However, HUISMAN *et al.* (1974) report that at lower pH (4.5) ATP increases such release of hydrolases; this calls into question the viewpoint of HAYASHI *et al.* (1973) who found that addition of ATP to liver homogenates at low pH increased the breakdown of protein and speculated that this was due to an ATP-mediated incorporation of proteins into lysosomes [2]. Furthermore, some nucleotides that lack high energy bonds can influence the stability of lysosomes in suspension. Ignarro's laboratory (IGNARRO and COLUMBO 1973, IGNARRO *et al.* 1974) claims that incubation of leukocyte granules with cyclic GMP leads to enhanced release of enzymes whereas exposure to cyclic AMP has the opposite result. But, according to MEGO (1973 a) no effects of ordinary AMP on lysosome breakage are detectable in his system.

Mechanisms for acidification based on metabolic inputs alternative to ATP driven proton-pumps have also been discussed. For example, TAPPEL (1968) and BARRETT (1972) mention the possibility of oxidation-reduction pumps, although there is no hard evidence for the requisite machinery. Several authors suggest proposals based upon passage into lysosomes of lactic acid or CO_2; the impetus for this derives from findings such as the demonstration that glycolysis is stimulated during phagocytosis in some leukocytes (HIRSCH 1972, KLEBANOFF 1971) and the observation that the acidification of human PMN leukocyte vacuoles containing indicator-dye stained *Candida* is prevented by fluoride or iodoacetic acid (MANDELL 1970). The amounts of lactate or other components produced during phagocytosis are small and may be inadequate to account for all of the pH changes, and one still must seek explanations for the selectivity of acidification of lysosomes, as opposed to the rest of the cytoplasm. But we really do not know how much acid is needed and there might well be several mechanisms that contribute to maintaining a low pH in lysosomes. It would certainly be premature to discard any plausible proposal. (See also page 212.)

---

[2] GOLDSPINK and GOLDBERG 1973, have identified another artifact that would affect a few published studies, although not those done with radioactive labels such as Mego's and the others discussed above: When ATP is added to tissue homogenates an increase in ninhydrin positive material is noted that might be interpreted as due to release of amino acids, but may actually reflect deamination of nucleotides.

II.2. Hydrolase Transport in Cells Other than Phagocytes

II.2.1. GERL

In most cell types, lysosomes play a less prominent role than they do in phagocytes, and hydrolase transport and packaging are correspondingly more difficult to study. However, even though distinctive-looking primary lysosomes often are not seen, cytochemical and morphological studies have suggested strongly that lysosome formation in the Golgi region is a very widespread phenomenon. A. B. NOVIKOFF (1960, 1963, 1967 b, 1973) has been central in developing and supporting this point of view. The data accumulated by him and by others can be summarized as follows: Some membrane-delimited sacs or tubules in the Golgi region share with the lysosomes the characteristic presence of cytochemically demonstrable acid phosphatase activity (Fig. 15). In many cell types, secondary lysosomes tend to be especially numerous near the Golgi apparatus. Configurations suggesting the budding of hydrolase-containing vesicles from Golgi-associated sacs or tubules are often encountered by microscopists and correspondingly there are appearances suggesting that these vesicles can fuse with digestion vacuoles in the Golgi region and elsewhere in the cell (see *e.g.*, FRIEND and FARQUHAR, 1967; obviously there are limits to the confidence with which one can extrapolate from static electron micrographs to such dynamic processes). In a few cases, particularly in injured cells forming many lysosomes (LANE and NOVIKOFF 1965, HOLTZMAN 1971, HOLTZMAN *et al.* 1973) acid phosphatase activity is also demonstrable in the rough endoplasmic reticulum. And, by autoradiography, BENNETT and LEBLOND (1971) have shown that labelled fucose (presumably entering glycoproteins) is present in duodenal and hepatocytic lysosomes very soon after it appears in the Golgi apparatus.

The simplest interpretation of these facts, especially when viewed in light of the more detailed information available for leukocytes, is that acid hydrolases, synthesized at the rough endoplasmic reticulum are conveyed via vesicles produced in the Golgi region to their ultimate sites of use. The movement of pinocytosis vacuoles and other bodies that acquire hydrolases to the Golgi region would facilitate interaction with lysosomes.

Fig. 15. Acid phosphatase preparations of neurons. *a* Shows part of a perikaryon from a cultured chick sympathetic ganglion. Reaction product is present in the nuclear envelope (*N*) and in cisternae of rough endoplasmic reticulum (*E*). A mitochondrion is seen at *M*. ×35,000. *b* Shows the Golgi region of a perikaryon from a culture similar to the one used for *a*. Reaction product is seen in vesicles, sacs and tubules associated with the Golgi apparatus (arrows). The configurations at *F* probably are face views of the same types of structure seen sectioned transversely at the arrows. ×35,000. *c* Portion of an axon from a cultured dorsal root ganglion 2 days after X-irradiation. Reaction product is present in elongate membrane delimited structures (arrows) that probably are part of a single continuous sac or tubule of axonal agranular reticulum. Schwann cells are indicated by *S*. ×10,000. (From TEICHBERG, S., and E. HOLTZMAN, 1973: J. Cell Biol. **57**, 88—108; and HOLTZMAN, E., S. TEICHBERG, S. J. ABRAHAMS, E. CITKOWITZ, S. N. CRAIN, N. KAWAI, and E. R. PETERSON, 1973: J. Histochem. Cytochem. **21**, 349—385.)

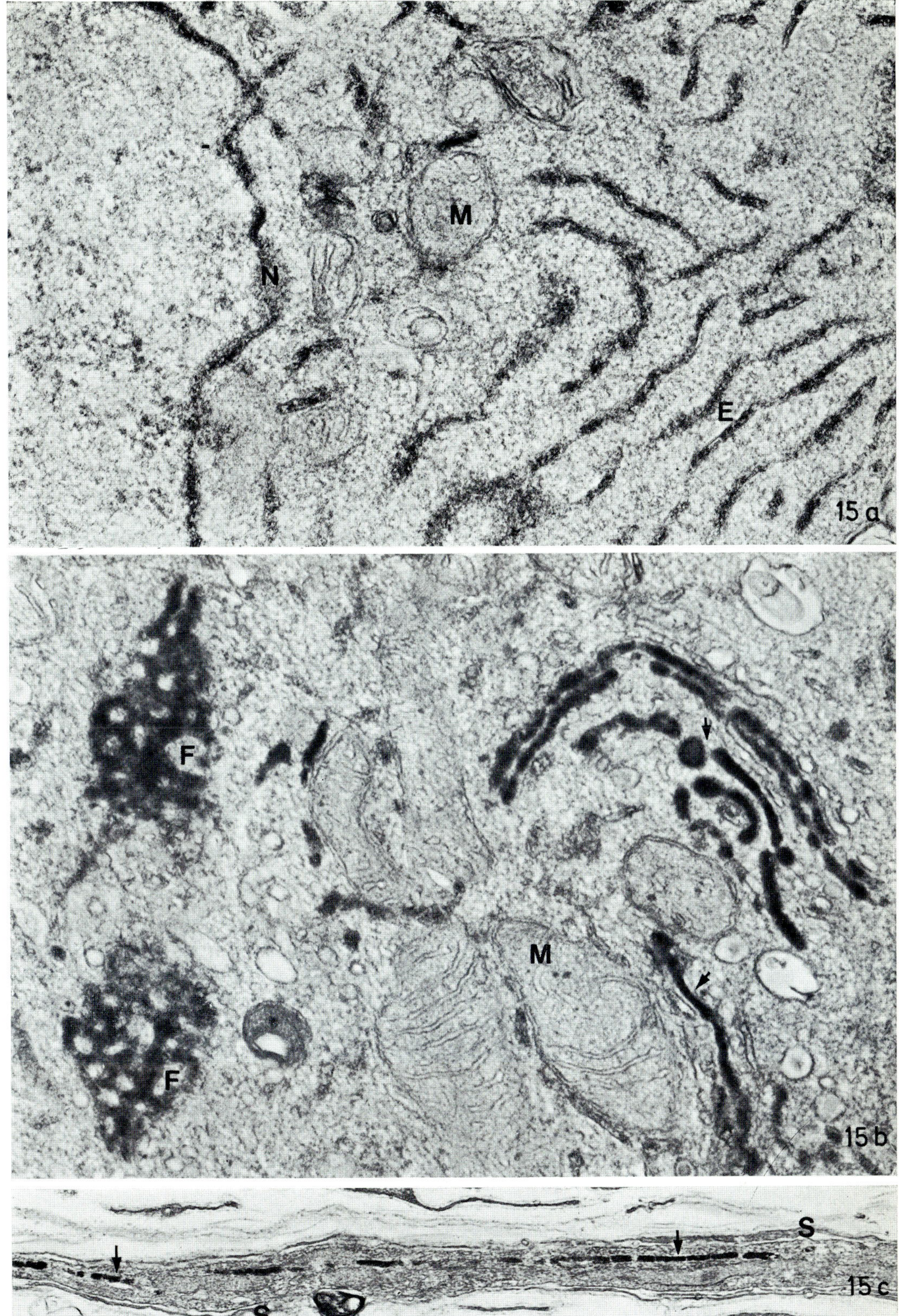

Figs. 15 *a–c*.

Novikoff and his co-workers (including myself) have carried this analysis a step further by contending that at least in some cell types there exists near the Golgi apparatus, a relatively direct route from rough endoplasmic reticulum to lysosomes [see also Brandes (1965) for his speculations along these lines]. The name GERL was coined to emphasize the observation that in lysosome formation by neurons of rodent dorsal root ganglia, the Golgi sacs apparently are by-passed with enzymes moving from the rough endoplasmic reticulum into smooth (agranular or "ribosome-free") ER that is continuous with the rough ER, and thence into lysosomes. (GERL = Golgi-associated Endoplasmic Reticulum from which Lysosomes form; see Fig. 16 for a 3 dimensional model and Novikoff 1967 a, b, 1973, and P. M. Novikoff et al. 1971 for details.) There is precedent for by-passing of the Golgi sacs in some secretory cells. For example, in the analysis by Jamieson and Palade (1967) the vesicle-mediated transport within cells of the exocrine pancreas can move secretory proteins directly from the ER to "condensing" vacuoles near the Golgi apparatus. In adrenal medulla, direct continuities have been seen between endoplasmic reticulum and secretion granules that appear to be forming near the Golgi apparatus (Holtzman et al. 1973) and there are claims that this is the case for a few other gland cells (Lazarus et al. 1966).

It is impossible at present to know how strongly to emphasize the apparent differences between situation in which GERL seems to generate lysosomes and those (e.g., the PMN leukocytes) in which Golgi sacs appear to do this, especially since the mechanisms by which hydrolases move from the ER to the Golgi apparatus of the leukocytes are still to be determined *. Many microscopists maintain that the sacs of the Golgi apparatus themselves derive from the ER or that there is a dynamic flow of membrane from the ER into the Golgi apparatus and then into structures formed by the apparatus (see e.g., Dubois 1972, Flickenger 1971, Holtzman 1971, Morre et al. 1971, Novikoff 1973, Teichberg and Holtzman 1973 for discussion and references). The appropriate morphology for this exists; for example ribosomes sometimes are seen along portions of sacs that are part of, or intimately abut upon the stacked sacs of the apparatus, as if the ER were in process of losing its ribosomes and transforming into a Golgi sac. And there must be a mechanism for replacing the membrane that passes from Golgi sacs or GERL to forming secretory structures, or lysosomes. But of course this is only suggestive of membrane flow, and definitive proof is still lacking (e.g., conceivably some new membrane of the Golgi apparatus is assembled locally from macromolecules, rather than being shipped in from the ER; Section III.4.3.3).

It may be that there are a number of equivalent routes and mechanisms for transport and packaging, with one or another being dominant in different

---

* Although they did not discuss the details of ER-Golgi relations, in a preliminary report, Bentfield and Bainton (1974) indicated that the lysosomes of megakaryocytes and platelets may originate from GERL. And, Essner and Haimes (1975) have presented evidence for the presence of GERL in normal and abnormal mouse macrophages; one micrograph they showed (taken by C. Oliver) also demonstrated continuity between rough ER and a lysosome with the appearance of a residual body. The macrophages' GERL sequestered particulate silver taken up by endocytosis.

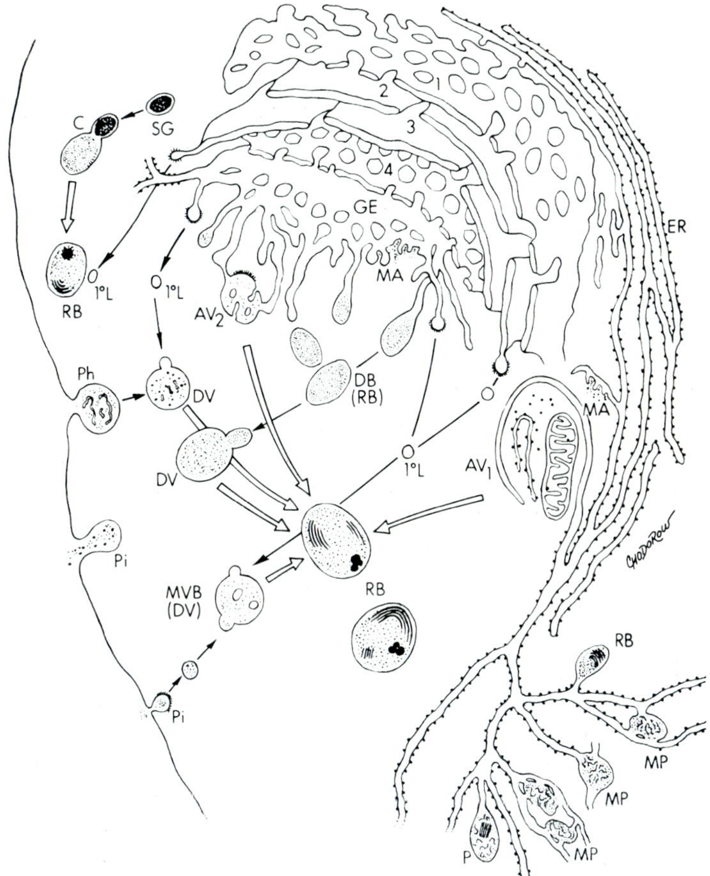

Fig. 16. Diagram of the Golgi region and part of the ER stressing some possibilities for the modes of formation of lysosomes and the interrelations of endoplasmic reticulum with other organelles; based partly on studies of neurons. Abbreviations: 1–4 are sacs of the Golgi apparatus; $1°L$ = primary lysosome; AV_1 and AV_2 are categories of autophagic vacuoles; C = designates the process of crinophagy (Sect. II.3.3); DB = dense body; DV = digestive vacuole; ER = endoplasmic reticulum; GE = GERL; MA = indicates the process of microautophagy (Sect. II.4.2.4); MP = microperoxisome; MVB = multivesicular body; Ph = phagocytic vacuole; Pi = pinocytic vacuole; RB = residual body; SG = secretory granule. (From NOVIKOFF, A. B., 1973: In: Lysosomes and storage diseases. [HERS, H. G., and F. VAN HOOF, eds.], pp. 1—41. New York: Academic Press; see also NOVIKOFF, P. M., A. B. NOVIKOFF, N. QUINTANA, and J.-J. HAUW, 1971: J. Cell Biol. **50**, 859—886.)

cells or circumstances or for different components in a given cell and with each producing a characteristic appearance of the Golgi region (HOLTZMAN 1971). This view is supportable by findings such as those of JAMIESON and PALADE (1971) that under varying experimental conditions secretory proteins may or may not be detectable in the Golgi sacs of exocrine pancreas cells. In addition the variation in topography of formation of specific granules as compared with azurophilic granules in PMN leukocytes (Section III.3.1) suggest that different routes through the Golgi apparatus may apply for different components [3]. Perhaps GERL is a variation on the same theme, with the Golgi sacs, in cells where GERL is present, functioning to package non-lysosomal components, and the situation in the PMN cells reflecting the central importance of hydrolase transport in the functions of these leukocytes.

GERL has been studied most carefully in neurons. Very likely, as Novikoff proposes (1971), comparable configurations are present in many other cell types (see Fig. 46, and the work on hepatocytes by Essner and Oliver 1974, Ma and Biempica 1971, and P. M. Novikoff et al. 1974, on thyroid by A. B. Novikoff et al. 1974 and on adrenal medulla by Abrahams and Holtzman 1973, Holtzman and Dominitz 1968, and Holtzman et al. 1973). However, the essential conceptual distinction between GERL-mediated packaging and Golgi-mediated packaging lies in the direct continuity between rough ER and GERL. Demonstrations of this continuity depend on use of serial sections, or reorientation of electron microscope sections through use of tilting stages (P. M. Novikoff et al. 1971) or upon rare fortuitous single sections (e.g., Holtzman 1971, Holtzman et al. 1973). In most cell types it is not yet clear whether the sacs or tubules found near the Golgi apparatus that contain cytochemically demonstrable acid phosphatase and show signs suggesting the production of lysosomes are best described as Golgi sacs or GERL. Thus we will use neutral terms like Golgi-associated membrane systems to refer to these configurations.

It should be reiterated (cf., Sections I.3.2.1; V.2.2), that the presence of acid phosphatase in a membrane-delimited structure does not unambiguously relate that structure to lysosomes. However, in neurons as in the PMN

---

[3] Many present views of the organization of the Golgi apparatus are strongly influenced by the tenet that materials are transported from one surface of the stack of sacs to the other. This certainly may be true for some substances or cell types. However observations of the type discussed here could prove difficult to reconcile with the conception that this is obligatory. Farquhar et al. (1974) have begun to delineate the biochemical differences among the sacs in a stack, and P. M. Novikoff et al. (1971) have described some of the key morphological and cytochemical differences, and detailed the close relations between GERL and Golgi. But we still have little detailed insight, for example, into how and where secretory cells, leukocytes or other cell types admix proteins made in the rough ER with polysaccharides that probably are synthesized by the Golgi apparatus and it is difficult to be confident about any available detailed picture of the functional anatomy of the Golgi region.

Fig. 17. Portions of neuronal perikarya from larval frog spinal cord. Tissue in *a* was incubated to demonstrate thiolacetic esterase activity and that in *b* and *c* was incubated to demonstrate aryl sulfatase activity. Reaction product is seen in structures associated with the Golgi apparatus (G). Arrows indicate configurations that probably are elements of GERL [cf., the interpretations by R. S. Decker (1974) who provided the pictures]. In the esterase preparation some Golgi sacs may also react. C indicates reaction product in what apparently is a coated vesicle in process of formation and L designates a large lysosome, perhaps a lipofuscin granule (Sect. II.5.2). The insert in *a* shows esterase-positive vesicles near the Golgi apparatus. *a:* from a degenerating neuron, ×85,000; insert, from a mature motor neuron, ×90,000. *b:* from a mature motor neuron, ×42,000. *c:* from a differentiating neuron, ×60,000. (Figures from Decker, R. S., 1974: J. Cell Biol. **61**, 599—612.)

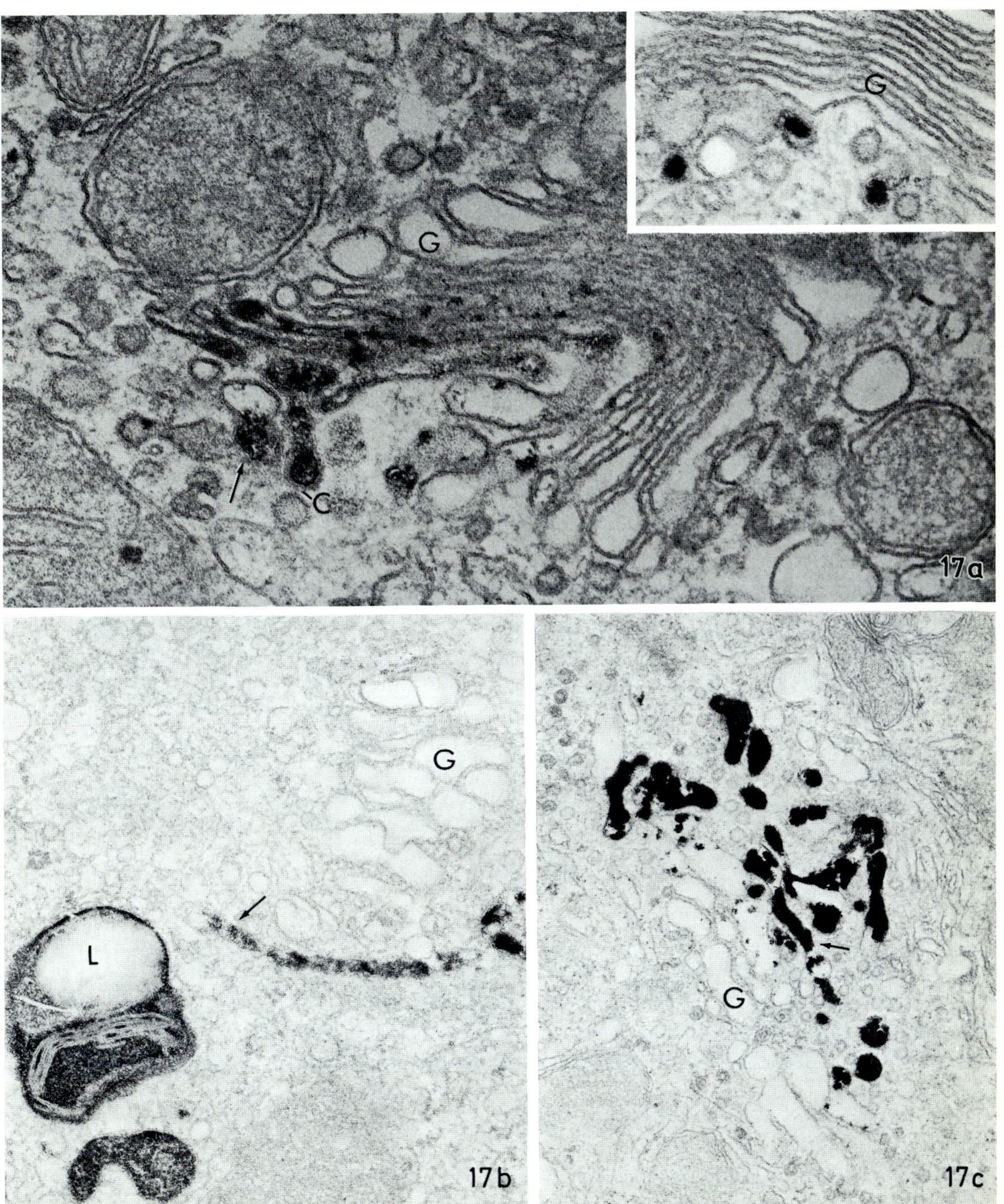

Figs. 17 *a–c*.

leukocytes and macrophages, employment of other cytochemical methods has substantiated the participation of Golgi-associated sacs or tubules in lysosome formation. In our own work on irradiated cultured neurons, aryl sulfatase as well as acid phosphatase was demonstrated in some sacs from which lysosomes form; these sacs appeared to derive from the ER (HOLTZMAN 1971). More detailed and extensive work by DECKER (1974) on frog dorsal root ganglion neurons showed the presence of acid phosphatase, esterase and aryl sulfatase activities in vesicles and sacs near the Golgi apparatus (Fig. 17); as DECKER suggests, most of the reactive structures probably correspond to elements of GERL. BRANDES *et al.* (1965) have also observed esterase as well as phosphatase in Golgi-associated sacs in cells of sebaceous glands. Still uncertain is the interpretation of situations in which both Golgi sacs and GERL-like configurations both contain cytochemically demonstrable acid phosphatase in the same cell (Fig. 15; BOUTRY and NOVIKOFF 1974). Perhaps this reflects some feature of ER-Golgi relations; related to lysosome formation. Or, possibly, the Golgi enzyme has a different role from the GERL phosphatase (Section V.2).

Some other interesting and potentially jarring possibilities arising within the framework presented in this section remain to be fully explored. In particular, NOVIKOFF (1973) has proposed that in some cases there may be no need to invoke primary lysosomes in accounting for hydrolase transport. The enzymes might pass directly into secondary lysosomes as the latter form from GERL. This view emerges from the frequent observations of connections between secondary lysosomes and smooth-surfaced, membrane-delimited sacs or tubules near the Golgi apparatus. Such continuities are commonly found with dense bodies, some of which have morphologies indicative of residual structures (*e.g.*, they may contain lamellae or electron dense grains) and also with autophagic vacuoles and multivesicular bodies (see Fig. 27 and *e.g.*, HOLTZMAN *et al.* 1967, NEHEMIAH and NOVIKOFF 1973, NOVIKOFF and SHIN 1964, NOVIKOFF *et al.* 1973). Mechanisms are known through which autophagic vacuoles and multivesicular bodies could arise from GERL (Sections II.3.2.1 and II.4 will discuss these in detail.) For heterophagic lysosomes one could think in terms of endocytic vesicles fusing with a dilated hydrolase-containing region of GERL, that retains continuity with the sac from which it formed.

We have stressed elsewhere (HOLTZMAN 1969) that the identification of smooth-surfaced sacs or tubules attached to lysosomes as derivatives of the ER or Golgi apparatus requires substantiation, since some configurations of this type probably arise through endocytosis (see Section II.4). But serial section studies coupled with cytochemical demonstration of acid phosphatase, and other work do suggest that attachment of secondary lysosomes to GERL may not be rare (see NOVIKOFF 1967 a, b, 1973). The extent to which elements of GERL are connected to the rest of the ER is not yet established but if these findings mean that extensive regions of ER are in long term or intermittent direct communication with lysosomes, one could conceive that not only hydrolases but also other macromolecules made by the ER move directly into the lysosomes, perhaps to be degraded there. Suggestions of this type have been made for lipids, a category of molecule known to be synthesized largely by ER enzymes (NEHEMIAH and NOVIKOFF 1973, NOVIKOFF *et al.* 1974; see also DIETERT and SCALLAN 1969). Perhaps more disconcerting is the possibility that lysosomal contents or the products of digestion move from the lysosomes into the ER; if so, there must be some constraints on such motion, since for example, indigestible tracers taken into the lysosomes through endocytosis are not seen in the rough endoplasmic reticulum or in ordinary smooth ER. (GERL may accumulate such tracers in macrophages [see the footnote at * on page 56] but this seems not to be the case in other cell types; *e.g.*, NOVIKOFF 1973.) A conservative view is that the sacs and tubules attached to lysosomes, chiefly represent interconnections among the lysosomes, and that continuities with the ER are of limited extent, duration, or frequency.

Crucially related to these issues is the lack of information about the morphology of primary lysosomes in most cell types. The available evidence suggesting the presence of such lysosomes, in many cases is solely the cytochemical demonstration of acid phosphatase within vesicles in the Golgi region. In a few cell types, such as Schwann cells (HOLTZMAN and NOVIKOFF 1965), neurons (HOLTZMAN 1969, 1971, and HOLTZMAN *et al.* 1967) and others (MOE *et al.* 1965, SELJELID 1966) small dense bodies that may be primary lysosomes have been observed; this is true for example, in injured neurons in which some of the dense bodies attached to Golgi associated membrane systems lack obvious digestive residues (see HOLTZMAN *et al.* 1967, HOLTZMAN 1971). Thus, candidates exist, but the dynamics are elusive—for the most part one can only guess that the propinquities between forming digestion vacuoles and dense bodies, small vesicles or other possible primary lysosomes seen in the electron microscope, might have been followed by fusion, had the cells been allowed to survive. And there is no simple way to decide about the relative importance of such fusions as compared with delivery of hydrolases from secondary lysosomes or by other mechanisms.

The literature also contains a number of cases in which bodies have been identified as primary lysosomes without adequate evidence to distinguish them from other organelles such as secondary lysosomes, variants of neuronal densecored vesicles, endocytic structures and perhaps even some "microperoxisomes" (see P. M. NOVIKOFF and A. B. NOVIKOFF 1973 for the characteristics of this last type of organelle).

Summarizing the several threads running through this section: 1. In various cell types, lysosomes, especially ones near the Golgi apparatus, are found to be attached to smooth-surfaced sacs or tubules. Sometimes, this seems simply to reflect lysosome origin by budding from the Golgi apparatus. In other cases, it is likely that the sacs and tubules are elements of a network, distinct from ordinary Golgi sacs. Some lysosomes may originate by budding from this network and it is possible that lysosomes also maintain or establish, longer-term, or intermittent, connections to it. This may mean that intracellular digestion sometimes occurs within a network of interconnected lysosomes, sacs and tubules, rather than solely within discrete lysosomes. 2. In a few cell types, continuities have been shown to exist between rough ER, and lysosomes or the smooth sacs and tubules attached to lysosomes. This suggests that hydrolases can move directly from the ER to lysosomes, perhaps to secondary types as well as to primary lysosomes. It also raises the possibility that there are other direct molecular interchanges between the endoplasmic reticulum and the intracellular digestive system.

II.2.2. Endoplasmic Reticulum and Lysosomes; Some Biochemical Findings

A number of acid hydrolases resembling the lysosomal enzymes are present in reasonably pure microsome fractions. This is true, for example, of some esterases of liver and of β-glucuronidase in kidney and elsewhere (BEAUFAY et al. 1974, CONCHIE et al. 1961, IDE and FISHMAN 1969, SWANK and PAIGEN 1973). Too little is known of the esterases to sustain an extensive discussion. But β-glucuronidase is highly interesting case. One might expect microsomal β-glucuronidase to be present as a result of formation of lysosomal hydrolases in the ER (cf., IDE and FISHMAN 1969, KATO et al. 1972, VAN LANKER and LENTZ 1970) and a proportion of the enzyme may have such a significance. However, particularly from the work of PAIGAN and his colleagues, it appears that much of the microsomal enzyme actually is a constitutive ER glucuronidase. Although a complete, direct analytical comparison of purified microsomal enzyme with lysosomal β-glucuronidase is not yet feasible (TOUSTER 1973), the fundamental enzymatic characteristics of the two are very similar, both share major immunological characteristics and both are affected by mutations at a single locus (DEAN 1974, LALLEY and SHOWS 1974, SWANK and PAIGEN 1973). However, important differences are observed in electrophoretic behavior and in size. SWANK and PAIGAN (1973; and others, see TOUSTER 1973) have marshalled evidence that the enzymatically active β-glucuronidase molecule is a tetramer of polypeptide chains common to both the lysosomes and microsomes; for the microsomal enzyme, Paigen's laboratory holds that this tetramer is joined to additional polypeptides that probably are non-enzymatic. At present the status of this notion is somewhere between a thoughtful working hypothesis and an unequivocally demonstrated fact. If it continues to hold up, it may have great significance for a number of general questions concerning organelle formation and interrelations and the control of intracellular transport of proteins; for instance one might hypothesize that the added polypeptides of the constitutive ER enzyme are related to the integration of the enzyme within a membrane. For the narrower purposes of the present monograph, the "β-glucuronidase situation" emphasizes the need for an open mind in studies of the intracellular distribution and transport of hydrolases. It also provides a concrete example of the sharing of "subunits" among proteins that are demonstrably not identical; as will be seen, this concept has also been used to help explain the complex effects of mutation upon hydrolases in certain genetic diseases (Section IV.1.2).

Future work on the relations of microsomal and lysosomal β-glucuronidase should be facilitated by the availability of a mutant mouse strain, that lacks the microsomal enzyme (TOMINO and PAIGEN 1975). It also has been claimed that the microsomal enzyme is converted to resemble the lysosomal form, by exposure, *in vitro,* to the other lysosomal hydrolases (OWENS *et al.* 1975); how this bears on events *in vivo* is not known, but the possibilities are intriguing. Does such conversion occur in hydrolase "packaging"?

From analyses based largely on biochemical work with partially purified fractions of renal tissues coupled with some autoradiographic investigations, GOLDSTONE and KOENIG (1972) maintain that acid hydrolases are handled by the cell more or less as are those glycoproteins destined for export to cell surfaces or extracellular channels; the polypeptide chain is completed on the ribosomes of the rough ER and then saccharide units are added, first by glycosyl transferases of the ER and then by the Golgi apparatus. This overall picture is a plausible one, given the morphological and cytochemical information outlined earlier, and GOLDSTONE and KOENIG have extended their studies to the point of isolating (as a partially purified sub-microsomal fraction of kidney) what they believe is a special region of ER involved in synthesis or accumulation of hydrolases (GOLDSTONE *et al.* 1973). Unfortunately little is known of the possible primary lysosomes of renal tissue and there still are important uncertainties and controversies that becloud this picture of hydrolase processing. TOUSTER (1973) points out that there is not much direct information about the saccharide side chains of lysosomal hydrolases, even though the presumption that most of the enzymes are glycoproteins has some support. He argues vigorously that interpretations based, as many are, upon neuraminidase-engendered changes in the electrophoretic properties of proteins present in complex mixtures must be extremely conservative, and require substantiation from more direct studies. (For example, some supposed neuraminidase effects on hydrolases may actually represent non-enzymatic effects of the incubations; see *e.g.,* Ann. Rev. Biochem. **44,** 357 [1975]). Nevertheless, the heterogeneity in electrophoretic behavior among molecules of a given type of hydrolase isolated from a tissue such as liver or kidney, probably does reflect to some degree, the simultaneous existence in the cells of enzymes varying primarily in their saccharide side chains (BARRETT 1972, GOLDSTONE and KOENIG 1974, IKONNE and ELLIS 1973). GOLDSTONE and KOENIG attach much weight to their hypothesis that many hydrolases within lysosomes are cationic and thus can bind tightly to the supposed anionic matrix of the organelle (Section I.3.2.3). Since in their work hydrolases that are newly completed seem to be anionic, they have introduced the proposal that upon entry into lysosomes the hydrolases are modified into more cationic form by removal of neuraminic acids mediated by the lysosomal neuraminidases [see also TOUSTER (1973) for his suggestion that mannosidases in the Golgi apparatus might cleave saccharides from newly completed glycoproteins, and MARSH *et al.* 1974]. What is badly needed now is thorough evaluation based upon extensive purification of enzymes and of cellular fractions, and careful considerations of possible sources of artifact such as lysosomal hydrolase-induced modifications of enzymes occurring isolation. Such a study should help in formulation of hypotheses that might explain how lysosomal enzymes are "directed" to the proper intracellular packaging sites and how hydrolases are kept from digesting components of the endoplasmic reticulum or other molecules they encounter while in transit from the ribosomes.

A recent publication by Noseworthy *et al.* (1975) reports that the per-oxidase-rich granules of guinea pig PMN leukocytes may be poor in glycoproteins.

II.3. Autophagy

The sequestration and digestion by cells of parts of their own cytoplasm is a very common phenomenon (see Ericsson 1969 a for review). Initial descriptions (*e.g.*, Ashford and Porter 1962, Novikoff and Essner 1962, Swift and Hruban 1964) focussed on stressed or pathological material (hence the proposal of the name "cytolysomes"; Novikoff and Essner 1962) but it was determined early that autophagic structures are neither very rare nor exceptional in normal situations. From our own experience with neurons, cells of the adrenal medulla, Schwann cells and so forth, we conclude that most normal cells of these types have one or more recognizable autophagic vacuoles at any given moment; this is congruent with the impressions of others for hepatocytes and other cell and tissue types Nonetheless, it is well-established that the frequency of autophagic vacuoles increases markedly under such a variety of abnormal conditions including starvation, exposure to numerous noxious agents and responses to injuries such as those inflicted by irradiation (*e.g.*, Arstila *et al.* 1974, Ericsson 1969 a, Holtzman 1971, Hugon and Borgers 1966, Lane and Novikoff, 1965, Swift and Hruban 1974). One convenient experimental method for inducing autophagy in mammalian hepatocytes is the administration of glucagon, which brings about a dramatic increase in vacuole frequency within one hour (Ashford and Porter 1962). After an injection of glucagon an appreciable fraction of one percent of hepatocyte cytoplasm may be autophagocytosed (Deter *et al.* 1967).

II.3.1. Basic Morphology

The fundamental characteristic by which autophagic vacuoles are recognized is the presence within a membrane-delimited structure of cytoplasmic material for which a phagocytic origin is ruled out (or at least very improbable). Bodies of this sort can be shown cytochemically to contain acid phosphatase (Fig. 18) and there also are reports of the presence of cytochemically

---

Fig. 18. From a cell of the proximal tubule epithelium of rat kidney 2 hours after administration of hemoglobin to the animal; incubated to demonstrate acid phosphatase activity. Autophagic vacuoles containing mitochondria (*M*) are seen at *A*; note the sprinkling of reaction product near their edges. A dense body (*D*) shows a somewhat heavier reaction. ×50,000. (From Miller, F., and G. E. Palade, 1964: J. Cell Biol. **23**, 519—552.)

Fig. 19. Autophagic vacuole forming in a cell from the fat body of the insect *Calpodes* during the preparations for the transition from larva to pupa. A mitochondrion (*M*) is enclosed within a sac (arrows). O indicates the outer membrane of the mitochondrion; it is clear that there are two additional membranes surrounding this, as expected for envelopment by a sac. ×85,000. (From Locke, M., and J. V. Collins, 1973: In: Pathological changes in cell membranes. [Trump, B., ed.] New York: Academic Press [in press].)

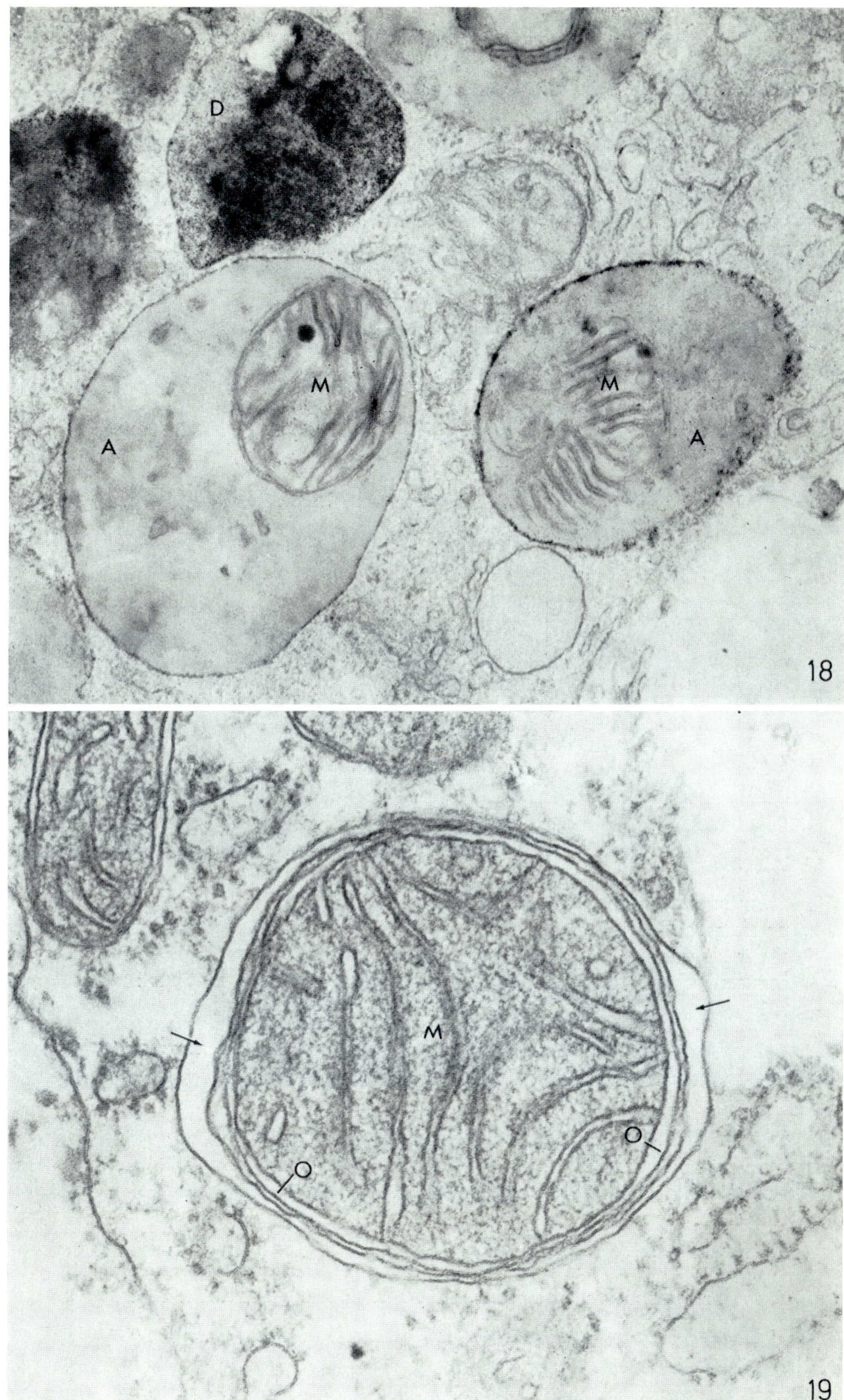

Figs. 18 and 19.

demonstrable esterase and aryl sulfatase within structures that probably are autophagic (*e.g.*, Houdry 1971, Topping and Trevis 1974). In addition, cell fractionation studies, chiefly by Deter and his colleagues, have provided biochemical backing for the identification of autophagic vacuoles as lysosomes (Deter and de Duve 1967, Deter *et al.* 1967, Deter 1971).

All of the common cytoplasmic organelles have been reported within autophagic lysosomes with the exception of recognizable centrioles, cilia and

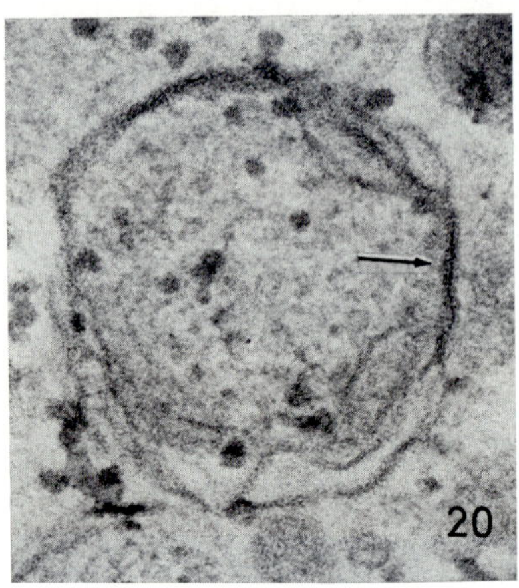

Fig. 20. Small autophagic vacuole from an epinephrine (adrenaline) cell of the rat adrenal medulla. The arrow indicates a thickened region of the membrane system delimiting the vacuole. Evidently this region comprises two closely apposed membranes since a dense line bisects it, and the thick membrane separates into two thinner ones at points above and below the zone indicated by the arrow. ×90,000 (approx.).

flagella. There even have been a few observations suggesting that smaller lysosomes may occasionally become sequestered within the vacuoles (Matthews 1973).

As indicated above, structures that are "intermediate" between autophagic vacuoles and residual bodies are commonly encountered. These frequently show signs of deterioration or digestion of the enclosed organelles and the accumulation of electron dense grains, lamellae and amorphous material.

II.3.2. Mode of Formation of Autophagic Vacuoles

II.3.2.1. Source of the Delimiting Membranes

At early stages of their formation, autophagic vacuoles often are delimited by a pair of membranes (Fig. 19) or by a thickened membrane that represents two closely apposed (compacted; Novikoff and Shin 1964) membranes (Fig. 20). Sometimes more complex arrays are found. These configurations are most simply explained as the result of the formation of the vacuoles by enwrapping of cytoplasm within structures derived from preexisting sacs. A major alternative viewpoint, that a membrane can form *de novo* around a cytoplasmic region, is difficult to support and might not be expected to result in pairs of delimiting membranes; there may however, be a "precedent" in the formation of the lipoprotein envelope of vaccinia virus (DALES and MOSSBACH 1968) which seems to arise *de novo*.

What is the source of the membranes that bound autophagic vacuoles? The two most obvious candidates are the endoplasmic reticulum (ERICSSON 1969 a, NOVIKOFF 1973) and Golgi apparatus, (BRANDES *et al.* 1964, LOCKE and COLLINS 1973) and as with the formation of other lysosomes it may well turn out that either one can be involved. It is very common to find cisternae of rough endoplasmic reticulum closely apposed to forming autophagic vacuoles, although the vacuole surfaces are almost invariably smooth (*i.e.*, they lack ribosomes) and direct continuities between rough ER and autophagic vacuoles are encountered rarely, if at all; perhaps this means that agranular ER is primarily involved or perhaps the delimiting membranes lose their ribosomes or their continuity with the rough ER early during sequestration processes. In cells such as neurons, autophagic vacuoles often (though hardly always) seem to form in the vicinity of the Golgi apparatus, and GERL is a strong candidate for the source of their delimiting membranes. A similar role may be played by the agranular ER of axons (HOLTZMAN 1969, MATTHEWS 1973). On the other hand macrophages, which have relatively little rough ER, can form numerous lysosomes thought to be autophagic (this happens *e.g.*, when the cells are exposed to chloroquine) and thus FEDORKO and co-workers (1968) suggest the Golgi apparatus may be of predominant importance in autophagy by these cells.

Efforts to trace the source of autophagic vacuole membranes through cytochemistry have produced disappointing results. ERICSSON's (1969 b) extensive studies have convinced him that endoplasmic reticulum is preeminently involved; part of his argument is based upon the cytochemical demonstration that glucose-6-phosphatase is split by an enzyme in the sacs surrounding forming autophagic vacuoles in hepatocytes (see also GRAY *et al.* 1974) Glucose-6-phosphatase activity is characteristically present in hepatocyte ER. A cytochemically demonstrable Golgi apparatus enzyme, thiamine pyrophosphatase (TPP'ase) is rarely evident in autophagic vacuoles; ERICSSON did find TPP'ase in some sacs surrounding newly forming hepatocyte vacuoles but the ER of the cells also reacted. Unfortunately, the glucose-6-phosphatase reaction product present around the reactive vacuoles in Ericsson's preparations is quite sparse and only occasional vacuoles show the activity; furthermore, acid phosphatase is also sometimes demonstrable in the sacs delimiting autophagic vacuoles (Fig. 21; see also HOLTZMAN 1971, WHITAKER and LaBELLA 1973) and this enzyme might be able to split glucose-6-phosphate (*cf.*, TEICHBERG and HOLTZMAN 1973 who also have noted configurations suggesting hydrolysis of glucose-6-phosphate by GERL). Thus the cytochemical findings can sustain only tentative conclusions with respect to relations of ER and autophagic vacuoles. These conclusions are somewhat strengthened by the observation (ERICSSON 1969 b) that newly formed autophagic structures are bounded by relatively thin membranes (50–60 Å) comparable to those that characteristically delimit the endoplasmic reticulum [4]. (Footnote is on p. 68.)

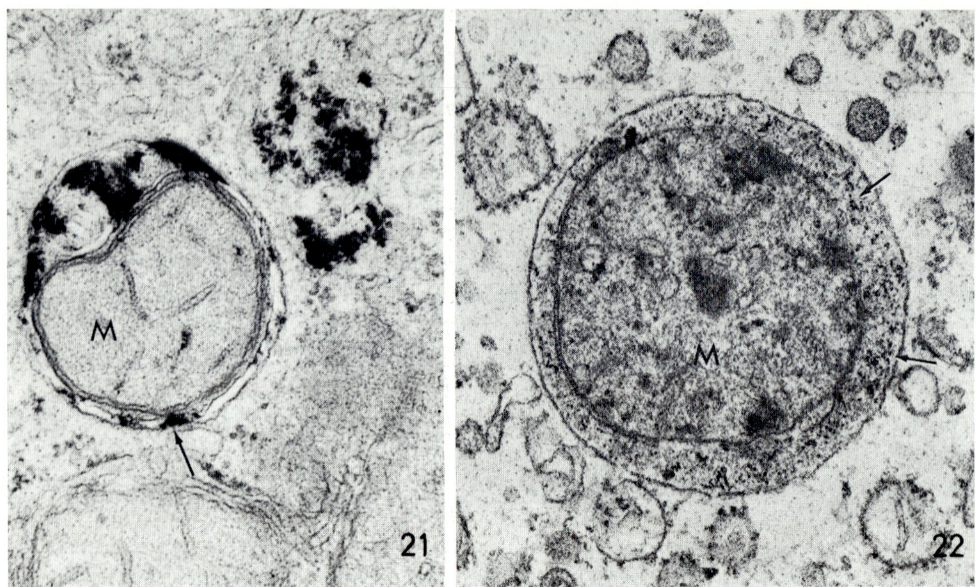

Figs. 21 and 22.

Fig. 21. Autophagic vacuole from an hepatocyte exposed to Triton WR-1339 and incubated to demonstrate acid phosphatase activity. Reaction product is seen within a membrane-delimited sac (arrow) surrounding a mitochondrion (*M*). ×64,000. (From Novikoff, A. B., 1973: In: Lysosomes and storage diseases. [Hers, H. G., and F. van Hoof, eds.], pp. 1—41. New York: Academic Press.)

Fig. 22. Autophagic vacuole isolated by centrifugation from the liver of a rat that had been injected repeatedly with the iron-complex, Jectofer for several days and with glucagon shortly before sacrifice. The vacuole contains a mitochondrion (*M*) and also numerous particles of the iron-complex (a few are seen near the tips of the arrows). Thus hetero-phagocytosed Jectofer has been incorporated in what is probably a newly forming auto-phagic vacuole, presumably through fusion of a heterophagic secondary lysosome with the vacuole. ×50,000. (Courtesy of R. Deter.)

Origin of the delimiting membranes of autophagic vacuole from the cell surface may occur, under some circumstances. The possibility has been raised that endocytic structures may, in the course of fusing with one another or with lysosomes, sometimes trap cytoplasm that there by winds up enveloped inside a forming lysosome (Miller and Palade 1964). Whether such "accidents" are quantitatively significant is not known. There is at least one

---

[4] Locke and Collins (1973) report that prolonged soaking of tissues in warm osmium tetroxide (Friend 1969) produces dense deposits in ER, Golgi vesicles and the sacs surrounding autophagic vacuoles. While the reliability of this technique for revealing membrane relations is not established, these observations could be cited as tentative "cytochemical" evidence for direct involvement of Golgi apparatus, ER, or both, in autophagy. With insect fat body, Locke and Sykes (1975) find osmium deposits in autophagic vacuoles and in Golgi-associated vesicles and sacs, but little in the ER. This, to them, suggests a preeminent role for Golgi elements in autophagic sequestration.

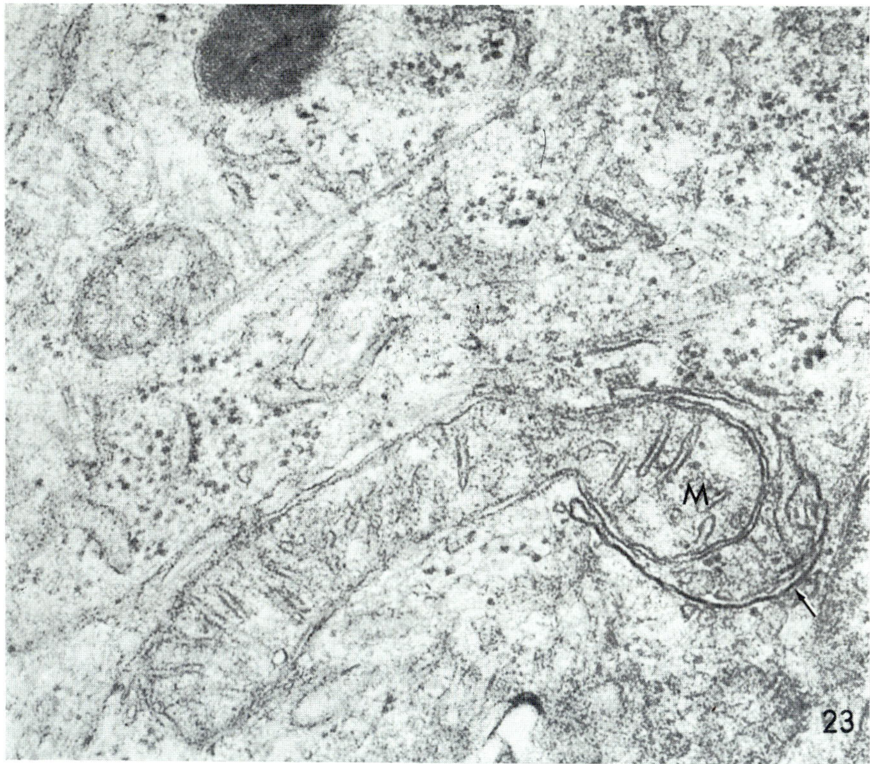

Fig. 23. From the perikaryon of a neuron of rat ganglion nodosum one day after the axon was cut (this sets off chromatolysis, which involves extensive autophagy in perikarya; HOLTZMAN *et al.* 1967). At the arrow, a thickened membrane (*cf.*, Fig. 20) partially surrounds the end of a mitochondrion (*M*). One interpretation is that this is an autophagic vacuole in early stages of its formation—perhaps only part of the mitochondrion was destined to be included in the vacuole (see Sect. II.3.1). ×50,000 (approx.).

case in which autophagic vacuoles that appear to originate with bounding membranes derived from the cell surface are of major importance. In the degradation of myelin during Wallerian degeneration, Schwann cells subdivide the myelin sheath segments they have previously formed and maintained, into smaller intracellular "globules" by intruding cytoplasm directly between the membranes of the myelin (HOLTZMAN and NOVIKOFF 1965). At least in their initial stages no membranes additional to those of the myelin separate the globules from the remainder of the Schwann cell cytoplasm. And since myelin is made of Schwann cell plasma membrane, plasma membrane is, in these circumstances, delimiting a forming autophagic vacuole. (Eventually the "globules" are transformed into lysosomal digestion vacuoles probably by fusion with dense bodies which often are found nearby.) When we discuss multivesicular bodies (Section II.4.2.2) we will take up another "special" autophagic mechanism in which membranes from the cell surface may be key participants.

Additional questions for which answers are still being sought include the following: 1. How is the delimiting system of two or more membranes that surrounds a newly formed autophagic vacuole "simplified" to give rise, eventually to a single membrane? Is it that the inner membranes break up into small vesicles, or disintegrate by some other route or are subtler mechanisms responsible? The membrane that delimits a "mature" autophagic vacuole may be fairly thick (ERICSSON reports thickness up to 100 Å) but at present there are few grounds for extrapolating from such a fact to the "outside possibility" that merger of two thin membranes to produce a thick one follows the "compaction" that is sometimes observed (Fig. 20). 2. Can an autophagic vacuole incorporate parts of organelles as well as whole organelles? Sometimes configurations are noted that suggest, for example, the pinching off of part of a mitochondrion within an engulfing membrane system (Fig. 23; see also, MATTHEWS and RAISMAN 1972, NOVIKOFF and SHIN 1969). While this may be a fanciful interpretation, there is no basis for firm conclusions one way or the other. A controversial concept receiving much current attention is that organelles usually thought of as discrete and separate actually may maintain extensive continuities with one another or with the ER (see *e.g.*, HOFFMAN and AVERS 1973, and RUBY *et al.* 1970 for mitochondria and Section III.4.3.3 for some additional comments and references). If this can be firmly supported, then pinching off of parts of larger structures will warrant especially careful evaluation. In general, the three-dimensional geometry of autophagic processes is still very poorly understood. (Unicellular organisms such as *Chlamydomonas* may contain only a single representative of organelles such as chloroplasts. Little is known of the steady-state turnover or of pertinent modulation properties of these organelles (see *e.g.*, IWANIJ 1975) but if autophagy is involved, then perhaps pinching off of portions of the structures, or some other kind of fragmentation takes place). 3. What are the metabolic prerequisites or controls for autophagy? Only very fragmentary data are at hand. Since glucagon acts via cyclic AMP, one might suspect that this nucleotide links the hormone to its induction of autophagic vacuoles. Perhaps the findings by SHELBURNE *et al.* (1973 a, b) that cyclic AMP itself is an inducer of autophagy will provide the opening wedge for a direct experimental attack upon this (as indicated at the end of Section II.1.4.4 there also has been a report by ZAHLTEN *et al.* (1972) claiming that glucagon enhances phosphorylation of membranes of lysosomes). TRUMP and BULGER (1965) have shown that isolated flounder kidney preparations which ordinarily exhibit much autophagy do not do so in the presence of cyanide, but in studies by FEDORKO *et al.* (1968) various inhibitors of energy metabolism present at concentrations that can supress pinocytosis did not prevent chloroquine from engendering autophagic vacuole production in macrophages.

Finally it is worth emphasizing that not every configuration seen by electron microscopy of thin sections as a sac surrounding a region of cytoplasm corresponds to a forming autophagic vacuole. During tissue preparation for electron microscopy curved portions of the Golgi apparatus or smooth ER often are sectioned so as to appear as though enclosing portions of cytoplasm. Comparable problems in interpretation hold for cases in which thin sections show islands of cytoplasm apparently segregated within dense bodies, but separated from the dense body contents by a membrane; perhaps there are situations, as has sometimes been suggested, in which preexisting dense bodies engulf cytoplasm by some kind of inpocketing of their surface (*cf.*, Section II.4.2). But it is very difficult to demonstrate such a mechanism particularly since it often is clear that the cytoplasmic island within a dense body really is continuous with the cytoplasm outside the body and appears isolated only because of the plane of section.

II.3.2.2. *Source of the Hydrolases*

Acid hydrolases of autophagic vacuoles can derive from preexisting secondary lysosomes that fuse with newly formed autophagic structures. ERICSSON and his co-workers (see ERICSSON 1969 a) have done studies, analogous to those described above for heterophagy (Section II.1.4.1), in which they labelled secondary lysosomes with endocytosed electron dense tracers and then induced extensive autophagy. The label appeared in newly formed autophagic vacuoles (Fig. 22) which indicates both that fusions with the secondary lysosomes had transpired and that autophagic and heterophagic bodies can intermingle their contents. Biochemical analyses by DETER and his colleagues (DETER and DE DUVE 1967, DETER *et al.* 1967, DETER 1971) have demonstrated that the glucagon-engendered autophagy in hepatocytes involves only small increases, if any, in the amounts of hydrolases present in the cells. Much more dramatic is the redistribution of hydrolases from dense bodies to autophagic vacuoles. This can be observed in part through use of Millipore filter isolation methods that aid in the enumeration of organelles of a particular size class. Roughly two dense bodies disappear for each autophagic vacuole that forms.

Involvement of primary lysosomes in hydrolase transport to autophagic vacuoles seems likely, from general considerations, (see *e.g.*, HUGON and BORGERS 1966) but it is difficult to prove since autophagy has been little studied in cells with large, readily recognizable primary lysosomes; the process is not prominent in PMN leukocytes, or at least it has not been the focus of much attention.

Another possible route for hydrolase transport is even more direct. As previously mentioned, cytochemically demonstrable acid phosphatase is occasionally present in the sacs that delimit newly forming autophagic structures [Fig. 21; *cf.*, ARSTILA and TRUMP 1968, BEAULATON 1967, HOLTZMAN 1971, NOVIKOFF 1973, WHITAKER and LaBELLA 1972, VORBRODT *et al.* 1971, but see also LOCKE and COLLINS (1973); for negative results with a system that is quite active in autophagy]. This has been taken as presumptive evidence that ER involved in the autophagic sequestration of cytoplasm might also contribute an initial charge of hydrolases (*cf.*, Section II.2.1). Presumably the hydrolases would mix with the vacuole contents during the changes that lead to the establishment of a single delimiting membrane. There are situations in which such a mechanism is particularly attractive. For example, in axons, acid hydrolases are known to be present normally [see *e.g.*, ORREGO'S

(1971) demonstration of proteases in axoplasm of giant axons] and the enzymes accumulate at experimentally induced interruptions or constrictions (HOLTZMAN and NOVIKOFF 1965, LAUDRON 1974). Autophagy occurs in uninjured axons and can be extensive after injury, but there are relatively few lysosomes present in normal axoplasm. And, from the quite limited information available, injury does not seem to provoke a special flow of lysosomes into axons from perikarya (see GORDON *et al.* 1968, HOLTZMAN 1969, HOLTZMAN *et al.* 1973 for reviews and references); so, lysosome fusion with forming axonal autophagic vacuoles probably would be infrequent at best. We (HOLTZMAN 1971, HOLTZMAN and NOVIKOFF 1965, HOLTZMAN *et al.* 1973) and others (WHITTAKER and LaBELLA 1972) have demonstrated acid hydrolases in the axonal (agranular) endoplasmic reticulum (Fig. 15) and have seen acid phosphatase-containing sacs apparently participating in autophagy in axons. However, despite such observations, one cannot easily rule out the possibility that the presence of acid hydrolases in a sac surrounding an autophagic vacuole is the result of early fusion of a primary or secondary lysosome with the outermost membrane of the vacuole. It is unfortunate that negative cytochemical results cannot be unambiguously interpreted, since the existence of autophagic vacuoles that do not show reaction product in cytochemically incubated tissues (see *e.g.*, PFEIFER 1969) might otherwise constitute proof at least that the provision of hydrolases is not an invariable function of the membrane systems that carry out the initial autophagic envelopment.

II.3.3. Crinophagy

SMITH and FARQUHAR (1966) determined that mammotrophic pituitary cells of rats accumulate numerous secretion granules within their lysosomes (multivesicular bodies and dense bodies) if secretory activities of the cells are suddenly discontinued, as takes place for example, when the pups are separated from lactating rats. Subsequently they found similar accumulations to occur in other categories of pituitary cells in analogous situations and noted also the occasional presence of secretion granules in lysosomes under "normal" conditions (FARQUHAR 1969, 1971, SMITH 1969). Since most of the granules in the lysosomes lacked their delimiting membranes it was suggested that the granules enter when their membranes fuse with the lysosomal membrane in a manner reminiscent of exocytosis; this suggestion was supported by the observation of the expected intermediate configurations (Fig. 24). Such fusion-based incorporation of secretory material within lysosomes is referred to as *crinophagy*. Apparently the process serves as an alternative fate to release from the cell, and can provide a mechanism for disposing of "excess" secretory material; the cells SMITH and FARQUHAR studied, eventually readjust their synthetic rates, but before this occurs they produce secretory material rendered superfluous by the sudden change in physiology. Fig. 25 summarizes the lysosomal events in crinophagy.

We have noted membraneless secretion granules within lysosomes of the adrenaline and nonadrenaline cells of the rat adrenal medulla (HOLTZMAN and DOMINITZ 1968). And, such bodies have been found in the islet cells of

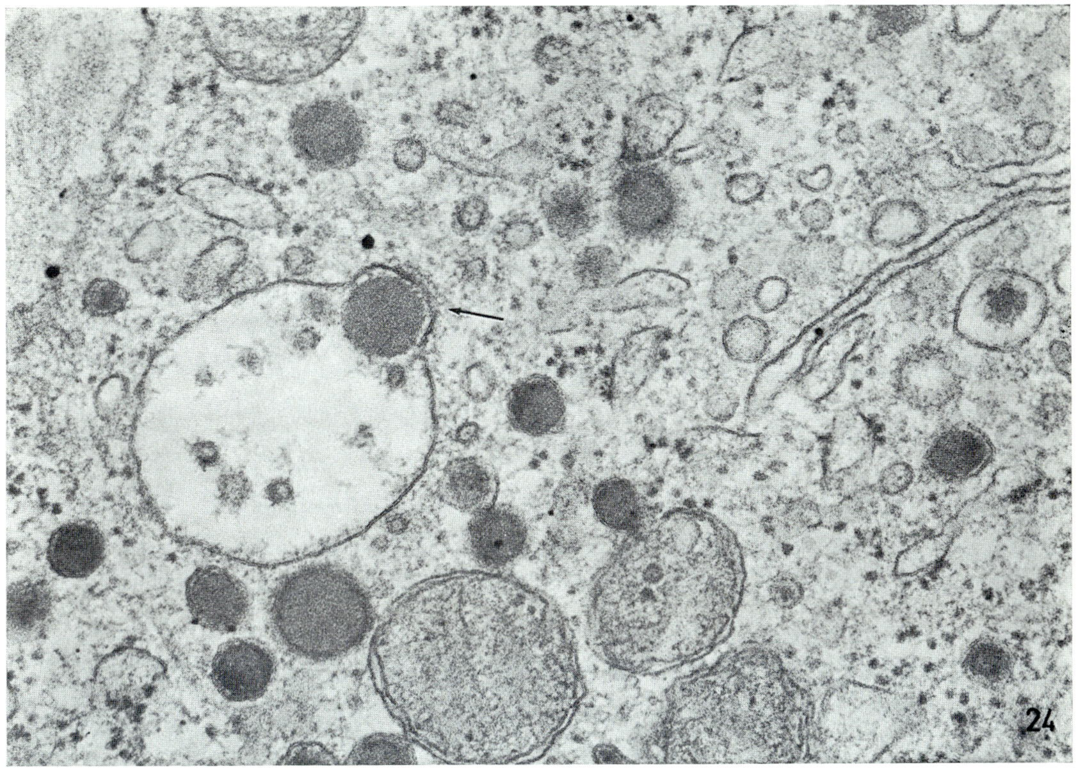

Fig. 24. Portion of a gonadotrophic cell of the pituitary of a castrated rat treated with oestrogen. The arrow-indicates a configuration that apparently represents a secretion granule which has just fused with a lysosome (multivesicular body). ×60,000. (From FARQUHAR, M. G., 1971: Mem. Soc. Endocrinol. **19**, 79—122.)

the pancreas (*e.g.*, the cells of hyperglycemic animals; ORCI *et al.* 1968, 1970). Interestingly, processes resembling crinophagy may also be involved in the degradation of some intracellular pigment granules, such as those in the retinula cells of invertebrate eyes (PERRELET *et al.* 1971, FAHRENBACK 1969) [5].

In various cells, including those of the rat pituitary and adrenal, secretion granules are found within autophagic vacuoles. As might be expected, "conventional" autophagy, results in intralysosomal secretion granules that are still clad in their delimiting membranes, at least for a time (see *e.g.*, DAVIES and KING 1972, HOLTZMAN and DOMINITZ 1967, HOLTZMAN 1971, MASUR and HOLTZMAN 1969, PALAY 1960, WHITAKER and LaBELLA 1972). Thus it is easy, in principle, to distinguish crinophagy from autophagy, although especially given the fact that structures can change considerably and rapidly

---

[5] DELLMAN (1972) has also speculated that the lysosomes which are moderately numerous in some of the perikarya and axon-like processes of neurosecretory cells (*e.g.*, in some categories of "Herring" bodies) might contribute to degrading excess neurosecretion (but see PICKUP and HOPE 1972, for some unresolved questions about the extent to which the cells actually make excess product).

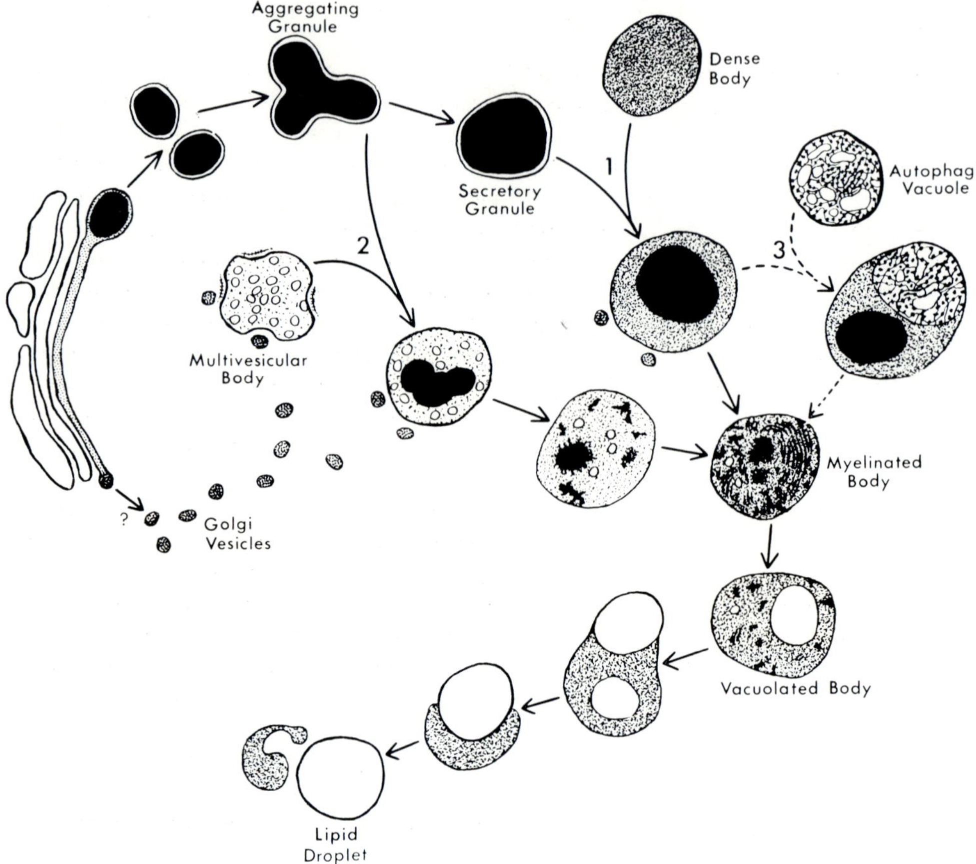

Fig. 25. Diagram of the events of crinophagy as studied in mammotrophic cells of the rat anterior pituitary gland. "Aggregating granules" are the ones that have recently formed, by fusion of Golgi-derived granules. The aggregating granules mature into secretory granules. In crinophagy, aggregating granules tend to fuse with multivesicular bodies and secretory granules tend to fuse with dense bodies. (From FARQUHAR, M. G., 1971: Mem. Soc. Endocrinol. **19**, 79—122.)

once inside a lysosome, caution is required in relying on such features as the presence or absence of a membrane. Increases in the frequency of secretion granule-containing autophagic vacuoles do take place, for example, when amphibian pituitary glands are transplanted from one individual to another; this transplantation results in enhanced secretion (MASUR and HOLTZMAN 1969). However, such changes in autophagy probably reflect general responses of cells to "stress" more than they do specific mechanisms for disposal of secretion materials. Still to be evaluated in such terms are some other observations, such as PALADE's (1972) finding that condensing vacuoles of the exocrine pancreas become incorporated in autophagic vacuoles when aspects of intracellular transport and secretory release are blocked by exposure to an N_2 atmosphere.

II.3.4. Some General Aspects of the Control and Specificity of Autophagy

As emphasized earlier, the frequency of autophagic vacuoles may increase markedly in cells experiencing sublethal injuries and in cells soon to die from a variety of causes. This is true for example for irradiated cells or for cells of many tissues destined for destruction during metamorphosis (Section III.1.2). Perhaps in some of these cases autophagy contributes to bringing about cell injury or death. But this is not self-evident and, in fact, proposals that lysosomes play such causal roles are among the most difficult to confirm conclusively (see Section IV). A strong case can be made for the alternative concept that increases in autophagic vacuole frequency in injured or stressed cells are manifestations of some kind of "defensive" mechanism promoting of facilitating adaptation of the cell to new circumstances, although since virtually nothing is known of the specific mechanisms by which rates of autophagy are controlled we must rely on circumstantial evidence, with all the attendant possibilities for error.

It is often noted that autophagic vacuoles are especially numerous in cells undergoing "remodelling" processes. A good example is provided by the cells of the fat body of the butterfly *Calpodes;* these cells survive during the metamorphosis of larvae into pupae but undergo dramatic changes in which their structure and biochemistry are considerably altered. LOCKE and co-workers have found that many of the organelles of the fat body cells are destroyed by autophagy and replaced by newly formed organelles; this holds for mitochondria, peroxisomes, endoplasmic reticulum and ribosomes. (Figs. 19 and 26; LOCKE and COLLINS 1968, 1973, PRICE 1973 a, SRIDHARA and LEVENBROOK 1974; Section III.1.2.) Autophagy is also dramatically increased in the tips of injured mammalian axons (and in the corresponding chromatolytic perikarya) during the stages in which the axons prepare to regenerate (see HOLTZMAN and NOVIKOFF 1965, HOLTZMAN *et al.* 1967, and HOLTZMAN 1969 for description and references and BUNGE 1973, DELLMANN 1973, MATTHEWS 1973, WETTSTEIN and SOTELO 1963 for work on various systems). In these cases one can plausibly speculate that autophagy contributes to removing structures that in some sense are superfluous or might even impede ongoing cell changes, but it scarcely needs emphasis that such teleological reasoning can prove misleading.

An explanation with a somewhat different focus might be more suitable for the increase in autophagy observed in hepatocytes of rats deprived of food or in a variety of unicellular organisms grown in nutrient poor media (ELLIOTT and BAK 1964, ERICSSON 1969 a, LEVY and ELLIOTT 1968, MALKOFF and BUETOW 1964, NILSSON 1970, STOLZE *et al.* 1969, TORO and VINAGH 1966). By analogy with turnover processes in procaryotes (Section III.4.2) it seems likely that in such situations, autophagy accomplishes degradation of macromolecules into low molecular weight materials necessary for continuation of energy metabolism or for other processes essential to cell survival. An interesting variant of this hypothesis suggests that under conditions of deficit in external sources of amino acids, cells might be able to utilize degradative routes to mobilize internal sources when induced to form new enzymes or otherwise change their enzyme complement. Some evidence pointing in this general direction has come from biochemical studies on unicellular organisms such as *Euglena*, in which maturation of chloroplasts in cells maintained in nutrient poor medium is prevented by inhibitors of proteases (ZELDIN *et al.* 1973; Section III.4.3.1).

There is an ongoing controversy about the significance of autophagy for normal cell function. *A priori*, one might expect autophagic vacuoles to be major participants in the degradative phases of normal intracellular turnover. They may well be, but as will emerge from our discussion of turnover (Section III.4), there are still critical gaps in information that preclude more than a very tentative assessment of the involvement of lysosomes. One fairly firm conclusion that will be amplified in later sections (Sections III.3 and III.4) merits mention here. Normal intracellular turnover under steady state conditions appears usually to occur at random with respect to molecule age. For example proteins within complex organelles such as mitochondria and peroxisomes, and proteins of the "soluble" fraction ("cytosol", "hyaloplasm") all seem to be degraded with first order (exponential) kinetics when the protein populations are pulse-labelled with radioactive amino acids and then followed in subsequent turnover. Characteristic and specific rate constants apply to different proteins but in each case the rate of loss of labelled protein depends chiefly on these constants and on the concentration of the labelled protein that is present at a given time.

These considerations bear on the fundamental question: does autophagy degrade cytoplasm at random, or are there selective mechanisms related to the "age" or "dysfunction" of the material that becomes sequestered? Especially when large autophagic vacuoles become numerous, as they do in stressed cells, one often finds a mixture of organelles in individual vacuoles *roughly* comparable to what might be expected for random "gulps" of cytoplasm. And it is difficult to think of mechanisms by which the vacuoles could be *completely* selective as to what becomes incorporated; after all, their formation does involve bulk envelopment of cytoplasm. Further, if autophagy is centrally involved in turnover, the information just cited concerning randomness of destruction of organelle macromolecules presumably implies some comparable randomness in autophagy.

Given the likelihood that there are aspects of randomness is there also evidence for selectivity in autophagic processes? The answer is a qualified yes. Crinophagy represents one situation in which intracellular structures are degraded selectively within lysosomes. Selectivity in more conventional autophagy has also been reported. For example, in the studies on *Calpodes* alluded to earlier, LOCKE and co-workers found that destruction of fat body cell organelles occurs through several sequential waves of autophagy each of which involves one type of organelle; first the peroxisomes (microbodies) are selectively sequestered, later one finds autophagic vacuoles that contain primarily mitochondria and still later, after a number of intervening events, there is autophagic destruction of endoplasmic reticulum (LOCKE and COLLINS 1973, LOCKE and McMAHON 1971). Less dramatic selectivity has also been reported on the basis of morphometric studies. For example, peroxisomes

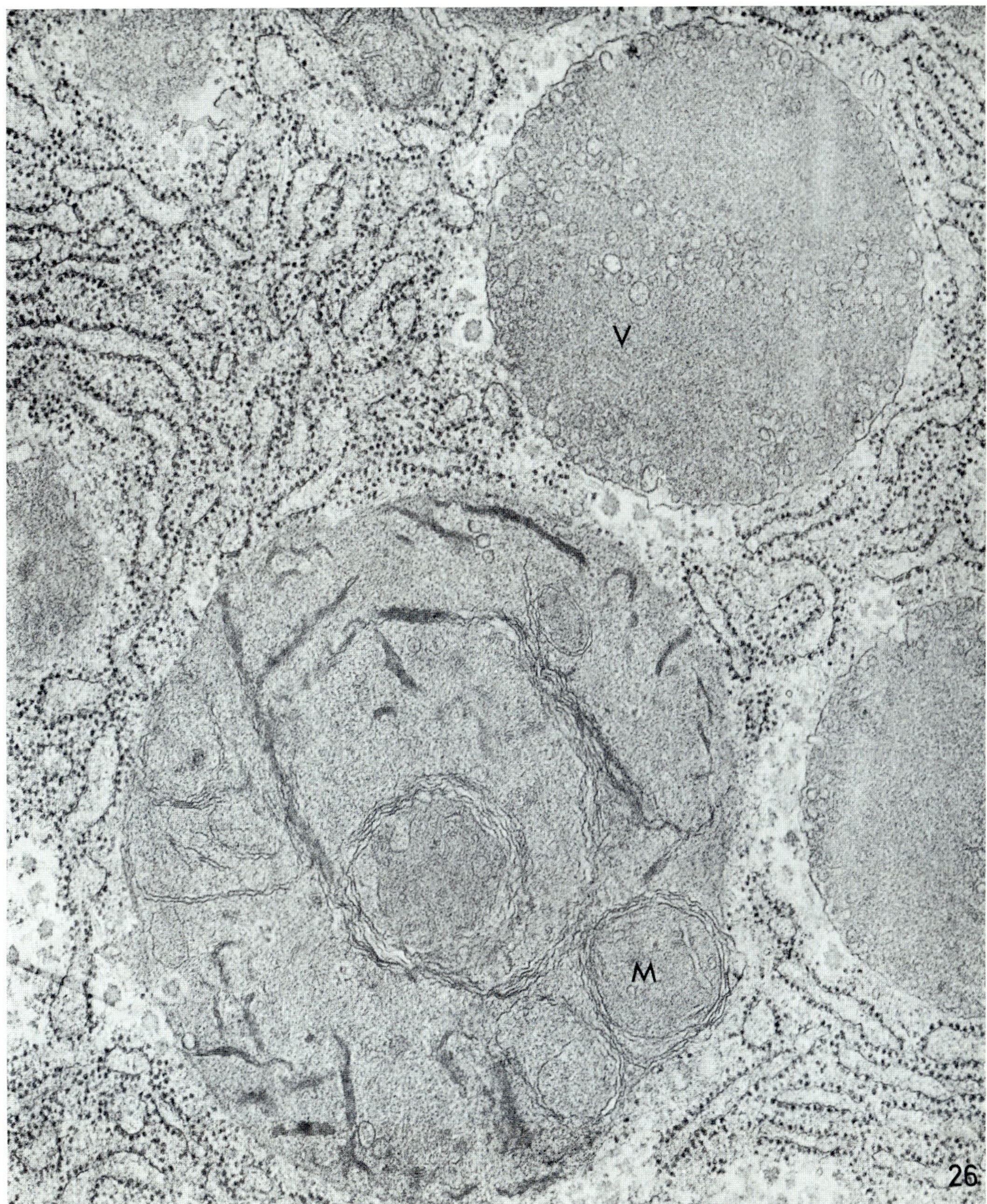

Fig. 26. Small portion of a cell from the fat body of *Calpodes* during the remodelling and protein storage processes that take place in the larva-to-pupa transition (Sections II.3.4 and V.3.2). Present in the cytoplasm is a large autophagic vacuole (note the mitochondrion within it at *M*) and a large multivesicular body (*V*); MVB's in this tissue participate in sequestering and storing blood-borne proteins. ×45,000. (From LOCKE, M., and J. V. COLLINS, 1965: J. Cell Biol. **26**, 857—884.)

are reportedly present in glucagon-induced autophagic vacuoles with a frequency much lower than that expected from their frequency in the cell (DETER 1971). And, during the recovery of hepatocytes from phenobarbital-induced hypertrophy of the smooth endoplasmic reticulum (BOLENDER and WEIBEL 1973) or during the regression of the prostate gland induced by castration (HELMINEN and ERICSSON 1971) ER is present with a disproportionately heightened frequency. To be sure, some of these observations may be affected by problems in recognizing organelles within the vacuoles, especially once digestion has altered their structure. And, for instance the fact that recognizable lysosomes are only rarely seen in autophagic vacuoles may reflect the abilities of lysosomes to fuse with the vacuoles rather than autophagic selectivity (DETER 1971). Nonetheless, it does seem that there can be non-random features to autophagy. One case that promises to provide some interesting information has recently been reported by MELMED, BENITEZ, and HOLT (1973). They have identified a small class of "intermediate" cells in the rat pancreas that seem, at least on morphological grounds, to contain both the characteristic exocrine secretion granules of the organ, and the endocrine secretion granules. In animals treated with alloxan, an agent that produces destruction of the (endocrine) β-cells, these intermediate cells survive, but they show what appears to be selective autophagy primarily of the endocrine types of granules. Another intriguing type of situation that might reward future studies is the destruction of specific organelles that sometimes occurs after gamete fusion or conjugation of unicellular organism (*e.g.*, there is a breakdown of one of the two chloroplasts in certain algal zygotes [BRATEN 1973] and of macronuclei in exconjugates of ciliates [BERGER 1973]); unfortunately, from present knowledge little can be said about possible involvement of autophagy.

What factors could account for autophagic selectivity? In some ways the most attractive possibilities are that special recognition devices or very localized changes in intracellular environments come into play; one could speculate about altered membrane surfaces, release of some "messenger" by injured or doomed organelles or even about interventions of microtubules or microfilaments [6]. Perhaps it will turn out that the signals are of a type that simultaneously affect many organelles of a given class. Such "class" signals could apply in the situations just cited. There are obvious technical problems in recognizing autophagy that is selective for individual organelles, if it does occur, and at present there is scant evidence to go on in evaluating possible special autophagic recognition mechanisms. It is worth noting that simpler considerations may also be germane. FARQUHAR and her colleagues stress that many of the secretion granules degraded through crinophagy are those that were very recently formed by the cell (FARQUHAR 1971). Since both secretion granules and lysosomes form in the vicinity of the Golgi apparatus, perhaps alterations in the rates of lysosome formation, coupled with changes in the rate at which new secretion granules move away from

---

[6] ARSTILA *et al.* (1974) find that intraperitoneal injection of vinblastine induces extensive autophagy in hepatocytes but there is no present reason to believe that this is due to effects on microtubules rather than to toxic influences.

the Golgi apparatus can increase the probabilities of an encounter between lysosomes and granules under circumstances propitious for fusion. In like manner, the topography and distribution of the membrane systems that carry out autophagic envelopment of organelles could influence the content of the vacuoles (see *e.g.*, LOCKE and COLLINS 1973). If, for example, ER is primarily responsible for processes of autophagic sequestration, then selectivity in autophagy might depend upon the geometry of the ER, the relations of the reticulum to other organelles (*e.g.*, possible continuities between them) plus the "signals" that induce a portion of the ER to undertake sequestration.

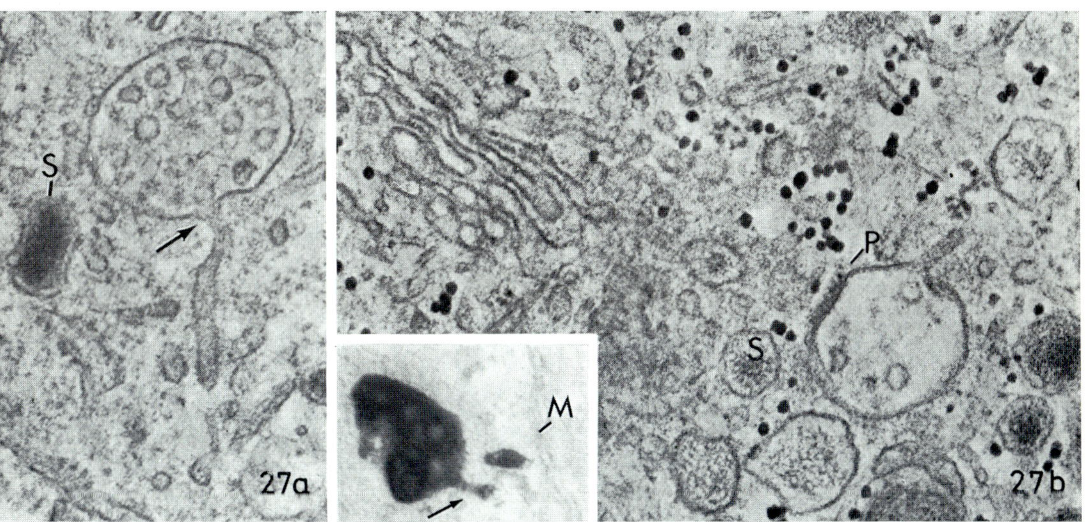

Fig. 27. Multivesicular bodies from rat adrenal medulla. *a* and *b* are from the epinephrine cells (note the secretion granules at *S*); the arrows point to smooth-surfaced membrane-delimited tubules attached to MVB's. A plaque (*P*) is present along part of the surface of the body in *b*. The insert is from a Schwann cell associated with one of the numerous axons present in the medulla; the preparation was exposed to horse-radish peroxidase and incubated to demonstrate peroxidase activity. Reaction product is present in an MVB; as is frequently the case the peroxidase-containing tubular tail of this body points towards the plasma membrane of the cell (*M*), which could reflect an endocytic origin of the structure (Sections II.4.1). *a* ×55,000; courtesy of S. ABRAHAMS. *b* ×50,000; from E. HOLTZMAN and R. DOMINITZ, 1968: J. Histochem. Cytochem. **16**, 320—336. Insert ×40,000. (From HOLTZMAN, E., 1969: In: Lysosomes in biology and pathology. [DINGLE, J. T., and H. B. FELL, eds.], Vol. **1**, pp. 192—216. Amsterdam: North-Holland Publishing.)

II.4. Multivesicular Bodies (MVB's)

To us, this term is best used on a morphological basis to designate bodies of the types illustrated in Figs. 26–30. The essential characteristic is the presence within the bodies of small vesicles (usually 50–75 nm in diameter) which constitute the primary recognizable intra-organelle structures. Often a portion of the multivesicular body surface appears as a plaque (Figs. 27 and 28) with organization different from the adjacent region. This may show radiating short projections reminiscent of vesicle "coating", or other less well-defined structure with no obvious resemblance to coating (it is not known if several types of plaques exist or whether the varied appearances reflect different microscopic views of the same type of region). In addition it is common to find portions of the MVB surface continuous with smooth-surfaced membrane-delimited tubules (Fig. 27).

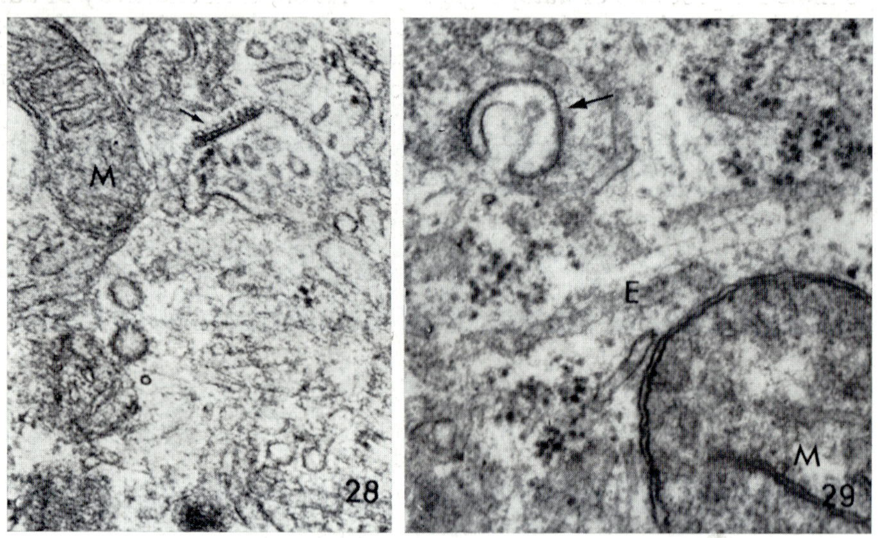

Figs. 28 and 29.

Fig. 28. Multivesicular body from the perikaryon of a chromatolytic neuron in the rat ganglion nodosum. The arrow points to a "plaque" of short projections along a region of the surface of the body. *M* indicates a mitochondrion. ×40,000. (From HOLTZMAN, E., A. B. NOVIKOFF, and H. VILLAVERDE, 1967: J. Cell Biol. **33**, 419—436.)

Fig. 29. Portion of a perikaryon of a neuron from a cultured mouse dorsal root ganglion. The arrow indicates a cup-like body in which a vesicle seems to be forming by budding from the infolded membrane that delimits the body. A mitochondrion is seen at *M* and endoplasmic reticulum at *E*. ×58,000. (From HOLTZMAN, E., and E. R. PETERSON, 1969: J. Cell Biol. **40**, 863—869.)

Multivesicular bodies are fairly numerous (several to quite a few per cell) in most cell types we have examined. They are frequent in the vicinity of the Golgi apparatus but certainly are not limited to this region. In preparations incubated for the cytochemical demonstration of acid phosphatase activity (Fig. 30) or aryl sulfatase activity (HOLTZMAN 1969, 1971) reaction product is seen within some multivesicular bodies but unreactive MVB's are commonly also present within the same cell. Bodies in which internal vesicles are surrounded by a dense matrix (Fig. 34) probably correspond to stages in the transformation of multivesicular bodies into residual bodies.

II.4.1. Heterophagic Roles

For many cell types, multivesicular bodies represent the major "depot" to which endocytosed tracer molecules are delivered. A very frequent pattern involves the formation of coated pinocytosis vesicles from the cell surface

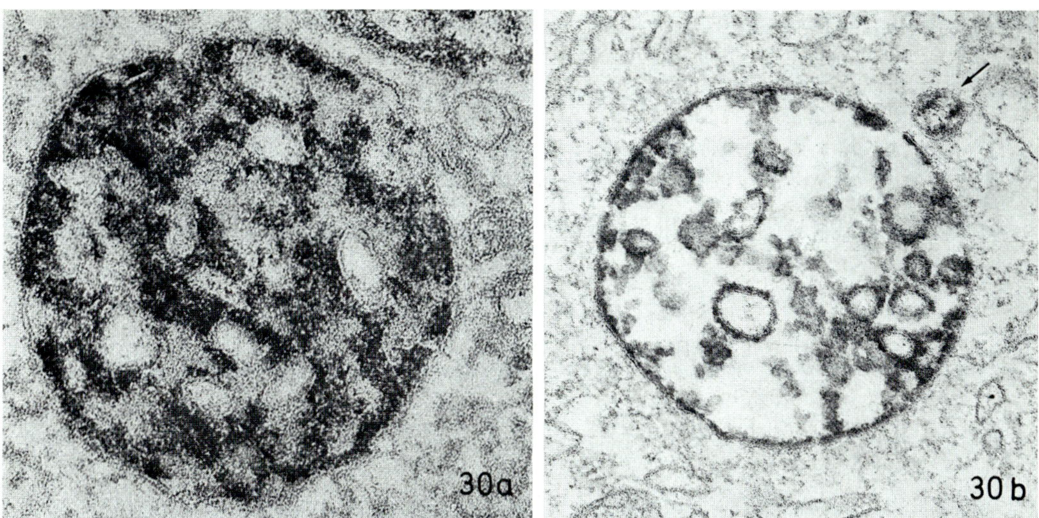

Fig. 30. Multivesicular bodies from the epithelium lining the rat epididymis. *a* From an acid phosphatase preparation. The electron dense reaction product is present in the spaces between the internal vesicles of the MVB. ×130,000. *b* From a preparation exposed to horse-radish peroxidase and incubated to demonstrate peroxidase activity. Reaction product is seen within a vesicle lying near the MVB (arrow) but product within the multivesicular body is present only on the outer surfaces of the internal vesicles and in the spaces between these vesicles. ×78,000. (From FRIEND, D. S., 1969: J. Cell Biol. **41**, 269—289.)

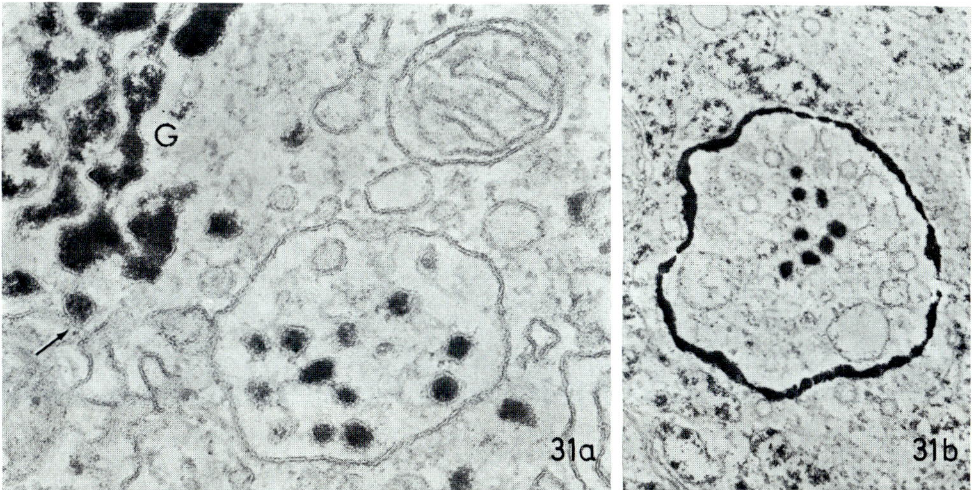

Fig. 31. From rat epididymis subjected to a prolonged soak in warm osmium tetroxide. *a* Osmium deposits are seen in sacs and vesicles (arrow) of the Golgi region (*G*) and within most of the internal vesicles of an MVB. ×70,000. *b* Osmium deposits are present in a sac surrounding a region of cytoplasm in which are found a few vesicles that also show deposits. Perhaps such configurations evolve into MVB's like the one in *a*. ×45,000. (From FRIEND, D. S., 1969: J. Cell Biol. **41**, 269—279.)

and the fusion of such vesicles with multivesicular bodies. The *dramatis personnae* are present in cells whether or not they have been exposed to exogenous tracers (*e.g.*, one can often find coated vesicles apparently in process of budding in from the surfaces of cells subjected to no special treatments). When tracers are present they usually accumulate outside of the vesicles within a MVB. From this, plus observations of the appropriate "inter-

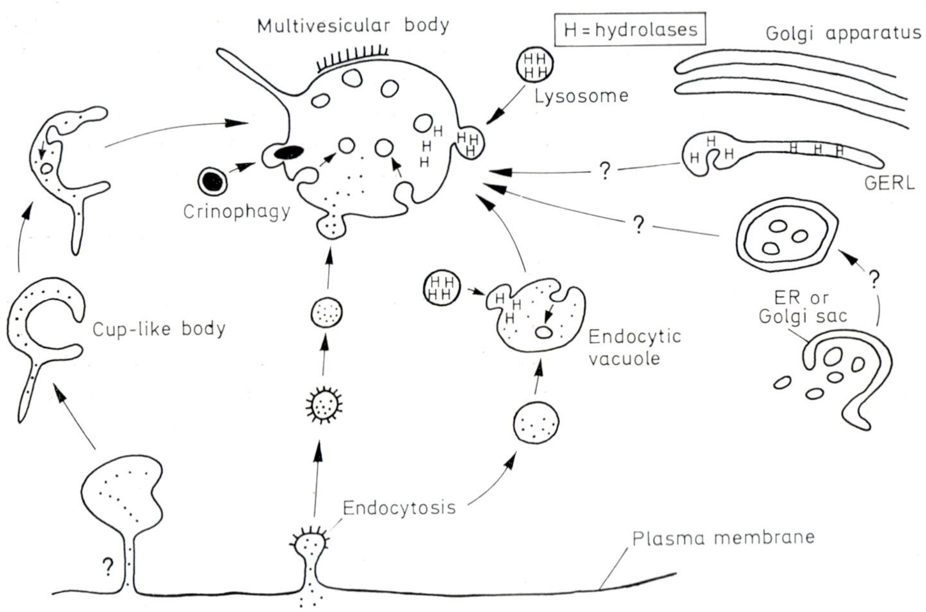

Fig. 32. Possible modes of formation of multivesicular bodies (Sections II.4.1–II.4.2). It is reasonably well established that MVB's can acquire hydrolases by fusion with lysosomes and that the internal vesicles can form by invagination and budding of the surface membranes. Membrane is added to the MVB surface from structures involved in endocytosis and probably in some cases also by fusion with secretion granules (crinophagy). That cup-like endocytic vacuoles transform into multivesicular bodies is likely but there is some uncertainty as to how the cup-like bodies initially form. Some MVB's may arise from dilated regions of GERL or Golgi sacs; not yet fully clear is whether this means that they can acquire hydrolases directly from GERL, and whether some of the bodies originate by autophagic-like sequestration of preexisting vesicles.

mediate" structures it is inferred that pinocytosis vesicles fuse with an MVB rather than somehow penetrating to the interior of the body. Hydrolases seemingly can enter the multivesicular body by fusion with lysosomes, although as with autophagic vacuoles this may be only one of the relevant mechanisms. The result of such fusions would be the presence, as is generally observed (Fig. 30), of the hydrolases in the same spaces as the exogenous materials and thus the intermingling of substrates and enzymes. Presumably multivesicular bodies that fail to react with cytochemical procedures for demonstration of hydrolase activities include some that have yet to fuse with lysosomes. (Fig. 32 summarizes key facets of MVB formation.)

Occasionally, preparations that have endocytosed horseradish peroxidase show cytochemical reaction product for peroxidase closely associated with the MVB vesicle membranes; the product may coat the external surfaces of the vesicles (as in Fig. 30) and sometimes this can create the impression that tracer is within the vesicles. There may also be cases in which a number of MVB vesicles actually do contain endocytosed molecule; some of LOCKE and COLLINS' (1968, 1973) tracer observations on *Calpodes* fat body merit follow-up along these lines. In addition, in a few studies, vesicles reactive for acid phosphatase have been noted within occasional MVB's (*e.g.*, HOLTZMAN *et al.* 1967, LOCKE and COLLINS 1973, LOCKE and SYKES 1975).

As will be evident as our discussion progresses it is possible that the internal vesicles of MVB's originate in several fashions. Studies of heterophagy have stressed one of these—configurations are frequently found that suggest the formation of the vesicles by invagination and budding of the membrane delimiting the MVB (Figs. 29 and 32; ARSTILA *et al.* 1969, FRIEND 1969, HIRSCH *et al.* 1968, HOLTZMAN *et al.* 1973, LOCKE and COLLINS 1968, NICANDER 1966). Vesicles formed in this manner would not be expected to contain either acid hydrolases or extracellular tracers. From our own studies on nervous tissue and the adrenal medulla we have proposed that multivesicular bodies can originate from cup-like structures (Fig. 29) whose inner membrane eventually vesiculates to produce the characteristic MVB morphology; cup-like bodies containing exogenous tracers and MVB's with large, deep infoldings of their surfaces (Fig. 34) are not uncommon in the tissues we have examined (HOLTZMAN 1969, 1971, HOLTZMAN and DOMINITZ 1968, HOLTZMAN and PETERSON 1969). Multivesicular bodies of the same tissues also show smaller invaginations of about the size expected if they were about to bud off and give rise directly to internal vesicles (HOLTZMAN *et al.* 1973). Probably the details of vesicle formation by invagination vary somewhat from cell type to cell type and perhaps also at different stages in the evolution of a given MVB.

For many tissues, small tubules are among the initial sites at which endocytosing cells accumulate tracers (see *e.g.*, ABRAHAMS and HOLTZMAN 1973, BECKER *et al.* 1968, HOLTZMAN and PETERSON 1969). Those tubules that are continuous with MVB's are among the ones accessible to endocytosed molecules (Fig. 27; HOLTZMAN 1969). Perhaps this sometimes means merely that material from the multivesicular body has diffused into the tubule. But our studies have led us to conclude that some MVB's may originate from dilatations of endocytic tubular infoldings of the plasma membrane (Fig. 32); what seem to be intermediates in this process are occasionally encountered (HOLTZMAN 1969, HOLTZMAN and DOMINITZ 1968), and ROSENBLUTH and WISSIG (1964) note that plasma membranes and MVB delimiting membranes are appropriately similar in thickness. Needless to say, the conceptual jumps from electron microscope images to dynamic process must be further supported but the notion does offer a simple explanation for MVB-tubule continuities: these might be persisting remnants of the original connections to the cell surface.

The significance of the MVB surface-plaques is not known. We have observed plaques on cup-like bodies (HOLTZMAN *et al.* 1967) and they have been described on other structures that probably are very early stages in MVB formation (ROSENBLUTH and WISSIG 1964). ROSENBLUTH and WISSIG maintain that each of the plaques that resembles the coating on endocytic vesicles occupies an area similar to the area of membrane bounding one such vesicle. But while the possibility that a plaque represents a region where a coated vesicle had just fused is attractive, some difficulties are created by the observations that endocytic vesicles generally seem to "lose" their coats soon after forming (Section I.4). Coating is present on some Golgi-associated sacs and vesicles (*e.g.,* FRIEND and FARQUHAR 1967, HOLTZMAN *et al.* 1967) as well as on endocytic structures. This triggers conjecture that its presence is a sign of molecular organization involved in such processes as the pinching of smaller structures from larger membrane surfaces, such as sacs or the plasma membrane. Perhaps the MVB plaques simply reflect the initiation of an MVB by a process of this type [7].

II.4.2. Participation in Autophagy

II.4.2.1. Incorporation of Secretory Material and of Intact Vesicles

Multivesicular bodies are prominent among the lysosomes responsible for crinophagy. This raises some tantalizing possibilities related to selectivity. Is there significance in the fact that the same types of structures (MVB's) can fuse with endocytosis vesicles derived from the plasma membrane and with secretion granules that normally fuse with that membrane during exocytosis? (It is not yet certain if an individual MVB can take part both in crinophagy and in heterophagy.) Or is a key to be found in the likelihood that some multivesicular bodies originate from the Golgi apparatus or GERL (see the next section) which are the systems that also give rise to the secretory bodies?

Multivesicular bodies in neurons occasionally are found to contain granules resembling the dense-cored vesicles or neurosecretory granules that the cells release by exocytosis (HOLTZMAN *et al.* 1967, HOLTZMAN 1971, PALAY 1960). Since some of these bodies in MVB's retain their delimiting membrane, they probably can enter by a process other than fusion with the MVB surface; mechanisms must also be sought to explain the occasional presence in MVB's of hydrolase or tracer-containing vesicles mentioned earlier. Were a cup-like precursor of an MVB to surround a membrane-delimited granule, the granule might eventually lie free within the MVB when the inner membrane of the cup broke up into vesicles or otherwise disappeared. Or perhaps some

---

[7] If suitably oblivious to the difficulties, one might conceive *a priori* that the plaques somehow generate the formation of vesicles from the MVB surface or even that they serve as binding sites for the endocytic-like incorporation of cytoplasm into multivesicular bodies that is envisaged in some discussions in the following sections. Unfortunately: the coating is on the wrong side for it to "produce" internal MVB vesicles in a manner analogous to endocytosis; except for protozoan food vacuoles, evidence is lacking for budding of vesicles in the opposite direction (into the adjacent cytoplasm); and coated vesicles or plaque-like regions are exceedingly rare within multivesicular bodies (HOLTZMAN *et al.* 1967).

other comparable sequestration processes take place. For example, in a number of cell types, configurations of the sorts shown in Fig. 31 have been found (FRIEND 1969, HOLTZMAN *et al.* 1967). These are the main evidence, such as it is, for the proposal that some multivesicular bodies can arise by segregation of pre-existing vesicles of the Golgi region, within a delimiting membrane that originates as an enwrapping sac, like that of conventional autophagic vacuoles. Sometimes sacs that might be carrying out such sequestration are coated over part or most of their surfaces (HOLTZMAN *et al.* 1967). FRIEND (1969) has determined that vesicles within some MVB's, vesicles lying free near the Golgi apparatus and the configurations that might represent vesicles in process of sequestration are all sites of heavy deposition of osmium after a prolonged high-temperature soak in osmium tetroxide (Fig. 31). This is suggestive support for the sequence under consideration, but it is not conclusive, especially since the basis of the deposition of osmium is not yet known.

II.4.2.2. Degradation of Membranes Participating in Endocytosis

This may be a major role of multivesicular bodies and one that accounts for the relatively high frequency of these bodies in many tissues (see also ARSTILA *et al.* 1971, LOCKE and COLLINS 1973). There are no available membrane "markers" adequate to demonstrate unequivocally where the vesicles within MVB's ultimately originate. However, particularly for the multivesicular structures derived from very large pinocytic vacuoles, such as those studied in macrophages by HIRSCH *et al.* (1968) and for situations in which large MVB's with very numerous internal vesicles participate in heterophagy (*e.g.*, Fig. 26; LOCKE and COLLINS 1973), it seems reasonable to advance the working hypothesis that much of the membrane internalized within MVB's as vesicles, derives from the plasma membrane. Probably endocytosis can deliver membrane to the MVB surface over an extended period, with the infolding and vesiculation of the surface providing both a means for controlling surface area and a mechanism by which a good deal of membrane can be incorporated within the lysosomes for ultimate degradation. [The amounts of membrane within a multivesicular body can be roughly estimated from the calculation that each internal vesicle of diameter 50–100 nm has a surface area on the order of 0.0025–0.01 μm^2. In large multivesicular bodies such as those of frog retinal photoreceptors (Fig. 42) or *Calpodes* fat body (Fig. 26; see also Fig. 34) there may be hundreds of vesicles.]

Endocytic-like processes participate not only in the cell's handling of exogenous macromolecules but also in the cycling of membrane between intracellular compartments and the cell surface. By now it is a fairly common observation that the addition of membrane to the cell surface, intrinsic to exocytic release of secretions and neurotransmitters, is followed by compensatory reduction of the surface area through endocytosis (Fig. 42; see review in HOLTZMAN *et al.* 1973 and for findings on a variety of cells, see AMSTERDAM *et al.* 1969, BUNT 1969, CECCARELLI *et al.* 1972, 1973, DOUGLAS *et al.* 1971, GEUZE and POORT 1973, HEUSER and REESE 1973, HOLTZMAN 1971, HOLTZMAN *et al.* 1971, McKANNA 1973 b, NAGASAWA and DOUGLAS 1972, ORCI *et al.* 1973, PELLETIER and PUVANI 1974, SCHACHER *et al.* 1974, TEICHBERG *et al.* 1974, 1975). Endocytic-like process also may serve to adjust plasma membrane area in fat cells as they undergo cycles of size changes related to their storage of fat (SLAVIN 1972, WASSERMAN and McDONALD 1960). Tracer studies indicate that at least some of the membrane removed from the surface by endocytosis accompanying exocytosis ultimately becomes incorporated in multivesicular bodies. For example, ABRAHAMS and HOLTZMAN (1973) have shown that when rat adrenal medulla cells are induced to release rapidly most of their adrenaline (epinephrine) stores, the cells subsequently show marked increases in endocytic activities and the accumulation of numerous large multivesicular bodies which ultimately transform into residual bodies (Fig. 34). If horseradish peroxidase is injected intravenously at appropriate times, the tracer accumulates in the multivesicular bodies, confirming their relationship to endocytosis. (The MVB's form whether or not peroxidase is present.) GEUZE and POORT (1973) have come to a similar view from their studies on the exocrine pancreas. But not all investigators agree that most of the membrane internalized by the cell during the events under discussion, is slated for lysosomal degradation—there is widespread belief that at least some is reutilized as intact membrane. As will be evident in a later segment of this book (Section III.4.3.3) this is part of a general controversy over the turnover of membranes, fueled by biochemical data of uncertain interpretation. More immediately germane to the present discussion are findings such as those of COPE and WILLIAMS (1973) whose morphometric measurements indicate that there is little change in lysosome size or in the volumes of other membrane-delimited intracellular compartments of the parotid gland after a period of intense secretory activity. They suggest that this might mean that membrane withdrawn from the surface somehow dissolves in the cytoplasm. However, LILLIE and HAN (1973) have made a seemingly contrary report—they do find increased numbers of lysosomes, perhaps due to their concentrating on different time intervals from COPE and WILLIAMS.

Coupled endocytic and exocytic phenomena may also be important for the normal turnover of cell surface material. Here again there is some evidence tieing in MVB's; in toad bladder epithelial cells induced by neurohypophyseal hormones or cyclic AMP to release surface coating material through exocytosis of Golgi-packaged granules, endocytosis is also greatly stimulated and endocytosed tracers are incorporated in MVB's (MASUR *et al.* 1971, 1972). Evidence along similar lines has been reported by HICKS (1966), from work on transitional epithelium of the rat urinary tract. In these cells, the characteristically thickened plasma membrane provides a natural "marker"; such membranes appear to be inserted in the cell surface through fusion of Golgi-derived structures, and after eventual withdrawal from the surface, at least some portions become incorporated within lysosomes (although many of these are not MVB's). PORTER *et al.* (1967) report similar findings.

In brief, a strong circumstantial case can be made for a role of MVB's as termini in biologically important phenomena of membrane circulation.

Still at issue however is the quantitative significance of this under varying conditions and in different cell types. Internalization of lysosome surface membrane within the organelles probably occurs with other categories of lysosomes as well as with MVB's and may contribute widely both to reduction of lysosome size and to aspects of membrane turnover. How and why membrane is internalized needs more study as does the fact that membranes within lysosomes are degraded whereas membranes at their surface are not (Section II.1.4.4). Perhaps the internalization process itself modifies the membrane. For MVB's one might also consider the possibility that the arrangement of lipids and other molecules at the surfaces of highly curved vesicles like those within the bodies, differs in some significant way from that at flatter surfaces (*cf.*, POSTE and ALLISON 1971); there may be alterations in molecular spacings, in charge distribution, or in other features that could influence the interactions of hydrolases and membranes.

II.4.2.3. Possible Roles in Degrading Other Types of Membrane

Apparent intermediates in MVB formation are frequent in the vicinity of the Golgi apparatus and some of the tubular extensions of the MVB surface may well correspond to portions of GERL or other Golgi associated systems (see *e.g.*, HOLTZMAN *et al.* 1967 and P. M. NOVIKOFF *et al.* 1971 for suggestive micrographs). The presence of acid phosphatase in some of the tubular "tails" of MVB's (*e.g.*, HOLTZMAN 1971) could reflect the direct entry of hydrolases from GERL or Golgi sacs as discussed in Section II.2.1. To be sure, some of the structures under consideration may have migrated in from the cell surface and their tubular tails may have an endocytic origin (Section II.4.1) but the opposite is true as well—there is no reason to believe that MVB's originating near the Golgi apparatus cannot travel toward the cell surface.

Assuming that MVB's can originate from membranes of GERL or Golgi apparatus what might be the consequences? Presumably the bodies would form by the infolding or sequestration mechanisms discussed earlier. Thus they could engage in a type of autophagy of Golgi associated membrane [8]. Present conceptions of the function of the Golgi apparatus and associated structures might imply that surplus membrane is generated in the functioning of the apparatus. This follows from the fact that proteins and other molecules to be packaged are transported to the Golgi region as the relatively dilute contents of membrane-delimited compartments (ER cisternae, vesicles and so forth) but they often leave the region in highly concentrated membrane-

---

[8] In view of the information in this and the previous section, we suspect that all MVB's engage in some kind of autophagy, and as yet there is no strong reason to believe that any MVB sub-population is excluded from heterophagy. Thus we hesitate to sub-categorize the organelles. Nonetheless it is worth noting that some "conventional" autophagic vacuoles containing fragments of ER can come to resemble multivesicular bodies. And NOVIKOFF (1973) has suggested the name "type-2 autophagic vacuole" for some Golgi-associated multivesicular structures. He feels that they are distinguishable from other MVB's on the basis of such morphological features as the tubular morphology of their internal "vesicles".

delimited structures with minimal (spherical) surface area. As already emphasized (Section II.2.1) much remains to be learned of the relevant transport and condensation processes. Maybe some of the apparently excess membrane shuttles back to the ER for renewed use in transport (JAMIESON and PALADE 1967). Or, perhaps the view that the membranes serve, in a sense, as "conveyor belts" is too oversimplified. But the possibility remains open that MVB-mediated autophagy comes into play here as one element in several that maintain the steady state.

During crinophagy MVB's probably incorporate and degrade secretion granule membranes (Fig. 32).

II.4.2.4. Microautophagy

The resemblance of infoldings of the MVB surface to more conventional endocytosis has not escaped the attention of "lysosomologists". Thus far we have stressed the internalization of membranes that results from the infolding but there also might be some internalization of molecules from the cytoplasm immediatedly adjacent to the MVB. In fact, the presence of membrane delimited granules within some MVB's seems to imply that one or another of the possible mechanisms by which the bodies form serves to incorporate cytoplasmic materials. Other observations point in the same direction. For example, from their electron-microscope studies FEDORKO and co-workers (1973) contend that some of the ferritin molecules which are seen within macrophage lysosomes after phagocytic degradation of erythrocytes, enter from the macrophage cytoplasm by inpocketing of the lysosome surface. (See also NOVIKOFF 1973 for his hypotheses relating to sequestration of glycogen and other molecules by agranular ER through invagination, or through coalescence of membranes around a region of cytoplasm.) This raises the interesting possibility that cells may continually autophagocytose cytoplasm a few molecules at a time ("microautophagy"; see DE DUVE and WATTIAUX 1965, POOLE and DE DUVE 1973). If cytoplasmic materials enter MVB's (or other lysosomes, cf., POOLE and DE DUVE 1973) by surface inpocketing, then mechanisms must be sought by which the molecules are released from the vesicles in which they enter. Given the degradative environment within the lysosome, proposals are not difficult to generate.

II.5. The Fate of Lysosomes

The "functional lifespan" of a lysosome might be defined as the period during which it continues to incorporate material for degradation, or perhaps the time it takes to convert into a residual body. One can only guess at meaningful numbers and presumably these vary depending on factors such as the digestibility of the lysosome contents and the levels of hydrolases in the cell. The evolution of protozoan food vacuoles takes place over the course of a number of hours (KITCHING 1956) and this seems to be the order of magnitude for the time-course of changes in other lysosomes as well (the "half-lives" of the digestible contents of hepatic and intestinal lysosomes have been estimated to be in the vicinity of one day: JAMES *et al.* 1971, SEGAL *et al.* 1974 a, b). When cells are induced to undergo enhanced autophagy, it often takes several hours for the contents of the new autophagic vacuoles to transform largely into residual materials (*e.g.*, ERICSSON 1969 a). "Professional" heterophagic cells such as macrophages, can require from three or four hours to an appreciable fraction of a day to complete digestion of exogenous proteins taken in during an initial period of extensive endocytosis (*e.g.*, PARK and COHN 1973). When one introduces microscopically visible tracers into an endocytosing neuron or other cell type, the timing with which tracer appears in different categories of bodies suggests that an MVB may survive as such for an hour or a few hours (*e.g.*, ABRAHAMS and HOLTZMAN 1973, GORDON *et al.* 1965); congruent with this are the observations that once tracers are incorporated into MVB's at nerve terminals it may be one to several hours before the labelled multivesicular bodies have disappeared from the terminals (which they do by intra-axonal transport to the perikarya and ultimate evolution into residual bodies; BIRKS *et al.* 1972, TEICHBERG *et al.* 1974, 1975).

II.5.1. Release vs. Retention

A priori one can readily think of two general possibilities for the eventual fate of a lysosome formed through autophagy or heterophagy. Either it could be released from the cell or otherwise disappear or disintegrate, or in some sense it could persist for the life of the cell. There is reason to believe that mechanisms approaching both of these "extremes" occur.

The contents of secondary lysosomes can be released from cells via processes resembling exocytosis, but this is thoroughly established only for protozoa. In many of these organisms such egestion of residual food vacuole materials is common—it represents a normal phenomenon of food-processing and takes place either at special points on the cell surface or at more random locations, depending upon the organism (see *e.g.*, ALLEN and WOLF 1974, DUBOWSKY 1974, KITCHING 1956, ROTHSTEIN and BLUM 1974 a). Comparable events occur during special phases of the life cycle—*e.g.*, *Acanthameba* is found to release previously phagocytosed latex spheres during encystment (STEWART and WEISMAN 1972). KITCHING (1956) advances the interesting idea that the ability to rid themselves of digestive residues and other indigestible substances permits protozoa to be somewhat nonselective in endocytic feeding and thus conveys evolutionary advantages in terms of seeking food.

In contrast to other protozoa, the suctorian *Tokophyra* seemingly lacks a defecation mechanism and accumulates increasing number of residual bodies with time. This organism reproduces by the budding of small offspring from a parent cell, which fills with digestive residues and eventually dies (RUDZINSKA 1966, 1970). ALLEN *et al.* (1966) have also claimed that the flagellate *Peranema* can be induced to "overeat" latex beads so that the organism becomes fatally engorged; apparently it cannot get rid of the beads fast enough.

The situation in metazoa is much more confusing. Some cells of metazoans utilize processes resembling exocytosis to release acid hydrolases to extracellular spaces as part of their normal functioning, or under abnormal conditions. For instance, cells of the prostate gland normally secrete acid hydrolases into forming seminal fluid, and under some circumstances cartilage cells or phagocytes such as PMN leukocytes and macrophages can release lysosomal enzymes in considerable amounts. At appropriate spots in the succeeding chapters we will consider these special cases in detail. For the moment, the important point is that these seem not to represent examples of cells ridding themselves of digestive residues. Suggestive evidence for such defecation by some cell types does exist, but it is not yet fully compelling. For example, the liver secretes hydrolases into bile at a rate approximating 5% of the total hepatic hydrolase content per day *; bile also contains various other components, such as copper and other metals, that could come from the lysosomes. Hydrolases also are present in urine and the circulation (see *e.g.*, GREBNER and TUCKER 1973, SLOAN *et al.* 1971). When perfused adrenal medulla preparations are stimulated to release catecholamines, there is a simultaneous release of a number of acid hydrolases into the perfusion fluid, as if some lysosomes had come to "exocytose" their contents along with the secretion granules (SCHNEIDER 1970, SCHNEIDER *et al.* 1967, SMITH 1970). This situation is potentially complicated by the presence in the secretion granules themselves of at least one acid hydrolase activity (acid phosphatase, demonstrable cytochemically; HOLTZMAN and DOMINITZ 1968) but this activity may not signify the presence in the granules of the other enzymes found to be released in the perfusion experiments (Section V.2.2 will discuss hydrolase activities in secretion granules).

Unfortunately, in neither the liver nor the adrenal has there been an unambiguous demonstration of the cellular or the intracellular sources of the released enzymes or of the release mechanisms. The fact that disrupted organelles and other debris are sometimes seen in bile canaliculi and in renal tubules (GLINSMANN and ERICSSON 1966, MAUNSBACH 1966) might reflect egestion of incompletely degraded contents of autophagic vacuoles from the hepatocytes and tubule epithelial cells. But, the very real possibility that the organelles actually are released through artifactual disruption of nearby cells during tissue processing (NOVIKOFF 1973) has yet to be ruled out. In some circumstances where many supposedly exocytosed residual structures are seen extracellularly (*e.g.*, in liver after albitocin treatment; KERR 1970, 1973), a detailed evaluation of cell death is also needed to strengthen the case. Similarly the observation that secondary lysosomes within hepatocytes tend to be found near the bile canaliculi is consistent with participation of these bodies in exocytosis but it is not decisive evidence; lysosomes commonly cluster near the Golgi apparatus, and the apparatus in hepatocytes tends to be adjacent to the canaliculi. One system that may be exploitable for resolving some of these ambiguities is the maturing red blood cell. Elimination of cytoplasmic organelles in these cells involves at least some intervention of lysosomes and evidence is available that can be tentatively interpreted in terms of exocytic release of lysosome contents (Section III.1.2.1).

---

* DEDUVE and WATTIAUX 1966. HOLDSWORTH and COLEMAN (1975) find very little acid phosphatase in bile of various mammals, but aryl sulfatases (and esterases) are present.

A few studies of the fate of tracer molecules that accumulate in lysosomes also suggest that defecation takes place. For example, MORGAN (1968) reports that in mouse exocrine pancreas, neutral red globules from the acinar cells are eliminated into the lumen of the acinus and from there pass into the pancreatic ducts. Triton WR-1339 after entering hepatic lysosomes gradually disappears from the liver (WATTIAUX *et al.* 1963, HERS *et al.* 1973); it takes many days for this to occur. BRADFORD *et al.* (1969) find that when iron-containing tracer molecules are administered to the liver via the blood stream, they (or derivatives) subsequently turn up in the bile; electron microscopy shows the presence in the bile canaliculi of bodies that contain the tracer and resemble the contents of hepatocyte residual bodies. GALLE (1974) has made similar claims for kidney based on his studies of uranium salts. In some of these experiments, toxic effects of the tracers may come into play; the implicit assumption, however, is that were tracer release to be due primarily to cell death, signs of such death would be much more prominent than they are.

There is, however, still need for circumspection. The fact that blood-borne tracer molecules enter in the bile is susceptible of more than one interpretation. Despite the fact that normally the bile canaliculi and blood stream are separated by highly restrictive cell junctions, it is possible in some cases that limited or abnormal extracellular routes between blood and bile are available (*e.g.*, CREEMERS and JACQUES 1971). Even if endocytic processes intervene, there is no firm reason to implicate lysosomes in the subsequent release of endocytosed materials into the bile: DAEMS *et al.* (1972) claim that some macromolecular tracers enter the bile from the blood much faster than expected for a process mediated by residual bodies, and others point out that by analogy to transport across the capillary wall one might expect some endocytic vesicles in hepatocytes to fuse directly with the plasma membrane bordering bile canaliculi rather than with lysosomes (GRAHAM *et al.* 1969, MA *et al.* 1974).

There is a body of information that points away from exocytic elimination. For example with macrophages, both in culture and *in vivo*, tracers or other non-digestible materials taken into the lysosomes persist there for periods of weeks or months—perhaps as long as the cell lives (CARR 1973, COHN and FEDERKO 1969, DAVIES *et al.* 1973 a, EHRENREICH and COHN 1967, VAN FURTH 1970, VERNON-ROBERTS 1972). Interestingly, COHN (1970; COHN and BENSON 1965) has demonstrated that the cellular content of hydrolases in cultivated macrophages can decrease markedly after a round of extensive heterophagy and that such decrease takes place without the release to the medium, of detectable hydrolases or of indigestible gold tracers previously incorporated by endocytosis. Presumably intralysosomal inactivation and degradation of hydrolases is responsible (COHN and FEDORKO 1969). Perhaps macrophages are not a good model system for our present considerations since their key functions are related to sequestration of potentially injurious material, which may make them particularly prone to retaining lysosome contents. However, observations of long-term retention of indigestible tracers have also been made with some slowly dividing and non-dividing tissue culture cells (*e.g.*, ALLISON and MALLUCCI 1964, BRUCK 1973, NOVIKOFF 1973) and some other cell types (*e.g.*, MAUNSBACK 1969). Furthermore, in storage diseases (Section IV.1) in which lysosome enzyme activities are missing, cells of many types show very heavy accumulations of lipid or polysaccharide-containing lysosomes; this hints strongly at severe limitations in the ability of cells to eliminate lysosomal contents.

We have already made the point that secondary lysosomes can fuse with new heterophagic or autophagic vacuoles and we have outlined direct evidence that endocytosed materials in lysosomes can be admixed with autophago-cytosed material. In addition, as discussed above, lysosomal volumes can decrease as digestion proceeds and there are mechanisms that result in diminution of surface area. Apparently, in the cell types discussed in the last paragraph, such retention and recycling phenomena predominate over release of lysosome contents. For such cells one might expect that the long-term consequences of their apparently limited abilities at defecation would depend in part upon the rates at which the cells divide, since a growing and dividing cell population presumably can dilute digestive residues among newer materials to a much greater extent than can a non-dividing one. Factors of these types probably accounts in some degree for the particular susceptibility of neurons to severe manifestations of lysosome malfunction (Section IV.1) and to the "normal" accumulation of increasingly evident deposits of residual materials as the cell ages (see next section). It is also commonly observed with those tissue culture cells that show a sharp decline in division rates after a period in culture (or upon "contact inhibition") that residual bodies become markedly more numerous when division slows (BROCK and HAY 1971, BRUNK *et al.* 1973, LIPETZ and CRISTOFALO 1972, ROBBINS *et al.* 1970).

One "special" fate of lysosomes may apply to some pathological circum-stances. During the response of axons to ligation or cutting, focal accumula-tions of lysosomal dense bodies are observed in several regions of the injured axons, including the portions destined to regenerate; as reorganization and regeneration proceed these dense bodies disappear. Some, at least, may be phagocytosed (along with other portions of the axons) by Schwann cells or invading phagocytes such as macrophages (see HOLTZMAN and NOVIKOFF 1965, MATTHEWS 1973). KERR (1973) claims that Kupffer cells can phagocytose lysosome contents released by hepatocytes after albitocin treatment. Cycles of release of lysosome contents and other material by dying cells and sub-sequent uptake of some of this by phagocytes, may be widespread at sites of inflammation (Section IV.3); often these phagocytes themselves die with their contents being phagocytosed anew. Such processes seem unlikely to contribute much to normal lysosome turnover in most tissues. But there are some special cases, such as the phagocytosis by Sertoli cells, of lysosome-laden cytoplasmic droplets shed by maturing sperm (DIETERT 1966).

Finally the behavior of one type of lysosomal residue merits some comment. Structures whose composition is not known in detail but with the appearance of lipid droplets, commonly are seen in lysosomes as they transform into residual bodies (*e.g.*, Figs. 25 and 34). This might be the consequence of the "special problems" in lipid degradation alluded to at several points above, but conceivably, newly synthesized lipid from the ER could also contribute, under some circumstances (Sections II.2.1 and II.5.2). In a number of cases, subsequent to the accumulation of lipid in residual bodies, lipid droplets have been found "free" in the cytoplasm. Microscopic observations suggest that this can result from extrusion of the droplets from the bodies (*e.g.*, Fig. 25; SMITH and FARQUHAR 1966) or by a process resembling transformation of a lysosome into a lipid droplet (HOLTZMAN and NOVIKOFF 1965, WAKE 1974). Presumably the lipid can in some sense "dissolve" or incorporate the lysosome membrane since many of the "free" droplets are not delimited by a visible membrane. Eventually the droplets disappear from view, probably by dispersal into the cytoplasm or use in metabolism.

On balance, despite the many caveats, and the paucity of reliable information, it seems more likely than not that some cell types of mammals and other metazoa do engage in extensive release of digestive residua, through exocytic defecation. But, it should be evident from the matters raised in this section, that such release is not the only "interesting" thing that can happen to residual materials. For some cell types, release may occur to only a minor extent, if at all.

II.5.2. "Telolysosomes" and Lipofuscin

Are there categories of lysosomes that no longer contribute to degrading either endogenous or exogenous materials and thus might be thought of as "telolysosomes"—terminal, nonfunctional types? No thoroughly detailed study of this possibility has yet been one. However, there are a few suggestive situations in which some of the lysosomes in a given cell appear to reach a state in which they no longer are accessible to exogenous tracers that enter many of the other lysosomes in the same cell. (ABRAHAM *et al.* 1973, CARR 1973, DAEMS *et al.* 1969, 1972, DAVIES 1973, MAUNSBACH 1969, NOVIKOFF 1973). DAEMS *et al.* (1972) cite as an apposite example, their observations that the large residual bodies present in mouse spleen macrophages do not accumulate Thorotrast, although the cells do endocytose this colloidal tracer and incorporate it in other lysosomes. These workers also are impressed by the fact that when hepatocytes are exposed repeatedly to intravenous dextran, the cells do not form increasingly large dextran containing lysosomes, but rather form increased numbers of digestive bodies which maintain a "rather narrow range of dimensions"; this, they assert could reflect some "maximal loading capacity" of an individual lysosome. However, SCHELLENS (1974) also reports that mouse hepatocyte residual bodies, initially loaded with thorotrast, are subsequently still able to accumulate other endocytosed tracers, such as gold, over a prolonged period. More work along such lines is needed, to better define the pertinent dynamics of the lysosome population within a given cell.

One type of lysosome that is often thought of as possibly nonfunctional is the lipofuscin pigment granule. These bodies are especially frequent in hepatocytes, muscle cells, neurons, and some other cell types of aging animals. Lipofuscin may occupy as much as $5^0/0$ to $10–20^0/0$ or more of the cell volume in different cells or circumstances (*e.g.*, BRIZZEE *et al.* 1969, REICHEL 1968).

The granules appear as yellow-brown bodies that have a characteristic morphology (Fig. 33) and distinctive staining, cytochemical and autofluorescence properties. The presence of acid hydrolases can be demonstrated both cytochemically (ESSNER and NOVIKOFF 1960, FRANK and CHRISTENSEN 1973, GOLDFISCHER *et al.* 1966, SAMORAJSKI *et al.* 1964) and biochemically in appropriate cell fractions (HENDLEY and STREHLER 1965). TAPPEL (1973)

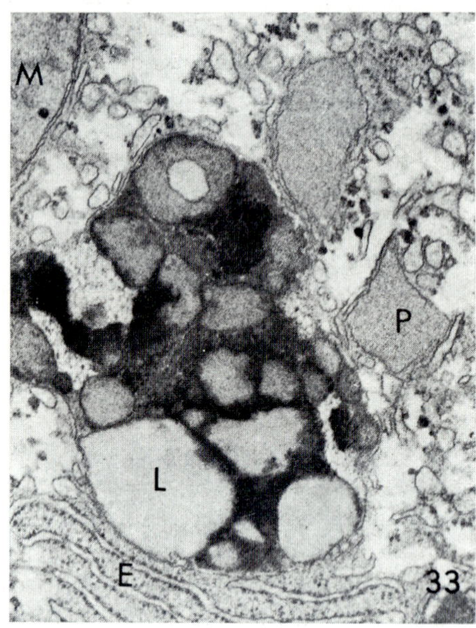

Fig. 33. Lipofuscin granule from a human hepatocyte showing the characteristic heterogeneous content of such bodies. Globules with the electron microscopic appearance of lipid (*L*) are interspersed among areas of lighter and darker electron density. At suitable magnifications the darker areas are often found to contain small "ferritin"-like grains. *E* indicates endoplasmic reticulum, *P*, peroxisomes (microperoxisomes) and *M*, the edge of a mitochondrion. (Fixed in glutaraldehyde followed by osmium tetroxide.) ×30,000 (approx.). (From NOVIKOFF, A. B., P. M. NOVIKOFF, N. QUINTANA, and C. DAVIS, 1973: J. Histochem. Cytochem. **21**, 1010—1020.)

and his co-workers argue from fluorescence and staining studies that lipofuscin bodies are especially rich in insoluble, polymeric molecules derived from unsaturated lipids and from proteins, through peroxidation reactions—material with properties comparable to lipofuscin components, can be generated in the test tube by incubation of cell fractions containing mitochondria or microsomes under conditions promoting such reactions (ARSTILA *et al.* 1972, CHIO *et al.* 1966). On the basis of cytochemical studies, GOLDFISCHER *et al.* (1966) stress the presence of iron-containing molecules (probably groups resembling hemes) and some copper compounds in lipofuscin, and they point out that such materials could catalyze a variety of non-enzymatic oxidation or peroxidation reactions (see also NOVIKOFF *et al.* 1971).

Although lipofuscin has traditionally been thought of as an "aging pigment" or a "wear and tear" pigment, this clearly is an over-simplification. Bodies of quite similar type are numerous in adrenal cortex and some other tissues of non-aged animals, comparable structures (rich in copper) have been noted in hepatic cells of new-born humans (GOLDFISCHER and BERNSTEIN 1969) and increases in lipofuscin-like bodies accompany some pathological conditions (BIAVA and WEST 1965, ZEMAN and SIAKOTOS 1973). In part since residual bodies in several cell types sometimes resemble lipofuscin in morphology (see *e.g.*, Fig. 34 and BRUNK and ERICSSON 1972), it is widely assumed that lipofuscin granules are a type of residual body; presumably their distinctive content reflects the abundance, in certain cells, of particular metabolic materials. Groups such as hemes and components such as unsaturated lipids abound in many organelles (mitochondria, ER, etc.) so it may not be surprising if the materials that accumulate in residual bodies (especially those derived from autophagic vacuoles) come to be modified by interactions between molecules of these classes. What is not yet understood is why particular cells accumulate lipofuscin in particular situations. The relations with aging although certainly not absolute, are still intriguing. Perhaps lipofuscin can represent a category of "ultimate residual-body material" that builds up slowly as a result of the repeated recycling of lysosomal constituents through fusions of old lysosomes with newer bodies, coupled to the gradual cross-linking of residues into insoluble polymers. As noted in the last section, such material could become prominent in cells whose abilities to release lysosomal components extracellularly are inadequate or impaired, and those that do not divide often enough to "dilute" the granules sufficiently with "newer" material. There also has been interesting speculation that vitamin E might influence rates of lipofuscin formation, based upon the supposed antioxidant properties of the vitamin and the fact that experimental vitamin E deficiencies of animals can involve a variety of changes in lysosome numbers and properties (HOCHSCHILD 1971, TAPPEL 1968 b, TOTH 1968, ZEMAN and SIAKOTOS 1973). The observation that steroid producing cells, such as those of adrenal cortex, often contain lipofuscin-like bodies, might reflect the high concentration, in these cells, of lipids that are prone to the pertinent modifications. (*E.g.*, FRANK and CHRISTENSEN 1968, feel that autoxidation of unsaturated lipids may give rise to lipofuscin in interstitial cells of the testis.)

There are a few other types of "pigments" such as ceroid and hemofuscin that resemble lipofuscin in containing altered lipids and metal and may have similar significance (see *e.g.*, NISHCOKA *et al.* 1968). A preliminary report by ARMSTRONG *et al.* (1974) claims that certain diseases in which pigments of these types accumulate ("ceroid-lipofuscinoses") are characterized by deficiencies in peroxidases in leukocytes although what this means for lysosomes of other cell types is not clear. It is too early to even speculate about how such a peroxidase lack might affect the peroxidations presumed to generate lipofuscin.

That lipofuscin may be a type of "telolysosome" is suggested by the fact that its content of lytic enzymes is reportedly meagre—cathepsins are prominent (HENDLEY and STREHLER 1965) but most other enzymes are present only in low concentrations. However, NOVIKOFF *et al.* (1973) contend that lipofuscin granules in guinea pig liver are attached to smooth surfaced tubules. If these are agranular endoplasmic reticulum then it could turn out, for example, that the tubules contribute to the movement of lipids or other components to or from the lipofuscin granules (*cf.*, Section II.2.1). In other words, it may be too early to count lipofuscin granules out as active participants in cellular metabolism. This relates to the unresolved issue: does the presence of lipofuscin affect the cell's function in some deleterious fashion that might be of importance for aging or dysfunction? RUDZINSKA (1962) has shown that the protozoan *Tokophyra* (see Section II.5.1) dies young if overfed; overfeeding produces a more rapid accumulation of residual bodies to the level normally attained only in "old" individuals. One could conceive for example that the presence of many residual bodies somehow interferes with normal cell functioning—perhaps the probabilities of fusions of younger lysosomes with new heterophagic or autophagic vacuoles are diminished by "competition" from the older ones *. However, as will be discussed more fully when we consider storage diseases (Section IV.1) distinguishing cause from effect is rarely possible in the situations where "abnormal" lysosomes are present. Theories of aging have also been advanced in which the accumulation of lipofuscin is regarded as a secondary consequence of the piling up in cells of macromolecules resistant to degradation; this might occur if errors in synthesis become more frequent (ORGEL 1973) or if reactions such as lipid peroxidations engender increasingly damaged membranes (PACKER *et al.* 1967, SHELDRAKE 1974, TAPPEL 1968). In contrast to the scheme implicit in the discussion above, the abnormal materials would be generated primarily outside the lysosomes (although further modifications could occur subsequent to autophagy). As they become more abundant, the abnormal components could interfere directly with the functions served by their normal counterparts.

The perplexing, enexplained observation has been made that certain drugs (centophenoxane; dimethylaminoethyl-p-chlorophenoxy acetate) can diminish the amounts of lipofuscin present in guinea pig neurons (HASAN *et al.* 1974, NANDY and BOURNE 1966).

II.6. Lysosome Heterogeneity

That lipofuscin granules are relatively rich in cathepsins and poor in other hydrolases is one of a collection of findings suggesting that there may be important quantitative differences in hydrolase content among lysosomes. There is little doubt that the lysosome populations of different cell types and different organisms vary in the relative proportions of given hydrolases and probably in other properties as well. This can be shown through bio-

---

* Although they do not explicitly discuss lysosomes, SMITH-SONNEBORN and KLASS (1974) mention the possibility that "aging" in clones of *Paramecium* may involve changes in the efficiency with which key precursors enter metabolic pools from endocytosed sources. These authors are thinking specifically about the utilization of nucleic acid precursors obtained from phagocytosed bacteria.

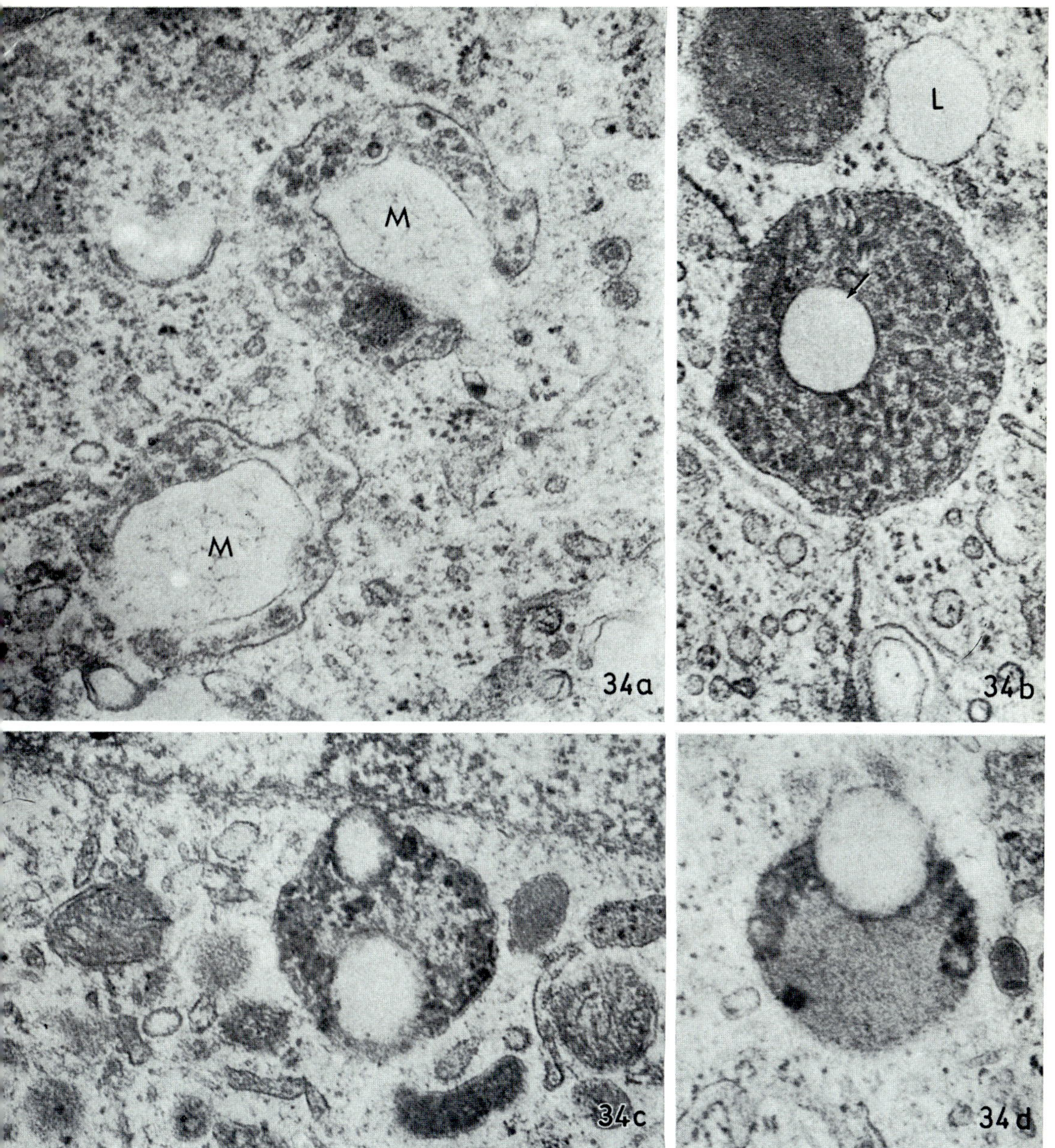

Fig. 34. Residual bodies. These micrographs are from epinephrine cells of rat adrenal medulla during the period of membrane degradation that follows insulin-induced depletion of secretory stores (ABRAHAMS and HOLTZMAN 1973; the membranes degraded probably are those that initially delimited secretion granules and have been endocytically "retrieved" from the cell surface subsequent to exocytic release of their contents). The bodies labelled *M* in *a* are cup-like stages in the formation of multivesicular bodies. Eventually the MVB's evolve into structures of the type shown in *b*: a dense matrix surrounds the internal vesicles and globules with the appearance of lipid accumulate (arrow). Note also the lipid globule lying free in the cytoplasm (*L*). Subsequently the MVB's transform into residual bodies like that shown in *c*; these resemble lipofuscin granules (Fig. 33). FARQUHAR and co-workers propose that residual bodies of the type shown in *d* may be in process of releasing their lipid inclusions into the cytoplasm (*cf.*, Fig. 25). ×50,000 (approx.).

chemical studies (for representative results see ARBORGH *et al.* 1973, 1974, BADENOCH-JONES and BAUM 1974, HOOK *et al.* 1973, SINHA and ROSE 1973) and also has been held responsible for differences in the intensity of reaction observed in cytochemical preparations (see *e.g.*, GOLDFISCHER 1965, NOVIKOFF 1973). Much less clear, however, is the extent to which the lysosomes *within* a given cell differ and especially the situation at the level of primary lysosomes. MILSON and WYNN (1973) report heterogeneity in the lysosome population isolable from what seems to be a "homogeneous" tissue culture line, but even here the possibilities of cell to cell variations complicate conclusions about intracellular heterogeneity. Advantage has been taken of the large size of some heterophagic secondary lysosomes for demonstration, by sequential cytochemical incubations, of the presence of more that one type of acid hydrolase within the same body (HOLT 1963, NOVIKOFF 1976 b). A similar inference can be drawn for PMN leukocyte primary lysosomes from comparisons among cells incubated for one or another cytochemically demonstrable activity. It is true that in acid phosphatase or aryl sulfatase preparations there often are both reactive and unreactive azurophilic granules within a given PMN cell (the uniformity of reaction seems better with peroxidase incubation [BAINTON *et al.* 1971]; as Section I.3.2.1 outlined, some kind of cytochemical "latency" is observed with the hydrolases). But given the number of granules present in the cells and the number of enzymes present in lysosomes it is virtually certain that a single azurophilic granule contains at least several different types of hydrolases. With eosinophils, acid phosphatase (and peroxidase) can be demonstrated in most of the granules of suitably prepared cells, so this hydrolase must overlap with the others (SEEMAN and PALADE 1967). In other words, the extreme condition—one lysosome, one hydrolase—seems very unlikely.

Less dramatic possibilities for heterogeneity are still to be evaluated. For example, STOSSEL *et al.* (1971) have studied the relative levels of enzymes present in phagocytic vacuoles isolated from guinea pig PMN leukocytes at different intervals after the induction of phagocytosis. They find that acid phosphatase continues to enter the vacuoles for some time after peroxidase and β-glucuronidase have largely finished accumulating. This could be interpreted as indicating heterogeneity in enzyme content of the lysosomes responsible for delivering the enzymes, but a more detailed characterization of the enzymes involved and study of additional hydrolases is needed to put this interpretation on solid ground (*e.g.*: Is the acid phosphatase exclusively lysosomal? Perhaps not, since BRETZ and BAGGIOLINI 1974 have found a nonlysosomal acid nitrophenyl phosphatase along with alkaline phosphatase in a membrane-containing fraction of human PMN leukocytes; see also Section V.2.2). ZEYA and LAZLO (1973) claim for myeloblasts of leukemic humans that the particles containing acid phosphatase (assayed with p-nitrophenylphosphate), those rich in β-glucuronidase and those containing peroxidase are separable to a substantial degree by density gradient centrifugation. This, they suggest, might mean that the enzymes are packaged separately during the initial phases of cell maturation; the enzymes seem to enter common particles later (in myelocytes), presumably through fusion of the different packages. Much more information about the cells and the particles must be gathered before this unorthodox view can be decisively tested *.

It is common in gradient ultracentrifugation studies of many cell types, to find some variations in the relative content of different hydrolases among lysosomes at different regions of the gradient. One example is illustrated in Fig. 3. Sometimes it is possible to identify in this way, distinct subcategories of lysosomes varying markedly in enzyme content. This is true for some protozoa, where intracellular heterogeneity seems a likely explanation. A number of protozoa release acid hydrolases to the extracellular medium (ARONSON 1973, EECKHOUT 1973, MÜLLER 1971, 1972, ROTHSTEIN and BLUM 1974 a, b, SOMMER and BLUM 1965)[9]. For *Tetrahymena* the "secreted" enzyme complement differs from the average intracellular complement in the relative proportions of various enzymes (*e.g.*, in the extracellular medium, protease activities are disproportionately low). This appears to reflect the selective involvement in release, of a class of intracellular lysosomes of appropriate "special" hydrolase content and with a buoyant density greater than that of the remainder of the cells' lysosomes (MÜLLER 1972, ROTHSTEIN and BLUM 1974 a). These high density particles disappear from the cell when hydrolases are secreted under starvation conditions; in these circumstances new hydrolase synthesis does not occur so the release can be unambiguously attributed to preexisting bodies.

A few observations of special subtypes of lysosomes have been made for metazoan tissues, but distinguishing between cellular and intracellular variation is more difficult. One dramatic example is the class of cathepsin D-rich organelles obtained from rat spleen and other lymphoid tissues by BOWERS (1969, 1970, 1973). Do these correspond to a special category of lysosomes involved in the unique functions of lymphocytes? Or is it merely that cathepsin D survives within residual bodies for a longer time than other enzymes due presumably to evolutionary factors that stabilize the enzyme against autodegradation? Differential enzyme survival could produce a spectrum of hydrolase contents in secondary lysosomes of virtually any cell. One could investigate these possibilities by introducing lysosome extracts into other

---

* ULLYOT and BAINTON (1974) report that the blood neutrophils obtainable from leukemic humans may include some interesting aberrant PMN types; in some, alkaline phosphatase or peroxidase may be missing from the granules, while in others, the entire specific granule or azurophilic granule population seems to be absent. Such heterogeneity of cell types could prove to be a nuisance for interpretations of studies on isolated granules. However, the aberrant cells may also be useful for future analysis of granule formation.

[9] Conventional wisdom has it that the enzymes can aid in nutrition, but MÜLLER (1972) points out that under usual conditions they would be enormously diluted soon after release. The extracellular acid phosphase induced in *Euglena* by phosphate starvation seems to be associated with the cell surface (SOMMER and BLUM 1965) and thus its effects might be sufficiently localized to "aid" the cell, but this may not be a lysosomal enzyme. [BLUM (1965) reports that it differs from the constitutive intracellular phosphatase in its responses to inhibitors and other properties.] ROTHSTEIN and BLUM (1974 a) have found that pharmacological agents such as catecholamine antagonists that enhance enzyme secretion from *Tetrahymena* also enhance egestion of digestive residues, which may mean either that the secretion and defectation mechanisms are identical or that they are under similar cellular controls.

lysosomes via endocytosis (Section IV.1) or perhaps by following the evolution of macrophage lysosomes after a bout of heterophagy. Such approaches also could clarify influences of the relative prominence of particular classes of macromolecules in the substrate population (see *e.g.*, MAACK 1967), possible intervention of inhibitory components (some of which might be formed as degradation products within the lysosome), and a number of other factors which must be examined for any detailed determination of the significance of variations among secondary lysosomes.

III. Lysosomes in Turnover and Modulation

A balanced discussion of the turnover of cells and of macromolecules under normal conditions, would take us far afield of our major topic and into questions of the regulation of macromolecule synthesis, cell division and so forth. Such broad ranging perspectives are reflected in some recent publications concerning turnover at one or another level of cell organization (see *e.g.*, CAMERON and TRASHER 1971, GOLDBERG and DICE 1974, POOLE 1971, RECHCIGL 1971 a, SCHIMKE 1973). We will restrict ourselves to degradative aspects of turnover and particularly to determining how lysosomes might fit in. One might expect lysosomes to play an essential role in turnover since these organelles represent the major intracellular sites of the known appropriate enzymes (POOLE and DE DUVE 1973). This is not to imply that non-lysosomal enzymes are not at all involved. But the specificities of the few relevant non-lysosomal suspects, such as the dipeptidase that is present in the "cystol" fraction of cells (BARRETT and DINGLE 1971, MARKS and LAJTHA 1970, McDONALD *et al.* 1971, PETERS 1970) suggest that their roles are ancillary [10]. Despite such considerations, direct demonstration of lysosomal participation in turnover has been possible chiefly for materials that function in extracellular locations and are destroyed through endocytosis. Even for these components we know much less than might be expected, especially about the mechanisms through which molecules come to be "selected" for destruction. And, the involvement of lysosomes in normal breakdown of *intracellular* materials is still a contentious matter.

III.1. Turnover of Cells and Tissues of Higher Animals and Features of Developmental Remodelling

III.1.1. Background and an Example: The Red Blood Cell Life History

Hydrolase levels reportedly do not change much in the epithelial cells of the intestinal mucosa once the cells have left the crypts and begun to migrate

---

[10] There are some "special function" enzymes such as the nucleases involved in intranuclear processing of RNA which almost certainly have little to do with the lysosomes; some others such as the protease responsible for processing of proinsulin (Section V.2.2) must be studied further in this regard. Also of uncertain status are such enzymes as the proteases reported to be present in mitochondria (D'MONTE *et al.* 1970, GEAR *et al.* 1974, LOVAS 1974); skepticism has been expressed about these in light of the possibilities of lysosomal contamination of mitochondrial fractions.

toward the tips of villi (DE BOTH *et al.* 1974) and there is little evidence implicating lysosomes as causative factors in the turnover of the epithelium. Degradation of the cells presumably follows upon their being sloughed off into the lumen where there are abundant digestive enzymes. Loss of cells by sloughing on a regular basis is seen in the epidermis of skin. Passage of cells slated for death across lining epithelia and into lumens probably occurs to some extent in a number of locations such as the respiratory tract. Holocrine secretion by sebaceous glands also results in loss of cell contents to the body exterior.

Beyond these cases, little is known of the destructive routes involved in steady state turnover of cells (CAMERON 1971, GOSS 1970). The major exceptions are the blood cells and their relatives, such as the macrophages. These cell types have restricted life-spans that range from a few days for some mature granulocytes to about 120 days for mature human erythrocytes and weeks or months for many macrophages. We will focus here on erythrocytes since their turnover under normal conditions has been studied most thoroughly. There are relatively few cases of well characterized materials being destroyed after a fixed life span. Red blood cells are a useful model both for looking into the kinds of factors that might contribute to this type of turnover, and for obtaining clues as the controls of degradation of other circulating materials. Further, they illustrate nicely some possible intracellular roles of lysosomes in a special "extreme" developmental context.

III.1.1.1. Lysosomes in the Maturation of Red Blood Cells

The loss of organelles, including, for mammals, the departure of the nucleus, is a striking feature of erythrocyte maturation. It might be anticipated that autophagy would be extensive in maturing red blood cells and at least in some species one does find fair numbers of autophagic vacuoles and probable residual bodies present at appropriate stages of cellular differentiation. This has been most thoroughly studied in amphibians (GRASSO 1973, TOOZE and DAVIES 1965) but similar bodies have been identified in mammals, including man (*e.g.*, KENT *et al.* 1966). Since micrographs convey little information about rates, it is hard to know if the frequency of such vacuoles is great enough to account for most of the loss of cytoplasmic organelles.

Mitochondria are readily identifiable within erythrocyte autophagic vacuoles; sometimes they seem to show abnormal morphology prior to their sequestration (GASKO and DANON 1972; GRASSO 1973). This might not be surprising in light of the unusual intracellular environment conditioned by the high concentrations of hemoglobin; or, perhaps the reactions of the cells to fixatives used for microscopy are to blame. But it does raise the possibility that the organelles "atrophy" via processes not initiated by the lysosomes. As an extension of this it has been argued occasionally that it is the changes in the mitochondria that provoke their autophagy, or alternatively that altered mitochondria and other structures can be extruded directly from maturing red cells. A proposal related to the latter suggestion is that small buds packed with mitochondria, ribosomes or other organelles might pinch off the cells.

Microscopic observations of mitochondria or cytoplasmic fragments lying free near maturing cells have been made but they are fraught with possibilities for artifact and cannot yet be reliably interpreted (they could be due, for example, to rupture or fragmentation of a few cells). It should also be remembered that a mechanism must be sought for disappearance of autophagic vacuoles since such vacuoles are not at all frequent in mature erythrocytes; exocytic extrusion of vacuole contents at the cell surface could account for appearances resembling direct loss of organelles, and images suggestive of this fusion process have been observed (SIMPSON and KLING 1968, 1970). Interestingly, the loss of vacuoles from maturing red blood cells is markedly diminished in splenectomized individuals (HOLROYDE and GARDNER 1970, KENT et al. 1966). As outlined in the next section, the spleen is able to remove some types of abnormal inclusions from red cells; apparently this ability extends also to autophagic vacuoles (B. T. SCHNITZER et al. 1971).

Numerous polysomes persist until relatively late in erythrocyte maturation. Usual schemes attribute their degradation to RNAses, but relatively little attention has been paid to the lysosomes in this regard. Ribosomes are seen in membrane-delimited vacuoles in erythroblasts (CAMPBELL 1972).

The nucleus of the maturing cell clearly is not autophagocytosed. Rather (in mammals) it is included within a surrounding thin cytoplasmic rim which detaches from the erythroblast and in this form it is phagocytosed and degraded in phagocytic vacuoles by the macrophages that abound in hematopoietic tissue (CAMPBELL 1968). Before separating from the maturing cell, the nucleus becomes condensed and eccentric in position. The suggestion has been made that the actual separation in bone marrow, may occur as the cell squeezes through the vascular wall to enter the circulation; this notion derives by analogy from the studies on splenic degradation of damaged erythrocytes (see below). However, SKUTELSKY and DANON (1967) and TAVASCALI and CROSBY (1973) have found that engulfment of the "extruded" nucleus by macrophages in hematopoietic tissues commences before separation from the red cell is completed and it is unlikely that "mechanical screening" by the vascular wall is obligatory for the loss of the nucleus. Nevertheless, such screening might conceivably serve as a kind of filtering that can minimize the entry of nucleated red cells into the circulation. (But, recall that leukocytes maturing in the marrow can enter the circulation with their nuclei present.)

III.1.1.2. The Destruction of Red Blood Cells

Under most circumstances, erythrocytes are destroyed through phagocytosis by the macrophages of the "reticuloendothelial system". Although there may be some hemolysis attendant upon such uptake, widescale intravascular hemolysis is prominent in erythrocyte destruction primarily in pathological conditions. Erythrophagocytosis occurs in liver, spleen, bone marrow and other sites.

Macrophages *in vitro* can carry out phagocytosis of whole red cells or fragments and it is not difficult to demonstrate the influence of antibodies, aging of the red cells and other factors on the rates of erythrocyte uptake (ESSNER 1960, MARUTA *et al.* 1973, VAUGHAN and BOYDEN 1964). Correspondingly, *in vivo* experiments have made clear that badly damaged cells, erythrocytes from a donor of one blood type introduced into the circulation of a recipient with a different blood type, and other types of severely abnormal red cells are cleared from the blood stream and destroyed rapidly (see *e.g.*, STIFFEL *et al.* 1970). Unfortunately, the relevance of this to normal red blood cell turnover is still to be established. Despite the availability of techniques such as the use of Cr^{51} to label red blood cells with little disruption, it is not even certain how the spleen, liver, marrow and other sites compare and interact in terms of their relative contributions to normal turnover. The liver reputedly contains the largest number of macrophage-like phagocytes (the Kupffer cells) of the body organs (VERNON-ROBERTS 1972) and secretes bile pigments derived from red cell breakdown (see *e.g.*, ARIAS 1972, BISSELL *et al.* 1972); Kupffer cells frequently are implicated in erythrophagocytosis of severely injured cells (RIFKIND 1966). The spleen, which has the highest mass of macrophages per unit weight of organ (VERNON-ROBERTS 1972) is held to be preeminent in destruction of mildly injured cells (KEENE and JANDL 1959, RIFKIND 1966) and probably in normal turnover as well. Marrow also has many phagocytes but rarely is thought of as a major site of red blood destruction except in abnormal circumstances (KEENE and JANDL 1965). However, one "product" of erythrocyte digestion, iron liberated from hemoglobin can rapidly accumulate in the marrow for use in erythropoiesis.

The central question yet to be answered is how cells normally come to survive in the circulation for a fixed time (and why cells aged without special mistreatment outside the animal show corresponding reductions in lifespan when reintroduced into the blood stream; RABINOVITCH 1970). Such behaviour argues for some threshold effect. Hypotheses to explain it have cited changes in the cell surface, changes in intracellular biochemistry and changes in mechanical properties.

Aged red cells do seem to differ in biochemical capacities from younger ones. For example, there is a decline in the activity of enzymes such as glucose-6-phosphate dehydrogenase and IMP-pyrophosphorylase (MARKS *et al.* 1958, RUBIN *et al.* 1969). Cellular "deformability", size, shape and resistance to osmotic rupture may also alter; thus old erythrocytes are often described as more "fragile" or prone to greater rigidity than younger cells (see *e.g.*, BUNN 1972 and many others). DANON and co-workers (DANON *et al.* 1971, MARIKOVSKY and DANON 1969) contend that younger red cells have a higher negative surface charge than do older ones—this is reflected in the electrophoretic behaviour of the cells (see also YAARI 1968) and in their binding of colloidal iron particles visible in the electron microscope—and they assert that the change in surface charge beyond some key point is intrinsic to the signal for destruction. The nucleated cell fragments produced during erythrocyte maturation also seem to bind relatively less iron than do the newly formed anucleate cells, which may help account for the differences in fates (DANON *et al.* 1971, SKUTELSKY and DANON 1969). A loss of charge-bearing saccharides (sialic acid and galactosamine) has been reported by BAL-

DUNI *et al.* (1974), for aging red cells. It could come about by shedding of surface coats or even as a result of the presence of low levels of neuraminidases or other enzymes in the serum. According to GREGORIADIS *et al.* (1974) introduction of neuraminidases into the circulation results in a lowering of red blood cell lifespans and JANCIK and SCHAUER (1974) have confirmed that neuraminidase-treated red cells are rapidly destroyed if reintroduced into the circulation. Alternatively, the crucial changes in the cell surface might involve the membrane lipids (COOPER and JANKL 1969, DAWSON and SWEELEY 1960) and reflect the presence of surface-active agents in the circulation, or there might even be a gradual accumulation in the membrane of some proteins of the red cell cytoplasm (KADLUBOWSKI and HARRIS 1974). The essential thing that must be specified however, is how any such changes can lead to cell death. Implicit in most hypotheses is the notion that some system recognizes the alterations and destroys the erythrocytes, rather than the red cell first undergoing some "autodestructive" process such as hemolysis, agglutination or fragmentation.

Despite the fact that splenectomized individuals turn over their erythrocyte population at rates that are not vastly altered (RIFKIND 1966) the spleen has been the focus of most attention as the probable normal major graveyard for red blood cells (PEARSALL and WEISER 1970) and for platelets (ASTER 1969). Apparently other organs, especially the liver, can increase their contribution to compensate for loss of the spleen. An attractive proposal for splenic function emerges from observations on moderately damaged red cells, such as those exposed to phenylhydrazine under conditions that result in the presence of "Heinz bodies" (RIFKIND 1966). These are intracellular aggregates visible in the microscope and thought to represent collections of denatured hemoglobin. It is known that some of the blood passing through the spleen normally is shunted into a special circulatory route in which the cells leave the arterial circulation, pass through a cellular meshwork (the splenic "cords") rich in macrophages and then reenter the venous circulation by moving through slit-like spaces in the vascular walls (CHEN and WEISS 1972, 1973). The slits in venous sinuses through which the cells must pass are less than 1 micron in diameter and thus, the cells are considerably deformed (Fig. 35); CHEN and WEISS argue that bands of microfilaments keep the slits at a restricted diameter. Heinz bodies, or other inclusions (*e.g.,* malarial parasites) that are relatively non-deformable will greatly retard passage of erythrocytes back into the circulation. If the inclusions are relatively small, "pitting" may occur—a fragment of the cell, in which inclusions are present, may be pinched off and left behind as the rest of the cell traverses the vascular wall (CHEN and WEISS 1973, CROSBY 1957, SCHNITGER *et al.* 1972). Fragments formed in this way are degraded by macrophages; the surviving portion of the cell may have a shortened life span (due, perhaps, to its altered geometry) but it can remain in the circulation for at least a time.

When inclusions within erythrocytes are very large or very numerous, the splenic cords show heavy accumulations of the abnormal cells and extensive erythrophagocytosis ensues.

From such observations it has been suggested that the spleen provides a

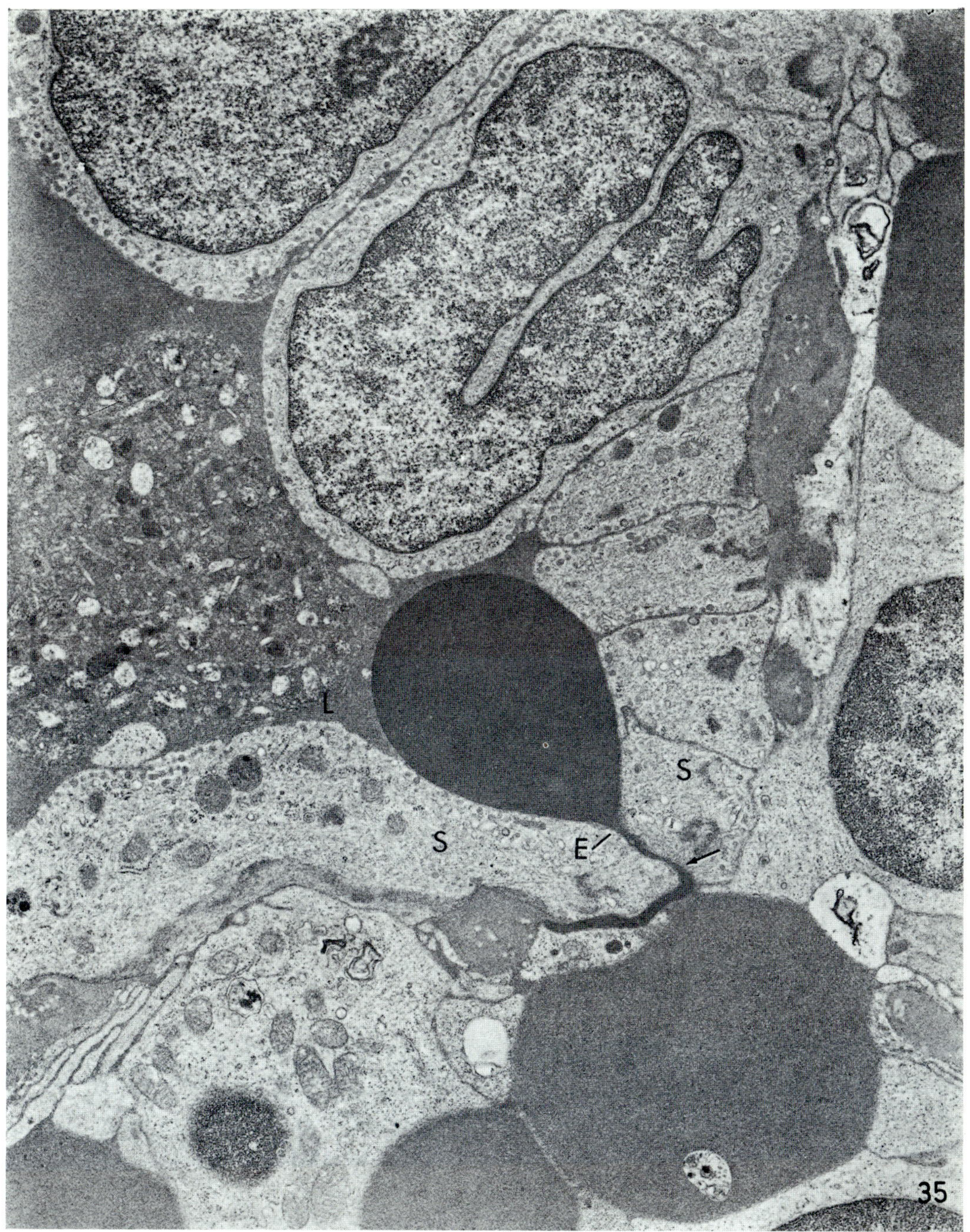

Fig. 35. From the red pulp region of human spleen. At *E* an erythrocyte is seen that apparently was in the process of crossing the wall of a sinus (*S*) to enter the lumen (*L*). Note the extreme distortion of the red blood cell (arrow) due, presumably, to its being forced to pass through a narrow aperture. ×10,000. (From CHEN, L. T., and L. WEISS, 1972: Amer. J. Anat. **134**, 425—458.)

kind of "testing ground" to which a given erythrocyte will be exposed numerous times in its lifespan. Local conditions in the splenic cords are likely to be quite stingent in terms of pH, oxygen and glucose concentrations etc. (RIFKIND 1966). And, the circulation in the cords is sluggish, which might permit more time for surface interactions between erythrocytes and potential phagocytes. Further work is needed to establish whether culling of aged cells reflects the detection of relatively minor surface or biochemical abnormalities or whether magnifications of osmotic or mechanical problems are more important. The fact that red cells low in surface sialic acids show altered aggregation properties (CHIEN et al. 1974) could prove of interest. It would also be useful to know if splenic phagocytes liberate enzymes or other factors into the extracellular environment which might contribute to erythrocyte alteration or destruction. Macrophages are known to be capable of releasing hydrolases (Section IV.3) and one could imagine that upon each passage through the splenic cords the red cells are modified by extracellular enzymes until finally a threshold is passed whereupon they are retained and destroyed.

III.1.1.3. Iron Storage and Release

The involvement of lysosomes in the cycling of iron liberated during the degradation of hemoglobin is incompletely understood. Even the form in which the iron leaves the digestive vacuoles of erythrophagocytes is not definitely identified. Perhaps ionic iron diffuses through the lysosome membrane and then, by complexing with apoferritin in the cytoplasm or with apotransferrin in the circulation, enters the cycles through which much iron can be stored as cytoplasmic ferritin or made available for erythropoiesis through transferrin mediated transport to bone marrow cells (see e.g., COOK et al. 1973, FEDORKO 1974, LIPSCHITZ et al. 1971, WYLLIE 1973). In addition to the extralysosomal storage and transport systems, membrane-delimited iron-containing bodies are found in erythrophagocytes and in erythroblastic cells. These vary in staining characteristics and electron microscopic appearance; hence the use of varying names such as "hemosiderin" and "siderosomes". Acid phosphatase is demonstrable in at least of these bodies.

None of the iron-containing bodies has been fully characterized. Those in macrophages seem to originate as residual bodies that form after erythrophagocytosis and come to contain small ferritin-like granules. FEDORKO and co-workers (1973; see also TRUMP et al. 1973) feel that many of these granules enter the bodies from the cytoplasm by a "microautophagic" inpocketing of the surface membrane. But it would be premature to rule out contributions by other mechanisms, such as the presence in the bodies of apoferritin or other iron-complexing molecules to which iron could bind directly upon degradation of hemoglobin. In either event, ferritin is resistant to digestion by lysosomal hydrolases (COFFEY and DE DUVE 1968, DRYSDALE and MUNRO 1966) and presumably could be stored for some time within lysosomes.

BESSIS and his co-workers (BESSIS and BRETON-GORIUS 1962) have called attention to endocytic phenomena in hematopoietic tissues wherein ferritin-like molecules are taken up by erythroblastic cells; some investigators have claimed that these molecules can become liberated into the cytoplasm by dissolution of the pinocytic vesicle membrane, but unequivocal demonstration of this through microscopy is virtually impossible (the relevant images are susceptible of several interpretations) and the likelihood is that the vesicles contribute their contents to membrane-bounded "siderosomes". The extent of erythroblast uptake of "ferritin" is quite dependent on species and on levels of circulating iron (TANAKA and BRECHER 1971).

It is not at all clear how the contents of the membrane-delimited bodies participate in iron cycling or in erythropoiesis. Present opinion strongly favors the concept that circulating iron, carried by transferrin, is delivered to cytoplasm of a maturing erythroid cell through dissociation of the iron from its carrier and transport across the plasma membrane (FIELDING and SPEYER 1974, MAZUR et al. 1960, MILLER and PERKINS 1969, WORKMAN and BATES 1974). At most, endocytosis and lysosomal degradation of transferrin might make a minor contribution (this may be more significant for iron uptake by non-erythroid cells; HENNAPLARDH and MORGAN 1974, MORGAN and APPLETON 1969, POOLE and DE DUVE 1973; see also EVANS 1973 for analogous considerations concerning copper cycling). Perhaps the bodies in erythroblasts sequester iron that in some sense is excess and eventually release it to non-erythroid cells as a kind of by-product of maturation (see BESSIS and BRETON-GORIUS 1962, SHEPP et al. 1972). The iron-rich structures in erythrophago-cytes are often thought of as storage devices although the details of this supposed function are still to be worked out. Heme groups from red blood cell hemoglobin eventually are metabolized into bile pigments [see ARIAS (1972) for a review]. This implies escape from the lysosomes of phagocytes but the pertinent routes and mechanisms are not known. According to CONWAY et al. (1975), the serum heme-transport protein, hemopexin, is catabolized rapidly by hepatocytes when heme is introduced into the blood stream; this "suicidal" functioning of the protein presumably involves the hepatic lyso-somes in heme disposal, at least under abnormal conditions.

III.1.2. Developmental "Remodelling"

In prior sections we have mentioned several situations, in addition to ery-throcyte maturation, in which autophagy seemingly contributes to develop-mental modulations of cells—this is true for example, in the fat body of metamorphosing insects in which much of the organelle complement of the cells is replaced but the cell themselves survive (Section II.3.4), and in regen-eration of injured axons (Section II.3.4). Here, we will take up examples of developmental situations in which cells die and lysosomal hydrolases participate in their degradation. Phenomena of this type are quite pronounced in insects, but numerous cases are encountered throughout the animal king-dom (CAMERON 1971, GOSS 1970, SAUNDERS 1966)—regression of the Mullerian duct and mesonephros in birds (SALZBERGER and WEBER 1966, SCHEIB 1963) and of mammary and uterine tissue in mammals (BRANDES and ANTON 1969, HELMINEN and ERICSSON 1968, 1969, WAESSNER 1969), resorp-tion of the tail in amphibia (FOX 1973, WEBER 1969) and localized death of nerve cells (O'CONNER and WIGTTENBACH 1974) and cells of developing limbs (SAUNDERS 1969) have been studied fairly extensively. An abundant literature documents the participation of hormones such as thyroxine (WEBER 1969) or steroids (SCHEIB 1963), neural activity, and other factors in controlling the timing of tissue destruction.

In most cases, what actually transpires at the level of the doomed cells is not clear. Commonly, lysosomes (especially autophagic vacuoles) increase in frequency in cells that are soon to die and once it has died, a cell's lysosomes probably rupture and contribute to autolysis (this seems especially prominent in plant development, Section V.1.1). But as stressed in Section II.3.4, causal roles for lysosomes in cell death are exceptionally difficult to establish; appreciable firm evidence for such roles is available only for a tiny handful of pathological conditions (Section IV.4). One cannot be sure whether an increase in autophagy is itself harmful to the cell, or reflects defensive efforts that finally are overwhelmed or even perhaps represents a mechanism whereby a cell in effect provides for its own autolysis by mobilizing enzymes that will act once it has died (in evolutionary terms such a mechanism could be advantageous since it might simplify or lessen the "demands" made on phagocytes mentioned in the next paragraph).

It was established early through biochemical studies that developmental breakdown of tissues often is accompanied by large increases in tissue acid hydrolase content. In usual situations, much of this is attributable to macrophages or other phagocytes whose number increases through influx from the circulation; sometimes this is accompanied by proliferation or maturation of phagocytes already present in the tissues. Probably the relevant "signals" for phagocyte accumulation resemble those that produce local collection of phagocytes in inflammatory responses and other pathological conditions (Section IV.3).

Some of the implications of the scavenging of dead cells by phagocytes may be interesting with respect to the economy of developing organisms—the reuse of degradation products from resorbed tissues for growth of surviving tissues is reportedly important for some invertebrates maintained under conditions of starvation (BROCK 1970, SAUNDERS 1966, STUART 1970) but presumably this is not essential under normal conditions of development of higher organisms.

III.1.2.1. Case Histories: Cell Destruction in Insect Metamorphosis

LOCKSHIN and colleagues have worked extensively on the lysis of intersegmental muscles of moths, that takes place shortly after the emergence of the adult from the pupal cocoon (LOCKSHIN 1969). SCHIN, CLEVER and coworkers (SCHIN and CLEVER 1968) have examined the programmed regression of salivary glands in Dipterans, chiefly *Chironomus*. In these cases, the stage is set by endocrine factors, such as ecdyson and, for the muscles, cessation of neuronal stimulation is an important trigger (LOCKSHIN and WILLIAMS 1965). Before the cells die there is an increase in biochemically demonstrable hydrolases not attributable to phagocyte influx and at about the time of death there is an increase in the size and frequency of lysosomes in the doomed cells (see also *e.g.*, RADFORD and MISCH 1971, and SCHARRER 1966 for other insect tissues in which prominent changes in the lysosomes accompany metamorphosis-associated cell death).

For the salivary glands, some evidence suggests that cell death is associated with release of acid hydrolases from the lysosomes into the rest of the cytoplasm; in cytochemical acid phosphatase preparations, reaction product in dead or dying cells is present in diffuse form throughout the cell (SCHIN and CLEVER 1965) and in cell-fractionation experiments there is a marked increase in the acid hydrolase activity that appears in the "soluble" fraction (HENRICKSON and CLEVER 1972). Elsewhere (Section IV.4) we will comment in detail upon the need for extreme caution in interpreting such findings—both diffuse cytochemical reactions and increases in "soluble" enzyme can reflect technical problems rather than genuine enzyme release. But the findings on the salivary glands do point, albeit tentatively, toward the conclusion that endogenous hydrolases participate in autolysis during metamorphosis. This, however, does not compel the further conclusion that hydrolase release is what kills the cell or even that it is an essential step in rendering relevant cell alterations irreversible.

In Lockshin's moths, the first physiological changes in the muscle fibers are in the electrical properties of the plasma membrane: the capacitance falls (LOCKSHIN 1973 a). Numerous autophagic vacuoles appear early in the subsequent degeneration but while many mitochondria seem to be degraded in these bodies, the visible disruption of the myfilaments occurs outside the lysosomes and without indications of massive hydrolase release into the cytoplasm (LOCKSHIN and BEAULATON 1974). Apparently the initial disruption of myofilaments does not involve extensive degradation of the constituent proteins (LOCKSHIN 1973 b) although at later stages of muscle breakdown such degradation does occur, almost certainly with the central involvement of lysosomal enzymes from phagocytic hemocytes (see also CROSSLEY 1966) and probably from the muscle itself (LOCKSHIN feels that autophagy of myofilament material may occur at advanced stages of disintegration and perhaps there is also some hydrolase release from the lysosomes of dead fibers) [11].

One promising line of investigation being pursued concerns the influences of inhibitors of macromolecule synthesis such as Actinomycin D and cycloheximide (HENRICKSON and CLEVER 1972, LOCKSHIN 1969 a, b, LOCKSHIN and BEAULATON 1974, SCHIN and LAUFER 1973). Both for the moth muscles and with the salivary glands such agents can retard or even prevent tissue breakdown under proper experimental conditions. The altered rates of breakdown are accompanied by effects on lysosome behaviour—in the salivary glands, hydrolase "liberation" into the cytoplasm is prevented and in the muscles, the increase in lysosome frequency does not occur. Although bio-

---

[11] Regression of vertebrate muscles after denervation or in pathological dystrophies has also been studied for possible lysosomal involvement (Section III.4.3.2 and POLLACK and BIRD 1968, SCHIAFFINO and HANZLIKO 1972 b, WEINSTOCK and IODICE 1969). While the matter is not settled, as with the insects there is no strong evidence in this work for an early involvement of lysosomes in myofibrillar deterioration.

chemical studies fully adequate for evaluation of the variety of possibilities have yet to be reported, there is suggestive evidence that the inhibitors are not affecting primarily the level of acid hydrolases present in the tissues. Rather some "activation" process dependent on protein synthesis is being sought. One focus of speculation is that the hydrolases themselves require activation in the sense of proenzyme to enzyme conversion; there is little precedent for this in the case of lysosomes (but HENRICKSON and CLEVER do claim that one of the proteases prominent in regressing salivary glands is activatable by trypsin). Another possibility is that cell death is initiated by a nonlysosomal enzyme or other protein synthesized shortly before use and that the lysosomal changes are secondary responses to events mediated by this protein.

The avilability of insect mutants with altered patterns of cell death has only begun to be exploited for work bearing on the matters discussed here. FRISTROM (1969) claims that mutation-dependent death in imaginal discs of *Drosophila* does not involve extensive autophagy. In his flies, an influx of specialized phagocytes may not be required to scavenge cellular debris in discs; cells already present seem able to carry out the task. There is also a mutant *Drosophila* strain in which fat body cells survive into the adult stages for much longer periods than normal (BUTTERWORTH and LaTENDRESSE 1973).

III.2. Turnover of Extracellular Materials

III.2.1. Connective Tissue Components

Extensive degradation of the special extracellular materials that abound in connective tissues is intrinsic to many early developmental processes. A tadpole cannot resorb its tail without breaking down a great deal of collagen. Proper growth of skeletal structures depends on extensive remodelling of bone and cartilage. And, there even has been some speculation that controlled degradation of intercellular materials such as hyaluronic acid may permit (or at least facilitate) developmentally significant cell-cell interactions (see *e.g.,* POLANSKY *et al.* 1974).

Turnover of connective tissue molecules occurs at an appreciable rate in adult organisms. Whereas for components such as collagen this may be relatively slow [*e.g.,* DINGLE (1973 b) gives the half life of rabbit articular collagen as several months to a year] estimates of the halflives of connective tissue polysaccharides are on the order of a few days or weeks (DINGLE 1973 b, MUIR 1973).

Lysosomal glycosidases, proteases, sulfatases, etc. can digest most connective tissue materials at acid pH. And there is morphological evidence for the participation of phagocytes in connective tissue breakdown during processes such as amphibian tail resorption. Although many details are lacking, it does seem likely that late stages of the degradation of connective tissue matrix material often, and perhaps usually occur intracellularly. But in the early stages, "problems" are posed by the geometry of the tissues plus the fact that many of the extracellular materials are in the form of very large

macromolecular complexes such as large fibers or the elaborate complexes
of covalently linked proteins and polysaccharides (proteoglycan) present in
cartilage. Initial extracellular processing of such materials would seem to be
necessary for their eventual uptake by cells.

III.2.1.1. Hydrolase Release in Cartilage

DINGLE, FELL, and BARRETT and their co-workers have been concerned with
the breakdown of cartilage matrix in circumstances such as the autolysis of
explanted pieces of cartilage (usually embryonic bone rudiments; for reviews,
see DINGLE 1969, 1973 b). In such preparations matrix degradation is stimulated or enhanced by vitamin A and other agents which also increase the
levels of acid hydrolases detectable biochemically in the culture medium. The
effects of vitamin A both on degradation and on hydrolase release are antagonized by cortisone. We will say more about vitamin A and cortisone in
Section IV.4. For the present, the important point is the parallelism between
levels of hydrolase release and the extent of matrix destruction; this is taken
as strong circumstantial evidence for central participation of lysosomes in
extracellular degradation processes. Since the cell populations in question do
not release much of the soluble cytoplasmic enzyme, lactic dehydrogenase
while liberating hydrolases, cell death does not seem a primary factor.
Instead, the phenomena are analyzed in terms of exocytic-like events (DINGLE
et al. 1965). Recently, immunocytochemical techniques have made possible
a direct demonstration that chrondrocytes and perichondrial cells of reasonably normal cartilage can be surrounded by extracellular hydrolases, confirming that cell death is not a prerequisite for enzyme release (DINGLE 1972,
POOLE et al. 1974, Phil. Trans. Roy. Soc. **B 271**, 233—410).

The roles in degradation, of cartilage cells at different stages of differentiation have yet to be sorted out. BARRATT (1973) suggests that the softer
tissues (perichondrium etc.) may initiate the breakdown since the presence of
such tissues seems important for autolytic responses of explants to agents such
as vitamin A. It is also not known if a given cell can simultaneously synthesize matrix materials and release hydrolases.

There are other important matters still to be settled. The pH optimum
for cartilage autolysis *in vitro* is about 5. At neutral pH, the acid hydrolases
can make only very limited attacks upon the extracellular materials of cartilage. During autolysis of cartilage explants, some acidification of the medium
occurs (DINGLE 1973 b) but whether this means that during less drastic
degradation, low pH zones can be established within the matrix still has to
be determined; further studies are needed on the possibilities that the sulfated
polysaccharides or the cartilage cells themselves contribute to favorable local
environments for acid hydrolases (*cf.*, DINGLE 1973 b). Matrix breakdown
in explants is antagonized by anti-cathepsin D antibodies and to some extent
by a fungal cathepsin inhibitor, pepstatin (DINGLE et al. 1972, WESTON and
POOLE 1973). These observations are obviously suggestive of a key role for
cathepsin D, an enzyme that is largely inactive at neutral pH. But there
are some difficulties to be finally disposed of. WOESSNER (1973; see also

SAPOLSKY *et al.* 1974) believes that "purified" cathepsin D tends to be contaminated with neural proteases, which in his view initiate the attack on cartilage matrix and also might be present in some preparations used to generate antibodies, producing misleading results. However, DINGLE, WESTON, POOLE and their co-workers report that the antisera they use are highly specific for cathepsin D, by the usual immunological criteria. The possibility that neutral proteases (or cathepsin B_1, which retains activity at pH 7) do participate in matrix breakdown is, of course, independent of the participation of cathepsin D. Proteases, with pH optima near neutrality are present in lysosomes of leukocytes (LoSPALLA *et al.* 1971, WEISSMANN *et al.* 1972, ORONSKY *et al.* 1973). And an enzyme of this sort is released to extracellular media from a presently unidentified intracellular source in cultured mouse fibroblasts (WERB and REYNOLDS 1974; since release of the fibroblast enzyme is not paralleled by secretion of acid hydrolases the lysosomes probably are not responsible). DINGLE (1973 b) mentions unpublished findings of a neutral protease in cartilage. Collagenase (Section III.2.1.3) also is likely to be present in the extracellular spaces where connective tissue breakdown is occurring.

The relative efficacy or abundance of different hydrolases (proteases, sulfatases, polysaccharides etc.) may largely account for the differences in turnover rates of various connective tissue components.

Little is known of the mechanisms of lysosomal hydrolase release from cartilage cells. Chondrocytes can be induced to take up sucrose by endocytosis; since they lack the enzymes needed to hydrolyze the sugar, the sucrose accumulates in intracellular vacuoles that presumably are secondary lysosomes (*cf.*, Fig. 5). If the cells are then placed in sucrose-free media, the intracellular sucrose gradually disappears. GLAUERT *et al.* (1969) maintain that this probably occurs via exocytic unloading of the vacuoles, although direct evidence is hard to obtain [12]. The disappearance of sucrose (and of dextrans when these have been taken up by endocytosis) can occur without large-scale release of hydrolases, if the cells are exposed to hydrocortisone, which inhibits the enzyme secretion; thus the suggestion has been made that primary, rather than secondary lysosomes are chiefly responsible for hydrolase release (DINGLE 1969, GLAUERT *et al.* 1969). This surmise requires additional support before it can be regarded as more than a working hypothesis. In any event, studies with cycloheximide suggest that new enzyme synthesis is not required for induced hydrolase release (HILLE *et al.* 1970); in this respect the system is similar to more conventional secretory cells (JAMIESON and PALADE 1968).

The occurrence in serum, of inhibitors of proteases and other enzymes (*e.g.*, CONTRACTOR and SHANE 1972; Sections III.2.1.3 and III.3.1) provides a mechanism whereby the organism can restrict the action of extracellular hydrolases. Serum inhibitors are large proteins that would diffuse into intact cartilage matrix slowly, if at all. But once extensive matrix breakdown occurs

---

[12] As discussed in Section II.5.1 the point is of sufficient general importance as to merit a very thorough study that might evaluate for example, possible contributions of slow intralysosomal hydrolysis, or of limited permeability of the vacuole membranes to sucrose.

further action of the extracellular hydrolases could be inhibited, thus preventing damage to adjacent tissues (DINGLE 1973 b).

The phenomena discussed in this section may also help explain the presence in blood stream and urine of fragments of molecules of the type expected to result from degradation of connective tissue matrixes. This can be quite marked in pathological conditions. The presumption is that once proteoglycans have been fragmented *in vivo* by initial extracellular attack by proteases and perhaps by other enzymes, the polysaccharides and peptides released are further degraded intracellularly; much of this may occur via local endocytosis. But passage of fragments of proteoglycan into the blood stream could also take place (DINGLE 1973; WATESON *et al.* 1972, WOESSNER 1973, WOOD *et al.* 1973); some of these products might subsequently be endocytosed in the liver or kidney.

Under some circumstances, apparently intact lysosomes are found extracellularly, in the matrix of cartilage undergoing breakdown or remodelling (*e.g.,* THYBERG *et al.* 1975). These may merely reflect persistence of the organelles after cell death, but one might guess that their enzymes or other components could have an effect on adjacent material.

III.2.1.2. Osteoclasts

VAES (1969) has summarized the biochemical and cytochemical evidence for substantial participation of lysosomal hydrolases in bone resorption. Large, multinucleated osteoclasts are agents in such resorption and these cells appear both to secrete acid hydrolases and to utilize endocytosis to continue digestion of molecules liberated from the bone matrix by the extracellular enzymes. As VAES pictures the process, hydrolases released by the osteoclasts accumulate where the cells abut upon resorbing bone. In this same region a low pH is maintained—one hypothesis proposes that lactic acid secreted by the osteoclasts is responsible for this. The acid conditions serve both in the demineralization of the matrix and in supporting hydrolase activity. As materials are released from the matrix they are endocytosed by the osteoclasts and degraded in phagocytic digestion vacuoles.

Essentially similar conclusions can be drawn from the work of LUCHT (1972 a–c). LUCHT (1972 a) make the interesting suggestion that phagocytosis vacuoles formed at the surface where an osteoclast is in contact with bone may acquire acid hydrolases from the extracellular medium along with the materials to be degraded. This circuitous route would not conflict with more conventional modes of hydrolase transport by the same cells.

III.2.1.3. Collagenases

Lysosomal enzymes such as cathepsin D have very limited effects on collagen especially at neutral pH. Cathepsin B 1 reportedly does do some degrading of collagen at pH 7 (BURLEIGH *et al.* 1974) but it and other lysosomal acid proteases are very much more effective at low pH (see *e.g.,* ETHERINGTON 1974) where the enzymes are "aided" by the acid-engendered alterations in the collagen fibers.

"Neutral collagenases" with pH optima near neutrality and characteristic modes of action and responses to molecules such as EDTA, have been isolated from several tissues studied under circumstances in which collagen degradation is extensive—for example, regressing amphibian tissues, synovial fluid from arthritic sources and involuting rat uterus (see LAZARUS 1973 for review and also EVANSON 1971, GROSS and LAPIERE 1962, WOESSNER 1971). Comparable enzymes are now being turned up in fibroblast cultures (WERB and BURLEIGH 1974) and other cell populations. At neutral pH, these enzymes split native collagen molecules into large fragments—at least some of them show little or restricted ability to catalyze more extensive degradation or hydrolyze other proteins. They are susceptible to inhibition by the α_2-macroglobulin components of serum (the PMN enzyme may be an exception; LAZARUS et al. 1972) and they tend to bind tightly to collagen fibers and perhaps to other materials (EISEN et al. 1971, NAGAI 1973, WERB et al. 1974); such factors make their detection, isolation and characterization quite difficult (e.g., see WOESSNER and RYAN 1973).

Little is known about the mechanisms for the intracellular mobilization and release of collagenases. In tissues such as tadpole tail, suspected enzymatically-inactive zymogens have been detected which seemingly can be activated outside the cell by limited proteolysis (HARPER et al. 1971, LAZARUS 1973, VAES 1972). NAGAI (1973) points out that inactive zymogens might be difficult to distinguish from collagenase, inactivated by complexing with serum inhibitors. But in at least some cases, supposed zymogens appear to differ only slightly in molecular weight from the active enzymes (HARPER et al. 1971); this would not be expected for complexes with serum antiprotease proteins. The existence of zymogens has led to suggestions for controls of collagenase activity such as the possibility that lysosomal hydrolases are responsible for their activation (VAES 1972). LAZARUS (1973) further speculates that the lysosomal enzymes might unmask collagen by degrading other components, and thereby providing access for the collagenase, and also that the hydrolases might eventually inactivate collagenase.

Only for rabbit PMN leukocyte collagenase has probable association with a known intracellular particle been demonstrated—the enzyme is reported to be present in the specific granules and not in the azurophilic granules (lysosomes). WERB and REYNOLDS (1974) surmise that the collagenase they study in cultured rat synovial fibroblasts is non-lysosomal. They observe that the cells "secrete" collagenase and the neutral protease mentioned in the previous section, at enhanced rates after phagocytosing latex; the increased release to the medium starts hours after phagocytosis has occurred and is not paralleled by similar changes in release of lysosomal hydrolases. DEWALD et al. (1975) point out that most of the known neutral proteases of PMN leukocytes are present in the azurophilic granules. Therefore, they suggest, the apparent localization of collagenase in the specific granules should be subjected to further experimental scrutiny. But WERB and REYNOLDS' findings, among others, do seem to establish a "precedent" for separate packaging of different proteases (for macrophage collagenase, elastase: J. Exp. Med. **142**, 346; 361).

The presence of neutral collagenase in macrophages is likely. A preliminary report by Fujiwara and co-workers (1973) claims that Kupffer cells isolated from liver have a higher collagease content per cell than do granulocytes. While Burleigh *et al.* (1974) assert that no such enzyme is present in alveolar macrophages, Werb and Reynolds (1974) mention that mouse peritoneal macrophages do possess a collagenase and Wahl *et al.* (1974) have found that guinea pig peritoneal macrophages release such an enzyme to their culture medium if stimulated by bacterial polysaccharide "endotoxins".

Thus, collagenases seem excellent candidates for participants in early steps of connective tissue breakdown. Subsequent to their action, lysosomal hydrolases could take over and complete the degradation of collagen. Endocytosis of partly degraded collagen fibers by macrophages has been observed (Fox 1972, Parakkal 1972, Schwarz and Galdner 1967, Usuku and Gross 1965). On the other hand, Evanson (1971) argues that extracellular sequences for degradation of some collagen to the level of dipeptides or tripeptides are also worth seeking since one can detect what seem to be low molecular weight collagen breakdown products in urine. The contribution to this of release from dying cells or of egestion of lysosome contents, is not known. The possibility that pertinent degradation mechanisms may show some complexities that are presently only vaguely evident is also suggested by the information that newly made collagen seems to have somewhat different turnover characteristics from older molecules.

An elastolytic enzyme activity is demonstrable in granulocyte lysosomes but breakdown of elastic fibrous tissue probably is of more interest for pathological conditions than for normal development and turnover (see p. 182; note footnote 24).

III.3. Turnover of Circulating Macromolecules

Disposal pathways are required for a large variety of macromolecules present in the circulation. The most abundant of these are the blood proteins—the albumins, globulins, fibrinogen etc. of human serum show steady state replacement rates that range for different proteins from a few percent per day to on the order of 50–75% or more. Mechanisms are also needed for inactivating and degrading circulating molecules such as hormones whose levels vary considerably depending upon physiological conditions (inactivation and degradation need not be simultaneous).

There is a luxurient literature on the clearance of injected colloids and macromolecules from the blood (see *e.g.*, Benacerraf 1964, Stiffel *et al.* 1970, Vernon-Roberts 1972). Early work with visible materials such as carbon particles and bacteria has been extensively supplemented through techniques such as the reintroduction into the circulation of proteins made radioactive through incorporation of radioiodine (I^{125}, I^{131}), or the estimation of turnover from steady state synthesis rates using radioactive amino acids. There are many pitfalls in use of these techniques—*e.g.*, the existence of extravascular pools, reutilization of label, denaturation during labelling, and effects of abnormal tracers upon cells (see *e.g.*, Heath and Barton 1973, Wolstenholme and O'Connor 1973, and Section III.4.3.1 for reviews). However, these affect primarily the finer details of interpretation. There

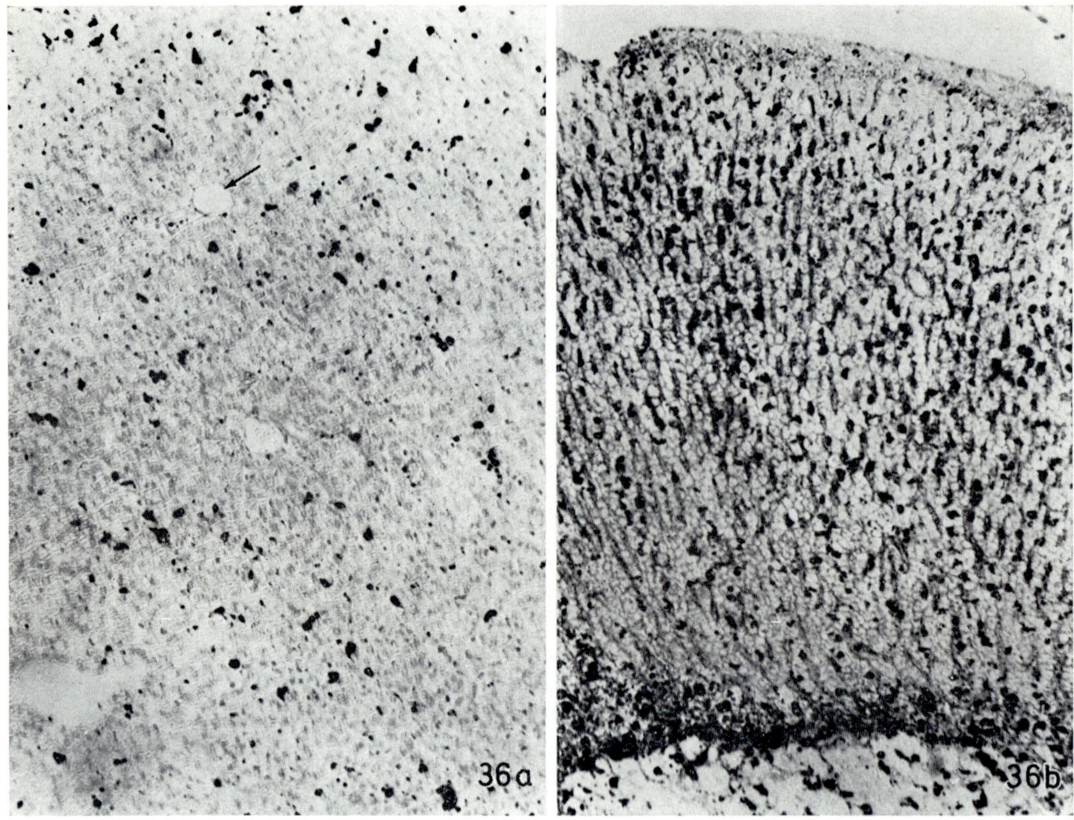

Fig. 36. Low power light micrographs of guinea pig tissues after intravenous administration of carbon. *a* Shows liver (note the probable central vein at the arrow) and *b*, adrenal cortex. The numerous Kupffer cells of the liver and the many macrophages lining the sinusoids of the adrenal cortex have taken up carbon and therefore appear black. (Courtesy of B. VERNON-ROBERTS.)

is little present basis for serious doubt that the major mechanism for the removal of many, and probably most macromolecular materials from the blood is endocytosis. This eventuates in extensive heterophagic lysosomal degradation with consequent reentry of low-molecular weight digestion products into local metabolic pools and into the circulation. Other routes, such as excretion in the urine are for the most part quantitatively unimportant, although in pathological conditions this often does not hold true.

Liver, and for some components, kidney are usually found to be the main sites for the clearance of labelled circulating macromolecules, although contribution can be made by various organs of the "reticulo-endothelial system". Endocytosis by Kupffer cells (Fig. 36), hepatocytes (see *e.g.*, DAEMS *et al.* 1972, GOLDFISCHER *et al.* 1970, GRAHAM *et al.* 1969, MA *et al.* 1974) and renal tubule cells (*e.g.*, ERICSSON 1965, MAUNSBACH 1969, MILLER and PALADE 1964, TRUMP 1965) seems straightforward. Important details are uncertain, such as the identity of possible primary lysosomes in the kidney or liver, or the extent to which endocytosed materials can escape lysosomal degradation. But investigations with a range of tracers indicate that endocytosis by the cells in question is capable of delivering large quantities of various exogenous materials for lysosomal degradation. This holds for materials that are exotic from the viewpoint of the organisms under investigation (*e.g.,* horseradish peroxidase), those that enter the circulation in large quantities only in abnormal circumstance (*e.g.,* hemoglobin; GOLDFISCHER *et al.* 1970) and molecules that are, or closely resemble normal blood constituents (*e.g.,* radioactively labelled serum proteins; MAUNSBACH 1969).

On the other hand, although the liver is usually considered to be preeminent in degradation of blood-borne macromolecules (see *e.g.,* VERNON-ROBERTS 1972) work with perfused organs has been interpreted as suggesting that normal hepatic clearance can account for only a modest fraction of albumin turnover (HOFFENBERG *et al.* 1970). Given the problems in maintaining perfused systems in normal states this may not be too significant a result but it does serve at least to remind one that the quantitative importance of different loci for turnover is known in detail only for a few conveniently studied materials, chiefly denatured or otherwise abnormal proteins and bacteria or other "colloidal" preparations.

In part from studies of storage diseases, in which many tissues show lipid deposits when the lysosomes lack esterases or other enzymes, the suspicion has grown that endocytosis and lysosomal "processing" may be an important mechanism for frezing blood-borne lipids from the proteins with which they are associated in the circulation, and making the lipids available for intracellular use (see Section IV.1.2 and PATRICK and LEAKE 1973).

III.3.1. Selectivity

Unlike the case of blood cells, circulating macromolecules seem generally not to be degraded with strongly age dependent kinetics. Under many circumstances, relevant tracers are cleared from the circulation with no evident appreciable lag and with kinetics that approximate first-order over much of the time course. In other words labelled macromolecules of a given type disappear at rates that are close to proportional to the relative concentrations of labelled vs. unlabelled macromolecules of that type. This is consistent with the operation of an unselective disposal mechanism such as one dependent upon bulk uptake of serum. However, when different macromolecules are compared, selectivity features become obvious—the rates at which different materials leave the blood do not depend simply upon their relative concentration (*e.g.,* JEFFAY and WINZLER 1958, MORRELL *et al.* 1971, SPECTOR 1971, WOLSTENHOLME and O'CONNOR 1973, TRASHER 1971). Why this is so is incompletely understood.

One factor is illustrated by the kidney. This organ provides an example of routing and filtering mechanisms whereby some circulating molecules are catabolized by cells to which others have little access. In mammals, only relatively small proteins (*e.g.,* L-chains of immunoglobulins, ribonuclease, lysozyme, "microglobulins") can enter the renal tubules, since the arrangement of capsule and glomerular cells (normally) restrains passage of larger components out of the circulation. The cells of the proximal convolution are especially active in endocytosis and can be shown to take up macromolecular tracers from the tubule lumen. Exogenous RNAse and endogenous L-chains are known to be cleared from the circulation fairly rapidly, perhaps reflecting the availability of the renal route (DAVIDSON 1973, DAVIDSON *et al.* 1971, STROBER *et al.* 1973). And in rats, the large heterophagic vacuoles that are prominent in the proximal tubule contain immunocytochemically demonstrable lysozyme that may enter the cells by endocytosis from the tubule lumen (KLOCKERS and OSSERMAN 1974).

There are large spaces in the walls of hepatic sinusoids and it seems unlikely that filtering on the basis of molecular size is an essential feature of the passage of macromolecules across these walls. In other locations the capillary walls and associated structures impose greater or lesser restraints on movement of macromolecules, which influences the rates and nature of exchanges between intravascular and extravascular compartments. But, except for the kidney there is little reason to ascribe major importance to filtering or special routing mechanisms as contributors to differential turnover of circulating molecules.

As outlined earlier, macrophages and other likely participants in the processes under consideration show important differences in the avidity with which they bind and take up different macromolecules. Obviously, specificities in the accumulation of molecules at appropriate cell surfaces, and selective endocytosis, could influence the turnover of many materials. The existence of highly specific cell-surface receptors can be invoked to help explain selective binding and uptake of "special" components such as antibody molecules. However, at present there is no firm basis for arguing that uptake of each recognizable component of the circulation depends upon a corresponding set of very specific cell surface binding sites. On the other hand, with massive introduction into the circulation of colloids or other materials one can obtain evidence for saturation of uptake and for some kind of competition among different components. It is true that in many cases such observations chiefly reflect gross effects upon the endocytosing cells such as exhaustion of their endocytic capacities. But they also may sometimes be attributable to saturation of binding sites "responsible" for clearance of different *classes* of molecules.

Do changes in the circulating molecules themselves provide some rate-limiting "signals" that are recognized by the cells carrying out their uptake and degradation? Several interesting possibilities along these lines warrant further investigation. Most obvious are cases where circulating macromolecules mediate the uptake of other materials. This is true for opsonization by serum immunoglobulins—complexes of antigens with antibodies usually are much more rapidly endocytosed then either antigens or immunoglobulins alone. Potentially toxic proteins, notably hemoglobin and proteases are rapidly cleared from the circulation through complexing with specific serum proteins (haptoglobin and the protease inhibitors such as α_2-macroglobulin); probably the endocytosis of hemoglobin by the liver which can be quite extensive (GOLDFISCHER *et al.* 1970) follows such complexing. Not known are the characteristics of the complexes that promote rapid uptake—perhaps upon associating with their "target" molecules, the serum factors undergo changes in conformation (*e.g.*, SASAZUKI *et al.* 1974) that expose sites which can bind to cell surface receptors. Or, perhaps features such as size or overall charge of a complex are important. (GANS *et al.* 1968 have shown that perfused liver takes up "active" fibrin and thrombin much faster than partly degraded and inactive forms of these proteins, probably as a result of effects analogous to those being considered.)

Various proteins denatured through exposure to heat, formaldehyde and so forth are cleared from the circulation far more rapidly than are the corresponding native forms (MEGO 1973 b). Although denaturation rates are influenced by "environmental" factors, such as ionic strength of the medium or the presence of substrate, at the level of individual molecules the process may be largely random—under given conditions a newly made molecule may have substantially the same probability of becoming denatured as does an old one. Denaturation can change the distribution of groups in proteins in ways that may promote association with membranes.

The notion of randomness in denaturation probably should not be carried too far, since, for example, there could be ways in which the "history" and "experiences" of a large protein sometimes contribute small conformational changes "step by step", which lead eventually to denaturation. But denaturation is only one suspect; the apparent lack of strong age-dependancy in clearance rates might in part reflect some other fairly simple rate-limiting changes occurring at random in the population of circulating molecules. Of major interest in this regard are the observations that several circulating glycoproteins shorn of their sialic acids by enzymatic means, are cleared from the circulation far more rapidly than are the intact glycoproteins (GREGORIADIS *et al.* 1970, MORELL *et al.* 1971). This holds true for ceruloplasmin, orosomucoid, gonadotrophic hormones and other glycoproteins (but turnover of transferrin seems insensitive to desialylation, which is interesting in light of the role of this protein, Section III.1.1.3). One can readily think of ways in which the charge or the conformation of saccharide side chains in glycoproteins might dramatically influence the interactions of the proteins with surfaces of cells.

As indicated earlier, changes in electrophoretic properties of some glycoproteins, that mimic those expected for enzymatic removal of charged saccharides, are reputed sometimes to take place under conditions (*e.g.*, very low temperature) where enzymatic removal seems unlikely (see *e.g.*, TOUSTER 1973). If they occur, such "spontaneous" alterations in proteins would be highly germane to our present considerations. Further, the possible presence at cell surfaces, of enzymes that cleave macromolecules (neuraminidases, lipases, etc.) has been reported (BRAIDMAN and ROBINSON 1973, SCHENGRUND *et al.* 1972, 1973, WAITE and SISSON 1973, VANNARELL and ARONSON 1973) although adsorption artifacts are very hard to exclude in the cell fractionation experiment on which such conclusions are based (DE PIERRE and KARNOVSKY 1973). In light of our later discussion of hydrolase release from phagocytes

(Section IV.3) one might also expect extracellular hydrolases to accumulate under normal circumstances at sites where phagocytosis is extensive, such as the spleen (*cf.*, the comments above concerning erythrocytes). Cell surface or extracellular neuraminidases clearly could play large roles in determining the fate of circulating glycoproteins.

When administered in sufficiently high concentration, different desialylated proteins appear to compete for a limited number of binding or uptake sites responsible for clearance; these sites are concentrated in the liver and autoradiographic evidence suggest that hepatocytes contribute most heavily to the clearance process (MORELL *et al.* 1971). Hepatocytic lysosomes degrade the asialoglycoproteins (in contrast, Kupffer cells are dominant in the breakdown of heat-denatured proteins, which might reflect some intrahepatic division of labor; GORDON 1973).

Perhaps the hepatocyte surfaces carries receptors with general affinity for asialoglycoproteins. In line with this, PRICER and ASHWELL (1971; see also RIORDAN *et al.* 1974) have reported that isolated hepatic plasma membranes bind neuraminidase-treated glycoproteins more avidly than they do intact ones, and HUDGIN *et al.* (1974) have isolated a membrane protein that may be responsible for this. Binding of desialylated proteins to the isolated membranes is inhibited if the membranes themselves are treated with neuraminidase. The purified protein shows no glycosyltransferase activity (HUDGIN and ASHWELL 1974), a finding which may dispose of one class of theories about the nature of possible receptors. Rather than their somehow detecting the absence of sialic acid, it is likely that the receptors recognize some group or feature common to those desialylated glycoproteins that are rapidly cleared; the best present guess is that key galactoses or other residues are exposed by the neuraminidase treatment (VAN DEN HAMER *et al.*, 1970; see also CONWAY *et al.* 1975).

If binding sites of such sorts control normal turnover, the likelihood is that concentrations of desialylated proteins would rarely, if ever under ordinary circumstances, reach levels where competition or saturation became important. Thus, the physiological implications of the fact that different asialoproteins compete with one another with greater or less success for clearance by the liver, may be less significant then the finding that when the concentrations of desialylated proteins introduced in the circulation are fairly low, proteins with very different apparent affinities for the ostensible receptors can be simultaneously cleared at rapid rates (MORELL *et al.* 1971).

One can also devise schemes for turnover of circulating molecules based upon possible "reversibility" of endocytosis. Perhaps only a proportion of the endocytic vesicles formed by a cell actually fuse with lysosomes (GORDON 1973) and the rest return their contents to extracellular spaces; one could hypothesize that endocytic vesicles with different contents have different fates. Plausible proposals of this type do require many assumptions that are difficult to support at present but, for example, some current thinking about transintestinal transport of antibodies is oriented in this general direction (see Section IV.5.3) and such hypotheses should not be totally discarded out of hand.

Among the desialylated proteins in the work described above, several were glycoprotein hormones. This raises the general question of the fate of hormones and of related "active" circulating molecules such as releasing factors. Lysosomal hydrolases can degrade various protein and peptide hormones such as insulin or angiotensin (ANSORGE *et al.* 1971; BOHLEY *et al.* 1971; MCDONALD *et al.* 1971, 1974) but the significance of this for events *in vivo* remains to be demonstrated. Intracellular locations are not identified for some hydrolases that may be pertinent, such as the "insulin specific proteases" of liver (BURGHEN *et al.* 1972, DUCKWORTH *et al.* 1972) or peptidases of pituitary (MARKS *et al.* 1974). Other potentially important enzymes are reputedly present at cell surfaces; this is reported for a pulmonary peptidase that can split angiotensin (for discussion and references see LANZILLO *et al.* 1973, SMITH and RYAN 1973, SOFFER *et al.* 1974), for an hepatic glucagon-inactivating system (POHL *et al.* 1972) and for hepatic glutathione-insulin transhydrogenase (ANSORGE *et al.* 1971; VARADANI 1973, 1974). Thus, hypothetical sequences can be constructed in which the overall levels of a hormone in the circulation are affected by endocytosis in the liver and kidney (many hormones are small enough to filter into the renal tubules) while the local levels at target cell surfaces are controlled initially by non-lysosomal inactivating systems; the inactivated hormones might then be fully degraded in the lysosomes. One topic of current debate bearing upon this is the relationship of the sites that mediate hormone action to the inactivation systems (see *e.g.*, DESBUQUOIS and CUTRECASAS 1972, FREYCHET *et al.* 1972 for initial experiments with "purified" membrane fractions in which binding and degradation processes can be dissociated).

III.4. Intracellular Turnover

Much data has been collected on the turnover of intracellular proteins under steady state conditions and during metabolic changes (see *e.g.*, Table 2; p. 130). Somewhat less information about degradation is available for lipids, nucleic acids and carbohydrates. Useful recent reviews on several aspects of intracellular turnover have been published by GOLDBERG and DICE (1974), PINE (1970), POOLE (1971 a), RECHCIGL (1971 a, b), SCHIMKE (1973), and SIEKEVITZ (1972 a, b).

The fact that proteins and other macromolecules do turn over at appreciable rates in non-growing tissues would seem to be the evolutionary outcome of a least three important factors. First, spontaneous denaturation and "wear-and-tear" phenomena eventually inactivate virtually all types of enzymes. During rapid growth, inactivated molecules might be diluted at a rate sufficient to render them unimportant but in cells that divide rarely, degradation or some other type of disposal is advantageous. Second, changes in metabolic patterns are facilitated if, in addition to synthesizing new mRNA's, enzymes and so forth, a cell can also remove "old" ones that may no longer be "appropriate". Finally, degradation of preexisting macromolecules provides small molecules that can serve as precursors for new synthesis or as metabolites for such processes as energy production. Under stress or starvation conditions the availability of intracellular sources of precursors or metab-olites could be essential for survival and there is also evidence that normal steady-state intracellular pools (*e.g.*, those of amino acids) include major contributions from intracellular degradation.

Considerations of these sorts do not place great constraints upon the mechanisms of degradation; in particular, they do not require that degradation be highly selective, even though this may be an aesthetically pleasing "solution". From current knowledge, one can barely guess at the balances of effects between random and selective inactivation and digestive processes, and between molecular stability and susceptibility to degradation, that might govern in the evolution of cells and macromolecules.

Present concepts of the mechanisms of the degradative phases of intracellular turnover are beclouded by numerous uncertainties. For eucaryotes, it seems inconceivable that autophagy plays no role since, as stressed earlier, autophagic vacuoles are part of the normal organelle complement of many—probably virtually all—eucaryotic cells. But the quantitative importance of this process as compared with the other possible mechanisms is very difficult to establish, especially for steady state phenomena. I am biased toward the belief that autophagy is of major importance for normal turnover.

III.4.1. Some Methodological Perspectives and Problems

The estimation of turnover rates usually has relied upon methods that are relatively simple in conception but problem-laden in execution (see POOLE 1971 a, b, and SCHIMKE 1973 for critical essays on methods). Most widely used are pulse-chase techniques in which radioactive precursors are introduced for a brief period and then the loss of label from a macromolecular population is followed during a period of maintenance of the cells in the absence of exogenous label. An important recent derivative of such approaches is the double-label method wherein a precursor such as ^{14}C-labelled leucine is administered at one time point, and then the same cells are exposed later to the same precursor labelled with a different isotope, *e.g.*, ^{3}H-leucine (ARIAS *et al.* 1959, GLASS and DOYLE 1972, SCHIMKE 1973). The second exposure is for a brief period shortly before disruption of the cells and analysis. Different proteins will show different $^{3}H/^{14}C$ ratios depending upon their turnover characteristics since, if timing is proper, those that turn over rapidly will have relatively less of the initial label (^{14}C) and relatively more of the second label (^{3}H) than those that turn over more slowly.

Sometimes evaluations of degradation rates are based upon determinations of steady state synthesis rates. For the steady state it is self-evident that this provides a measurement of molecular replacement. Occasionally replacement rates are estimated by inhibiting macromolecular synthesis pharmacologically or physiologically and determining the rate at which a component of interest disappears.

These various methods are affected by the usual difficulties relating to cellular and organelle heterogeneity, the presence of a few dying cells in an otherwise healthy population, the production by the cells of secreted macro-molecules as well as ones that remain within, difficulties in controlling tracer levels in intracellular pools, uncertainties as to proper choice of times for introducing tracers and taking samples, complexities in the effects of metabolic inhibitors and so forth. Reutilization of precursors has often proved vexing—thus, for example, it is commonly found that hepatocyte proteins show apparently slower turnover when leucine or uniformly-labelled arginine is used as the precursor than when guanidino-labelled arginine is utilized (Fig. 40; SWICK 1958). Each of these amino acids can be returned with its label to intracellular pools through protein degradation but the guanidino label is rapidly removed from the pools by the intervention of arginase, so reuse of label is minimized (although it is not entirely avoided; GLASS and DOYLE 1972). In addition, different portions of complex molecules may turn over at different rates—exchange reaction resulting in this behavior are evident with several categories of lipids and the result is that the degradation rates obtained depend partly upon the precursor used. While, especially for comparative studies, the double label method helps overcome many of these complications, even this procedure is prone to problems in timing of isotope administration and it is dependent on substantial purification of the components examined (POOLE 1971 b). Other special approaches employed particularly to avoid reutilization problems include carbonate labelling techniques (SWICK et al. 1968) and labeling of heme groups in proteins for which this is applicable (Fig. 40; DRUYAN et al. 1969).

Important difficulties are related to macromolecule identification. When one seeks to determine how fast label is lost from a population of macromolecules, the answer really is an operational one that can be strongly conditioned by the methods used to identify the molecules. Standard isolation procedures, chromatography, light absorption techniques, immunological precipitation approaches and so forth all depend upon relatively complex features of the molecules being studies. Thus, for example, the fact that a change occurs in the specific radioactivity of a given protein whose absolute amount in the cell or cell fraction is demonstrated to be constant through immunological inventory or measurements of enzyme activities, does not by itself imply that the radioactive molecules that are no longer detectable have been degraded to low molecular weight forms. Strictly speaking, all that has been shown is that a proportion of the relevant population of radioactive molecules no longer is immunologically identifiable, or enzymatically active. This can be misleading. For example, BALLARD et al. (1974) report data they interpret as demonstrating that incubation of a liver homogenate results in loss of enzymatic activity, of solubility, and of immunological specificity of the enzyme phosphoenolpyruvate—carboxy kinase before extensive proteolysis takes place. Additional problems arise in evaluating turnover of organelles—to what extent does the decline in radioactivity of the proteins in a mitochondrial or peroxisomal population imply degradation of the proteins as opposed to their departure intact from the organelles? This last point leads to another related one—the geometrical complexity of some cell types must be taken into account. For example, a number of authors have studied turnover in neuronal synaptic terminals by pulse-labelling brain and then following the decline in radioactivity in terminals ("synaptosomes") isolated at intervals thereafter (*e.g.*, DeLores Arnais *et al.* 1971, Morris *et al.* 1971). Unless suitable adjustments of interpretation are made this approach runs a high risk of generating misleading results since the decisively important pools for many synaptic macromolecules are actually not the low molecular procursors but rather the macromolecules that are synthesized in the perikarya and transported to the terminals (see Holtzman *et al.* 1973 for further discussion) [13].

Finally, the methodological problems outline in this section are exacerbated in studies of non-steady state events, such as the responses of cells to altered physiological conditions. It is rarely possible to predict precisely how precursor pools behave in such circumstances (see *e.g.*, Klevecz *et al.* 1971) and reuse even of guanidino-labelled arginine may become a significant problem (*e.g.*, Goldberg and Dice 1974). Even for steady state investigations, diurnal cycles, effects of diet and feeding schedules (Rechcigl 1971 b) and other experimental parameters probably should be factored into interpretations to a greater extent than is usually the case.

III.4.2. Turnover in Bacteria

Intracellular proteins and nucleic acids do turn over in bacteria, which implies that lysosomes as such are not essential for degradation of such molecules. However, relatively little is known of procaryote degradative mechanisms responsible for turnover and as will develop below, it may still turn out that some sort of sequestration sometimes is involved.

Bacteria elaborate many hydrolases but most of the ones that have been studied are exoenzymes by which the cells can hydrolyze macromolecules in extracellular spaces (for examples and references see footnote 14 [p. 125] and Berkeley *et al.* 1973, Corpe and Winters 1972, Hartmann *et al.* 1974, Scher and Dubnau 1972, Schwarz *et al.* 1969). De Duve and Wattiaux (1966) have pointed out that the secretion of exoenzymes can be meaningfully

---

[13] Geometry might also influence actual rates and mechanisms of turnover. For example, tracer experiments suggest that some synaptic molecules are transported back along axons to perikarya where they are degraded by the abundant lysosome population (Birks *et al.* 1972, Holtzman *et al.* 1973, La Vail 1974, Teichberg *et al.* 1974, 1975). The implications of this for the timing or details of the breakdown of macromolecules have not been studied. In striated muscle cells, lysosomes are relatively scarce and many of those that are present, cluster near the nuclei. This, plus the highly organized distribution of myofibrils, sarcoplasmic reticulum and so forth poses special "problems" for access of the lysosomes to potential intracellular substrates. [Hoffstein *et al.* (1974) have made a preliminary report that acid hydrolases may be transported in the sarcoplasmic reticulum of cardiac muscle; if so, some of the "problems" may be more apparent than real since such enzymes might rapidly enter lysosomes forming at many points in the cytoplasm; Essner *et al.* (1965) have also demonstrated acid hydrolase activity in sarcoplasmic reticulum. It is difficult to envisage how the extended arrays of filaments that make up the myofibrils, could be autophagocytosed without prior fragmentation, or other modification. Perhaps this is reflected in the lack of evident lysosomal involvement in the early stages of fibril degeneration (*cf.*, Section III.1.2.1).

analogized to heterophagic processes in eucaryotes [14]. Bacteria do not engage in endocytosis or exocytosis and as far as is known most exogenous proteins cannot cross their plasma membrane (although there are membrane associated peptide transfer systems that might facilitate entry of some fairly large molecules and perhaps there is similar entry of specific proteins; SUSSMAN and GILVARG 1971). The simplest explanation for the ability of procaryotes to translocate exoenzymes across their plasma membranes is that the proteins move directly from the corresponding ribosomes across the membranes and do not assume their folded conformation until such passage is well advanced or complete (BRAATZ and HEATH 1974, GLENN et al. 1973, GOULD et al. 1973, LAMPEN et al. 1971, MINDICH 1974): a similar explanation has been developed for entry of newly synthesized proteins into the endoplasmic reticulum of eucaryotes.

Turnover can be quite rapid for bacterial mRNAs under many conditions. For the other components, notably proteins and ribosomes, rates and characteristics of degradation vary greatly with growth conditions. Ribosomes are subject to little detectable degradation in log phase *E. coli* but when growth is stopped (*e.g.*, in nitrogen and amino acid poor medium) much breakdown of these organelles becomes evident (FAKHER and SCHLESSINGER 1966, KAPLAN and APIRION 1974, MARAVALDI et al. 1974). Pioneering pulse-chase experiments on protein turnover suggested that rapidly growing cells conserved almost all their proteins whereas starved ones showed degradation rates that ranged up to 5 to 10% per hour, and turnover also was found to be marked during sporulation (see reviews by MANDELSTAM 1960, WILLETTS 1967). The concept arose that bacteria starved for carbon or nitrogen sources can use amino acids and other molecules originating through degradation of intracellular material to support energy metabolism or induced enzyme synthesis, etc. More recent studies in which care has been taken to determine rates of label loss immediately after labelling, continue to support the idea that protein turnover differs substantially in growing bacteria as compared to non-growing ones but these studies also indicate that there is a small, rapidly turning-over fraction of proteins in growing cells (*e.g.*, NATH and KOCH 1970, PINE 1970). When cell growth is stopped, in some sense the population of proteins that is accessible to degradation expands (PINE 1970) although under the experimental conditions studied so far there remains a population that is degraded very slowly, if at all.

Relatively few investigations have concerned turnover of specific proteins so the members of the populations referred to in the preceding paragraph are for the most part not identified [*e.g.*, WILLETTS (1967) summarizes work on a few enzymes and discusses some hints of different behavior of membrane bound vs. soluble proteins; SUSSMAN and GILVARG (1971) comment on turn-

---

[14] The enzymes may sometimes serve nutritional functions by hydrolyzing exogenous macromolecules and they can participate in such processes as the penetration of bacteria into tissues of higher organisms. In addition, exoenzymes (plus bulk sloughing in some species) take part in the selective remodelling and turnover of cell walls that accompany bacterial growth, sporulation and other "developmental" events (*e.g.*, FIEDLER and GLASER 1973, SCHWARZ et al. 1969, SHOCKMAN et al. 1973).

over of a set of proteins that may be degraded at the time of cell division and KAHANE and RAZIN (1969) report that Mycoplasma membrane proteins turn over with half-lives of a few hours under conditions in which other labelled cytoplasmic macromolecules seem stable]. We also lack a clear conception of the contributions made to the observed patterns of degradation by a number of potentially important phenomena; for instance, proteolytic processing of large precursor proteins is thought to be involved in the "maturation" of some proteins of organisms ranging from viruses to mammals (Section V.2) and polypeptide fragments generated by this might appear as a rapidly turning over "protein" fraction. However, even though the overall picture is fragmentary there is a pregnant line of evidence indicating that degradation in bacteria can be selective. Under various growth conditions *E. coli* breaks down abnormal proteins rapidly [GOLDBERG (1972 a) provides a brief review of his work and others']. This has been demonstrated for proteins synthesized in the presence of puromycin or other drugs (HIPKISS and KAGUT 1973, KEMSHEAD and HIPKISS 1974), those that have incorporated amino-acid analogues (Fig. 37; KEMSHEAD and HIPKISS 1974) mutationally-engendered nonsense fragments (GOLDSCHMIDT 1970, LIN and ZABIN 1972) and abnormal repressor molecules. Among other considerations these striking findings have led to speculation that cells may have the ability to screen their output of proteins for features such as errors in translation.

The enzymatic machinery responsible for degradation of procaryotes' macromolecules is very poorly understood. Several of the known RNAse's are reasonably good candidates for participants in mRNA or rRNA turnover but the precise contribution of each, and the relevant control mechanisms are not known. Proteases and lipases are also readily detectable in bacteria; some of the proteases probably are intracellular, but the mechanisms that limit their activities are not known and in addition it often is difficult to establish definitely whether a given type of enzyme is located in the cells, bound to cell membranes or functions as an exoenzyme (see *e.g.*, ALBRIGHT *et al.* 1973; REGNIER and THANG 1972). For that matter, there is little unambiguous evidence that degradation of intracellular protein takes place within the cell. There have been reports that the intensive protein degradation that accompanies sporulation is inhibited by mutations affecting extracellular proteases (MANDELSTAM 1969, MANDELSTAM and WAITES 1968, PINE 1972) but the literature is complex and the significance of such findings is unclear. For example, SLAPIKOFF *et al.* (1971) assert that certain strains of *Bacillus brevis* can sporulate with very little protein turnover and very low protease levels.

The rapidity with which starvation initiates changes in patterns of proteins and ribosome turnover suggests that at least a few molecules of the enzymes responsible preexist in the cells (*e.g.*, SCHLESSINGER and BEN-HAMIDA 1966). This also is indicated by GOLDBERG's (1971 b) report that certain temperature-sensitive mutants can show marked enhancement of degradation when subjected to conditions in which they cannot charge specific tRNA's with amino acids and thus presumably cannot synthesize new proteins. (From this evidence and some other consideration GOLDBERG 1971 a, 1974 a and SCHLES-

SINGER and BEN-HAMIDA 1966 ascribe to the tRNA's an essential role in the control mechanisms governing protein degradation.) On the other hand, chloramphenicol, an inhibitor of protein synthesis, can affect the changes in protein and RNA degradation rates seen with altered growth conditions. Unfortunately the effects and the kinetics depend strongly on the experimental

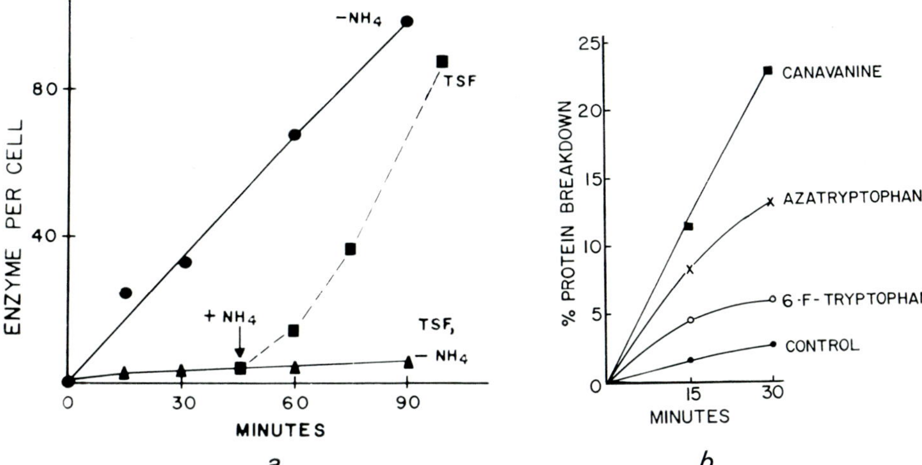

Fig. 37. Breakdown of bacterial proteins: *a* When thiogalactoside inducer is added, *E. coli* grown on a medium lacking a nitrogen source can synthesize β-galactosidase (curve labelled —NH₄). However, if toluene-sulfonyl-fluoride (TSF) is added, this ability is lost (TSF, —NH₄ curve); TSF is an inhibitor of proteases. The effects of TSF are reversed if ammonium is added (TSF, + NH₄). These results suggest that protein degradation can support synthesis of new enzyme molecules in cells starved for nitrogen. *b* An *E. coli* auxotroph for arginine and tryptophan was grown briefly in media containing tritiated leucine (³H-leucine) plus either other ordinary amino acids (control) or the arginine analogue, canavinine, or tryptophan analogues. The cells were then transferred to normal, non-radioactive media and the degradation of proteins made during the growth in labelled leucine was followed by determining the amounts of ³H-leucine released in TCA soluble form as compared to the initial levels of ³H-leucine in the proteins. The breakdown of proteins made in cells exposed to amino acid analogues along with the radioactive leucine is faster than than the breakdown in control cells. (Figures from GOLDBERG, A. L., 1972: Proc. nat. Acad. Sci. [U.S.] **69**, 422—426; GOLDBERG, A. L., 1971: Nature New Biol. **234**, 51—52; GOLDBERG *et al.*, Fed. Proc. 33, 1112—1120.)

design and interpretations are not clear-cut. However, in some experiments chloramphenicol seems to reduce rates of protein breakdown (MANDELSTAM 1970, SCHLESSINGER and BEN-HAMIDA 1966, WILLETTS 1967) which has fueled speculation about turnover of proteases, synthesis of new enzymes and involvement of protease inhibitors in the control of intracellular degradation. There is little direct concrete evidence.

Work with inhibitors of energy metabolism, such as cyanide or dinitrophenol has produced ambiguous findings. Turnover can generate metabolites for the production of energy but there also are indications from the inhibitor studies that under some circumstances degradative processes may themselves require energy (GOLDBERG 1972 a, PINE 1972, SCHECHER *et al.* 1973). At first glance this is somewhat surprising since proteolytic enzymes in general seem not to be energy dependent; perhaps, one is looking at indirect manifestations of perturbations in fairly complicated networks of processes and controls or perhaps there really are energy-dependent steps in degradation (one intriguing possibility, for which, however, no direct experimental support exists, is that energy might be required to sequester or transport molecules destined for destruction).

Very interesting results have emerged from work with protease inhibitors, notably the studies of GOLDBERG and his colleagues in which they use agents that affect "serine-proteases" (*e.g.*, p-toluene sulfonyl fluoride). This laboratory has reported that cells grown in the presence of such inhibitors can show fairly normal rates of "background" protein degradation (this is the rapid turnover of a small population of proteins that occurs both in growing and in non-growing cells) and they also can rapidly degrade abnormal proteins (GOLDBERG 1971 b). But, the cells are defective in their degradative responses to starvation (GOLDBERG 1971 b, PROUTY and GOLDBERG 1972 a). (This means, for instance, that when the inhibitors are present induction of β-galactosidase does not occur during starvation; normally, nitrogen-starved cells can synthesize this enzyme, apparently by using the amino acids liberated through proteolysis, Fig. 37.) The effects of protease inhibitors on cells may turn out to be less selective than might be hoped [see *e.g.*, SHECHER *et al.* (1973) for their claim that ATP levels are drastically reduced in circumstances comparable to those used by GOLDBERG's group and ROSSMAN *et al.* (1974), for evidence suggesting that chloromethyl protease inhibitors may directly alter macromolecules synthesis]. But while this does introduce cautionary aspects, it does not invalidate the proposal drawn from inhibitor studies that there may be two somewhat separate degradation systems—one responsible for "normal" turnover and the other for enhanced or "hard-times" degradation. The isolation by BUKARI and ZIPSER (1973) of mutants defective in their ability to degrade nonsense fragments suggests that a critical test of this concept may not be far off.

An important lead towards elucidation of cellular degradation mechanisms could lie in the observations that some bacterial proteins known to turn over relatively rapidly *in vivo* also are relatively more susceptible to proteolysis when exposed to enzymes such as pronase or trypsin *in vitro* (GOLDBERG 1972 b, 1973). This holds for abnormal polypeptides produced in the presence of puromycin or amino acid analogues and also for at least some normal proteins. It may thus be highly significant for analysis of *in vivo* degradation that since the pioneering work of LINDERSTROM-LANG (1950) it has been generally accepted that denatured proteins often are more readily attacked by proteases than are the corresponding native forms and that in a number of cases the presence of substrates or other ligands can help "protect" enzymes against denaturation, inactivation or degradation *in vitro* (*e.g.*, GRISOLIA 1964, MARKAS 1965). Hints of similar effects of functional states upon the "resistance" of ribosomes to RNAse attack have also been reported [NOLAN and HARTMAN (1973); see also LUZIKOV 1973 for eucaryotic organelles]. KAPLAN and APIRION (1975) speculate that starvation induces ribosome degradation via effects related to ribosome subunit conformation; a given conformation might be particularly susceptible to the endonucleolytic attack they believe is the primary event in ribosome breakdown.

Are these steps in procaryotic intracellular turnover that can realistically be compared to sequestration in lysosomes? The abnormal proteins containing amino acid analogues that are degraded rapidly *in vivo,* aggregate within bacteria to form distinct, microscopically visible inclusion bodies (PROUTY and GOLDBERG 1972 b, PROUTY *et al.* 1975). These are not membrane-delimited. Rather they seem to be due to some change in the solubility, ease of denaturation or in the affinity of the proteins for one another. KEMSTEAD and HIPKISS (1974) have reported that when *E. coli* cells are pretreated with puromycin and crude extracts are then made, the extracts, like the intact cells, show rapid breakdown of the puromycyl-peptides. But this does not hold for proteins containing the analogue, canavanine; these are degraded rapidly *in vivo* but not in extracts. One possible implication is that the full-length proteins that are synthesized with canavanine must undergo some processing step in the cell, that is not needed for the shorter puromycyl-peptides; since this step does not occur when the cells are disrupted, one might reasonably think in terms of some special structure-related events. In trying to relate findings of these types to degradation in eucaryotes it is probably best to keep an open mind—since we really do not understand well the mechanisms that contribute to intracellular turnover in any class of cells, choosing to emphasize procaryate-eucaryote similarities, rather than differences (or *vice versa*) is more a matter of faith than of analysis.

III.4.3. Turnover of Organelles and Intracellular Macromolecules in Eucaryotes

III.4.3.1. Some General Points and Some Experiments Paralleling those Done with Procaryotes

Normal eucaryotic cells show appreciable intracellular turnover under growing and non-growing conditions (EAGLE *et al.* 1959, RECHCIGL 1971, SCHIMKE 1970). Turnover is controlled at the levels both of synthesis and of degradation and such "dual control" can contribute to subtle modulations of cellular biochemistry as well as to more dramatic changes, such as the use-related atrophy and hypertrophy of muscle (*e.g.,* GOLDBERG *et al.* 1974 b, RABINOWITZ *et al.* 1973). From study of tissue culture cells under varying conditions it has been estimated that the steady state turnover of protein usually occurs at an overall rate of approximately 1–2% per hour (EAGLE *et al.* 1959, HERSHKO and TOMPKINS 1971, RECHCIGL 1971 b, PINE 1967). Estimates for hepatocytes are of the same order of magnitude (SCHIMKE 1973). Work on several cell types indicates that different proteins within a given cell have quite different half-lives; values of a few hours to a few days are common (Table 2) but there are examples that fall well outside this range.

In multicellular organisms under ordinary circumstances, different cell and tissue types may show distinctive rates of turnover of apparently identical

Table 2. *Reported Half-Lives for Selected Components of Rat Liver and Some Other Tissues*. Where available, data are for steady-state conditions and were obtained with methods designed to minimize problems of reutilization of precursors (exceptions to the latter include figures for intestinal brush border and brain tubulin). References are to articles in the reference list: (1) ARIAS *et al.* 1969; (2) ASCHENBRENNER *et al.* 1970; (3) BOCK *et al.* 1971; (4) DINGLE 1973 b; (5) DRUYAN *et al.* 1969; (6) FISCHER and MORELL 1974; (7) FRITZ *et al.* 1969; (8) GLASS and DOYLE 1972; (9) HIRSCH and HIATT 1966; (10) JAMES *et al.* 1971; (11) JOHNSON and KENNEY 1973; (12) JUNGALWALA 1974; (13) KAWASAKI and YAMASHINA 1971; (14) LINDY 1974; (15) LOEB *et al.* 1965; (16) MAJERUS and KILBURN 1969; (17) MELDOLESI 1974; (18) QUINCEY and WILSON 1969; (19) SABRI *et al.* 1974; (20) SCHIMKE 1973; (21) SMITH 1968; (22) SWICK *et al.* 1968; (23) TERJUNG *et al.* 1973; (24) TWETO *et al.* 1972; (25) WATESON *et al.* 1972; (26) WENTHOLD *et al.* 1974; (27) POOLE 1971; (28) DUTTON and BARONDES 1969; HEMMINKI 1973; (29) MARVER *et al.* 1966.

Rat liver components:	Half-life	References
Average protein in homogenate	3.3 day (d)	1
Average nuclear protein	5 d	20
Soluble (cytosol) proteins (average)	4–5 d	8, 20
Soluble enzymes (these tend to be inducible or responsive to diet, etc.):		
tyrosine amino transferase (basal rate)	2 hour (h)	11
acetyl CoA carboxylase (normal diet)	2 d	16
tryptophan oxygenase (pyrrolase) (basal rate)	2–4 h	20
arginase (several diets)	5 d	20
lactic dehydrogenase isozyme 5	disputed: 4–6 d; 16 d	8, 14; 7
Ferritin	1.3 d	8
Fatty acid synthetase (multienzyme complex)	3–4 d	24
Average peroxisomal protein	1.5–2 d	27
Plasma membrane proteins	1.5–2 d	13, 20
Average ER protein	2–2.5 d	5, 20
smooth microsomes	2 d	1
rough microsomes	2 d	1
NAD glycohydrolase of ER	18 d	3
Average mitochondrial protein	4–6 d	2, 5, 22
inner membrane (cytochromes c, b; heme a)	5–6 d	2, 5, 23
outer membrane (cytoch. b₅)	4.5 d	2, 5
inducible aminotransferases [1]	less than 1 d	22
δ-amino levulinate synthetase [2] (inducible; matrix)	1–2 h	2, 20, 29
Average ribosomal protein	5 d	9, 20
Ribosomal RNA's (5 S, 18 S, 28 S)	5 d	9, 15, 18
Examples of some other components:		
Rat heart lactic dehydrogenase isozyme 5	1.6 d	7
Rat skeletal muscle lactic dehydrog. isozyme 5	31 d	7
Rat intestine brush border proteins	18 h	10
Rat brain tubulins	4 d	28
Rat brain acetylcholinesterase	2.8 d	26

[1] Estimated from changes in enzyme activity after alteration in levels of dietary protein.

[2] Estimated from changes in enzyme activity after induction with allylisopropylacetamide followed by exposure to puromycin.

Table 2 (*continued*)

Rat central nervous system myelin:
    some proteins (many disputes)   35 d (?)   12, 19, 21
    phospholipids (some types may turn over faster) 41 d   20, 21

Mouse central nervous system myelin:
    different classes of proteins   40–95 d   6

Guinea pig pancreas:
    proteins of zymogen granule membranes (average) 4.5 d   17
    proteins of rough ER membranes (average) 5 d   17

Rat costal cartilage polysaccharide   8–16 d   4, 25
Rabbit cartilage collagen   more than 50 d 4

components, and differences in the patterns with which precursor molecules flow among extracellular pools, intracellular pools and macromolecules (RECHCIGL 1971 b, SCHIMKE 1973, SEGAL 1969, VESSELL and FRITZ 1971). There also are tissue characteristic responses of turnover to stress or to hormones such as insulin, that are important for the overall physiology of the organism. Thus, normal skeletal muscle derives roughly 30% of its amino acid pool from steady state intracellular degradation (GLAN and JEFFAY 1967); for hepatocytes the figure is about 50% (ARIAS *et al.* 1969, GAN and JEFFAY 1967). Under conditions of starvation intracellular sources initially account for as much as 90% of the hepatic amino acid pools and as the hepatocyte proteins contributing to this become depleted, degradation of muscle protein can be much enhanced as a major mechanism by which levels of amino acids in the blood are maintained and metabolites are provided, to the liver and other organs (see *e.g.,* FELIG and WAHREN 1974, GAN and JEFFAY 1967, GOLDBERG *et al.* 1974 a, b, JEGGERSON *et al.* 1974 for details, and also for some similar effects related to insulin) *. Regenerating liver shows little net loss of label incorporated in protein (SCORNIK 1972) whereas in the steady state, the measured halflife of total liver protein averages only a few days; differences in turnover and in reuse of precursors probably account for this.

Although few detailed studies have been done, it seems likely that in eucaryotes as in procaryotes, there are circumstances in which intracellular turnover can serve as the major source of metabolites for synthesis of new types of macromolecules, as well as for support of "maintenance activities". Some interesting preliminary work on unicellular organisms points in this direction. For instance, the protease inhibitor phenyl-methane-sulfonyl-fluoride (PMSF) can prevent the genesis of chloroplasts that normally occurs when dark-grown *Euglena* are placed in the light, even in nitrogen deficient media (ZELDIN and SKEN 1973, ZELDIN *et al.* 1973). One would like to think that the PMSF effect is on the proteolytic mobilization of amino acids (*cf.,* MANDELSTAM 1960) especially since the effect is partially overcome by

---

* The inhibition of protein degradation, produced in muscle by insulin, is accompanied by changes in the "latency" of lysosomes isolable from the tissue (RANNELS *et al.* 1975). Whether these changes are importantly related to the altered turnover is not known (*cf.,* Section IV.4.1).

supplementing the medium with amonium. However, the processes for generating chloroplast are complex and the inhibitor also interferes with cell division (even if ammonium is present) so the findings must be supplemented through more direct studies.

Studies on the mold, *Blastocladiella,* on slime molds, and on the alga *Chlamydomonas* have turned up some interesting changes in protein degradation rates, correlated with particular stages in the life cycles, or with responses to nutrient-poor growth media (JONES *et al.* 1971, LODI and SONNEBORN 1974). For example, extensive degradation of protein may take place during sporulation, or gamete formation. Little is known of the underlying mechanisms, but some of the phenomena are reminiscent of those observed with bacteria. In *Blastocladiella,* altered levels of acid protease activity accompany some of the changes in protein breakdown (LODI and SONNEBORN 1974).

When the "chymotrypsin inhibitor" TPCK is included in the growth medium, HeLa cells show a rise in the average molecular weight of newly synthesized proteins (TABER *et al.* 1973). The inhibitor had previously been shown to interfere with the proteolytic processing of large precursor proteins to form smaller ones, that is involved in virus formation. Thus it may eventually prove possible to evaluate the extent to which processing of this or related types contributes to the measured rates of turnover of proteins in eucaryotes (see also Section V.2.2. where we take up proteolysis of precursors of secretory proteins).

In the few eucaryotic cells studied thus far, there do not seem to be major populations of intracellular macromolecules that are not accessible to degradation under normal conditions, with the exception of nuclear DNA. This even applies to many, if not all the chromosomal proteins (*e.g.,* BORUN and STEIN 1972). In the steady state, loss of label from most of the intracellular macromolecular populations that have been analyzed takes place with time courses that resemble first-order kinetics (most studies have concerned proteins; for reviews see RECHCIGL 1971 b, SCHIMKE 1973). As outlined above, this suggests that the rate-limiting steps are not strongly dependent upon molecular age; there is currently a controversy as to whether or not this is generally true for mRNA's (COWAN and MILSHIN 1974, MATTHEWS 1973, PERRY and KELLY 1973, SINGER and PENMAN 1973) although it does seem to hold for rRNA's (HIATT and HIRSCH 1966).

While some efforts to increase rates of protein degradation by exposing cells to amino acid analogues or puromycin have not been strikingly successful (see *e.g.,* JOHNSON and KENNEY 1973, POOLE and WIBO 1973 a) it is claimed that peptides made in the presence of puromycin are degraded very rapidly in reticulocytes and hepatoma cells (MCILHINNEY and HOGAN 1974) *. RABINOWITZ and FISCHER (1964) report that reticulocytes rapidly degrade the aberrant protein they make in the presence of the valine analogue, threo-

---

* KNOWLES *et al.* (1975) have also found that exposure of hepatoma cells to fluorotryptophane, or to the arginine analogue, canavanine, results in enhanced rates of protein degradation.

amino-p-chlorobutyric acid and ADAM *et al.* (1972) have invoked selective degradation to explain why the genetically-engendered abnormal protein, hemoglobin Ann Arbor, seems not to accumulate to apprecible levels in the circulating population of red blood cells. MCILHENNY and HOGAN (1974) maintain that the enhanced degradation seen with puromycin does not occur in cell free-extracts (see also SCHIMKE *et al.* 1965 for comparable findings) raising the possibility that the procedures used to make the extracts disrupt some necessary "structural" feature of the cell.

CAPECCHI and co-workers (1974) have noted that L-cells rapidly degrade missense mutant forms of hypoxanthine-guanine phosphoryl transferase. Presumably, in this situation and in the other work on abnormal proteins outlined above, some conformational feature, or other aspect of protein structure, is disturbed, so as to make the molecule more prone to "attack" by the (unknown) systems that govern degradation. Perhaps, in its normal, native state, the molecule buries within its structure, certain bonds that are especially susceptible to enzymatic hydrolysis, or groups that can be "recognized" by the degradation system. Alteration of the structure might expose these groups or bonds. In any event, the ability to screen their protein populations for severe abnormalities, might be even more advantageous for eucaryotes than it is for procaryotes: Diploid cells possess recessive genes that code for abnormal proteins, which may be produced even when the phenotype is dominated by "normal" alleles present on the homologous chromosomes. One can readily imagine circumstances in which the presence of such abnormal molecules could be deleterious (*e.g.,* through competition), and their degradation, advantageous (see *e.g.,* CAPECCHI *et al.* 1974). (It does, however, appear most unlikely that all abnormal proteins are rapidly degraded, since some can be detected in cells [*e.g.,* Section IV.1.2]; a systematic study is needed to determine whether screening actually occurs on a significant scale.)

Turnover of several enzymes *in vivo* can be affected by the presence or absence of specific substrates or ligands. Thus SCHIMKE *et al.* (1965) have observed that hepatic tryptophan pyrollase (oxygenase) is less rapidly degraded if tryptophan is present, and the turnover of aspartate transcarbamylase in *Chlorella* may be controlled by the intracellular levels of UMP (VASSEF *et al.* 1973).

One can demonstrate steep declines in the rates of protein degradation when tissues or cultured cells are exposed to inhibitors of energy metabolism (BROSTROM and JEFFAY 1970, HERSHKO and TOMPKINS 1971, POOLE 1971 a, POOLE and WIKO 1973 a, SIMPSON 1973); the mechanisms accounting for this are not known. POOLE and WIBO (1973 a, b) note that turnover within cultured rat fibroblasts is inhibited more by fluoride than by iodoacetic acid despite the fact that the latter agent produces a more profound depression of ATP levels. This discrepancy might be partly a matter of direct inhibition by fluoride of cathepsin A or some other action additional to the inhibition of glycolysis but it also sounds a general cautionary note for interpretations of the effects of inhibitors.

Another finding by POOLE and WIBO is that the fibroblast proteins can be regarded as comprising two classes, ones that turn over rapidly and proteins that turn over slowly. Iodoacetic acid, fluoride and chloroquine inhibit degradation of both classes to approximately the same extent, but the addition of fresh medium to the culture has a selectively inhibitory effect on turnover of the slowly turning-over population. Perhaps, these results indicate that there is some common mechanism for degradation of all proteins (conceivably this involves inclusion in a lysosome) but that the two fibroblast protein populations also differ in some important regard (conceivably, mode of access to the degradation mechanism).

The fact that chloroquine affects turnover is reassuring if one ascribes general importance to the lysosomes for intracellular protein degradation since this drug is known to accumulate in lysosomes and can inhibit cathepsin B (WIBO and POOLE 1974). Other observations of interest within the same purview include those of NEELY and MORTIMORE (1974) who find that when liver proteins are labelled *in vivo* with radioactive amino acids and then the liver is perfused *in vitro*, the labelled proteins that undergo degradation accumulate in lysosomes. WOODSIDE *et al.* (1974) have also used perfused liver to establish that glucagon, the agent employed to stimulate autophagy, enhances protein degradation. See also Nature **257**, 414, for pepstatin effects.

III.4.3.2. Turnover of Macromolecules that Are Not Components of Membrane-Delimited Organelles

Since the organization and binding of molecules within membrane-delimited compartments might have special implications for turnover we will discuss the degradation of the membrane-delimited organelles in a separate section. Here we will be concerned with the "soluble" proteins of the hyaloplasm (the "soluble fraction", "cytosol" or "cell sap") and the components of such organelles as microfilaments, microtubules and ribosomes.

A. Soluble proteins and ribosomes: The different soluble proteins of a given cell appear to turn over at quite heterogeneous rates. (Some hypothetical mechanisms, based on totally non-selective degradation, would produce uniform turnover rates for all components.) The variation in rates does not seem to depend heavily upon relative intracellular concentrations, but there are few pertinent studies of specific identified proteins. SCHIMKE and his co-workers have utilized the double-label method on several cell types to compare the behavior of the numerous soluble proteins that are separable by gel electrophoresis or comparable methods. They report great heterogeneity in turnover rates and they also find that when one looks at the subunits obtainable by dissociation of the proteins there is a general trend toward faster turnover of the larger subunits (Fig. 38; DICE *et al.* 1973) [15].

Further, when the radioactively labelled proteins are exposed to pronase or trypsin *in vitro*, the ones that are degraded fastest are the ones with incor-

---

[15] KIEHN and HOLLAND (1970) believe there is an inverse relationship between polypeptide size and frequency in the cell; if so, this emphasizes even more the lack of simple relations between concentrations of different molecules and their relative rates of turnover.

poration characteristics suggesting relatively rapid *in vivo* turnover (Fig. 39) (see also BOND 1971, HUISMAN *et al.* 1974, SEGAL *et al.* 1974 for comparable results). One could argue that these correlations reflect some subtle properties of the proteins that independently influence their turnover *in vivo* and their behavior *in vitro*. But the observations are more simply

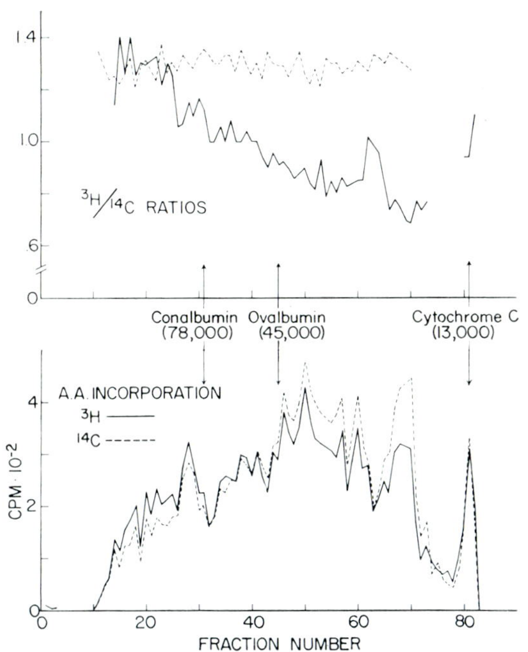

Fig. 38. A rat was injected intraperitoneally with ^{14}C-leucine. Four days later it received an injection of ^{3}H-leucine and four hours later a liver homogenate was prepared. This figure concerns the supernatant from a 100,000 g centrifugation—the soluble proteins of the cell. The proteins were separated according to size by SDS-acrylamide gel electrophoresis; protein subunits associated by non-covalent bonds dissociate and migrate independently. The positions of proteins of known molecular weights are indicated. The upper panel shows the $^3H/^{14}C$ ratios of the proteins separable by these techniques. Note the trend towards higher ratios (faster turnover; Section III.4.1) for proteins of higher molecular weights. The dotted line shows a control experiment in which ^{3}H-leucine and ^{14}C-leucine were given simultaneously to a rat. The lower panel presents the actual radioactivity data from which the ratios for the solid line above were calculated. (From DICE, J. F., P. J. DEHLINGER, and R. T. SCHIMKE, 1973: J. biol. Chem. **248**, 4220—4228.)

explicable if: 1. the events leading to protein degradation are initiated (or become irreversible) by proteolytic attack and 2. for proteins constructed of subunits, the attack is at the level of the subunits, probably when these are free, during cyclic dissociation and reassociation. The relations of turnover rate and size, though presently puzzling, could find an explanation in some factor like the relative ease with which proteases might have access to a susceptible link (*e.g.*, DICE *et al.* 1973, GOLDBERG and DICE 1974, SCHIMKE 1973, SIEKIVITZ 1972 b).

The components of such multienzyme complexes as fatty acid synthetase, of ribosomes, microtubules, and of microfilaments such as those of striated muscle, turn over at appreciable rates (*e.g.*, Table 2). Initial data on multienzyme complexes (HAYES and LARRABEE 1972, TWETO *et al.* 1971) and on ribosomes (HIRSCH and HIATT 1966) suggested that the rates of turnover of the several constituents of the structure (*e.g.*, the RNA and the proteins of ribosomes) were quite similar. If degradation via autophagy of entire organ-

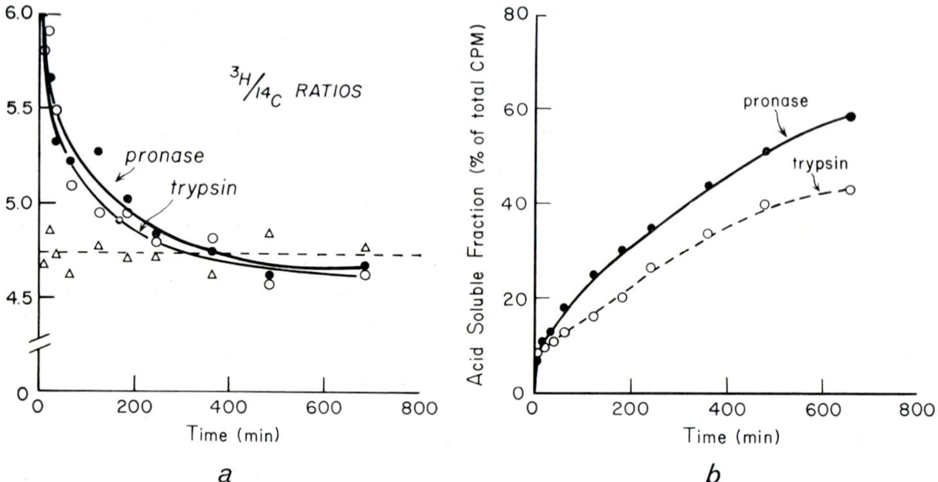

Fig. 39. Proteins in the soluble fraction of rat liver labelled sequentially with ^{14}C and ^{3}H as in Fig. 38 were exposed to pronase or to trypsin. The ^{3}H/^{14}C ratios of the acid soluble material liberated by the proteases were determined. The results (panel a) indicate that the proteins with high ^{3}H/^{14}C ratios are more rapidly hydrolyzed *in vitro* than are the proteins with lower ratios; hence there seems to be a correlation between turnover *in vivo* and susceptibility to proteolysis *in vitro*. The dotted line shows the result of a similar (control) experiment in which the cell fraction used was from a rat given [3]H- and ^{14}C-leucine simultaneously. Panel b indicates the percentage of the total radioactivity released by the proteases. (From DICE, J. F., P. J. DEHLINGER, and R. T. SCHIMKE, 1973: J. biol. Chem. **248**, 4220—4228.)

elles or some comparable bulk-processing mechanism were the only relevant phenomenon one might expect such correlations to obtain (*cf.*, FLETCHER and SANADI 1961) since, for example, inclusion of a ribosome within a lysosome will result in the simultaneous loss of all of its RNA's and proteins from the population of ribosomes isolable from the cytoplasm.

However, more recent studies of ribosome turnover using the double label approach (DICE and SCHIMKE 1972) have suggested that the rates of turnover of the ribosomal RNA's and the several ribosomal proteins may actually be quite diverse, with the proteins that have larger subunits turning over faster than the smaller ones. Heterogeneity has also turned up in the measured half-lives of fatty acid synthetase components (TWETO *et al.* 1972). While heterogeneity of rates does not rule out autophagy as a degradative mechanism, it does indicate at least that autophagy is not the sole relevant mechanism. Perhaps the fact that the two ribosomal subunits dissociate and reassociate cyclically as they participate in proteins synthesis, and the situation that some ribosomes associate with the endoplasmic reticulum while others do not, are important in this regard *. Additionally, a number of investigators maintain that individual ribosomal proteins may undergo extensive exchanges with soluble pools whose sizes and characteristics are poorly understood (HANOUNE 1970 but see Eur. J. Bioch. **57**, 49). Exchanges and dissociations of these types could lead to diversity in measured turnover rates for ribosome constituents even if the organelles themselves are degraded through autophagy. (Some *ad hoc* assumptions are needed for reasonable hypotheses along the latter lines, but this is true for any model of degradation.)

TSURUSI *et al.* (1974) hold that DICE and SCHIMKE's observations on ribosomes, may, in part, be attributable to the presence on the ribosomes, as DICE and SCHIMKE isolate them, of transiently associating proteins, such as elongation factor. TSURUSI and co-workers have obtained data on regenerating liver indicating that ribosomes freed of such "contaminants" show similar rates of turnover for RNA's and proteins when studied with the double labelling approach and with methyl labelled methionine (which should label both RNA's and proteins simultaneously). Whether this is really at variance with DICE and SCHIMKE's findings cannot easily be decided since hepatocytes of regenerating liver show notable increases in autophagy. (BECKER and LANE 1966. Regeneration may not affect RNA degradation; J. Exp. Med. **142**, 575.)

B. Microtubules and microfilaments: Pertinent data on turnover of microtubules and microfilaments is scanty. For skeletal muscle, LOW and GOLDBERG (1973) report that the lighter chains of myosin turn over faster than the heavier, which might prove to be a "violation" of the correlations between size and half-life stressed by SCHIMKE and associates. RABINOWITZ's (1973) data suggest, for cardiac muscle, that the myofibrillar proteins do not show much divergence in turnover rates; they discusses technical problems (pool complexities reutilization of precursors etc.) that could produce apparent heterogeneity. In studies by JEFFERSON *et al.* (1974) perfused muscle subjected to low insulin conditions shows extensive protein degradation and enhanced autophagy, but as in the work on regressing muscles (Section III.1.2.1) the implications for myofilament destruction are not clear—the vacuoles are not laden with filaments. Studies on regressing smooth muscles might be of use in clarifying the roles of lysosomes in myofilament turnover since there is some morphological evidence for participation of autophagic vacuoles in degrading the filaments (BRANDES and ANTON 1969).

Microtubule proteins turn over to amino acids at about 2% per hour in *Tetrahymena* (NELSON 1972, WILLIAMS and NELSON 1973). The two types of rat brain tubulin both have half-lives of 4 days (DUTTON and BARONDES

---

* Recent studies on tissue culture cells indicate that ribosomal-RNA turnover may be negligible during rapid growth, and also that the two ribosome subunits may turn over at different rates, under some growth conditions (ABELSON *et al.* 1974, KOLODNY 1974). Perhaps for ribosomes bound to the ER, it is important that only one of the subunits (the large one) is attached directly to the membrane (SABATINI *et al.* 1966).

1969, HEMMINKI 1973). Cycling of tubulins between organized structures such as the spindle or cilia, and cytoplasmic pools has been studied to some extent (see *e.g.*, HOLTZMAN 1974, RAFF *et al.* 1971, WILLIAMS and NELSON 1973) and there are many demonstrations of tubule depolymerization with no reason to suspect any participation by lysosomes. But the mechanisms for actual degradation of tubulins are obscure. Perhaps such breakdown usually occurs only after the tubulins have dissociated from organized structures. As indicated earlier, microtubules of unknown provenance and significance are occasionally seen in lysosomes (BOLER and ARHELGER 1966, JOURNEY 1964) and it is possible that upon further study, phenomena such as the ciliary resorption observed in dividing protozoa by ROTH and SHIGENAKA (1964) might turn out somehow to depend on lysosomes (the latter authors note that the cilia are retracted into a membrane-delimited system but the 3-dimensional arrangements of structures and the fates of the configurations need further analyzing). However, such observations are few and far between and they are more than matched by analogous situations in which tubule-rich structures or their relatives such as centrioles seem simply to disappear (*e.g.*, centrioles sometimes become "superflous" during gametogenesis or fertilization and there are reports for such cases that the organelles fall apart into their constituent substructures which subsequently are no longer detectable [WOOLEY and FOWCETT 1973]). (For a recent review, see BLOODGOOD 1974.)

C. Possible mechanisms: From this brief summary, it is apparent that if one assumes the lysosomes are likely to serve as the major sites at which soluble proteins, ribosomal constituents and so forth are ultimately degraded to the level of small molecules, then much of the analysis of intracellular turnover eventually will have to be articulated in terms of the mechanisms controlling entry of the relevant molecules into lysosomes. Perhaps it is here that metabolic energy comes into play, serving a sequestration mechanism that in some fashion is selective. To explain the sort of data we have been discussing, and especially the correlations between turnover *in vivo* and susceptibility to proteolysis *in vitro*, the interesting hypothesis has been advanced that molecules might enter autophagic vacuoles or other lysosomes by processes that are reversible, so that the molecules can be exposed temporarily to hydrolases and, if they survive, subsequently return to the outside. (HAIDER and SEGAL 1972 expound this view and SEGAL *et al.* 1974 b stress the correlations between turnover *in situ* in rat liver and rates of degradation of rat liver proteins exposed to lysosomal hydrolases *in vitro*.) The most straightforward schemes with this focus seem improbable since large molecules apparently cannot cross lysosome membranes, but some possibilities for this type of screening, perhaps involving membrane-mediated transport, are not completely implausible. The converse notion, that hydrolases are released into the cytoplasm (see *e.g.*, PANTREMOLI *et al.* 1973) and function there for a limited period (perhaps being stopped by inhibitors or other control devices) has some obviously unattractive features but there is no unequivocal evidence against it and it is consistent with some proposals that have been advanced (and disputed) for pathological events (Section IV.4).

Earlier, when we took up autophagy and multivesicular bodies (Section

II.4.2.4) it was pointed out that there do seem to be processes by which soluble molecules and small structures such as ribosomes can enter lysosomes through infolding and vesiculation of the lysosome surface. Perhaps part of the heterogeneity of degradation rates outlined above is attributable to events that influence such processes. Differential affinity of molecules for the lysosome surface could contribute. And, alterations might occur that predispose molecules to uptake. Spontaneous denaturation or modifications through "accidents" that might befall molecules or through encounters with enzymes, could very well affect their solubility and aggregation properties or their binding to membranes. Changes in solubility are common for denatured or partly degraded proteins (*e.g.*, BALLARD *et al.* 1974 see also BOHLEN *et al.* 1971) and one can think of ways in which, for instance, hydrophobic groups normally buried within a protein might be exposed on denaturation and contribute to an increased affinity for membranes.

ROBINSON *et al.* (1970) have put forth the imaginative suggestion that it is deamination of glutamine and asparagine that controls turnover since the authors are impressed by the rough correlations between half-life *in vivo* and the relative abundance of glutaminyl and asparaginyl residues in different proteins.

A slightly less speculative line of analysis which is gradually gaining adherents is that non-lysosomal enzymes with specificity for particular classes of proteins, or for particular proteins, initiate breakdown, with the lysosomes finishing the job. Through studies *in vitro,* certain partially characterized proteases have been shown to make limited attacks on other enzymes, inactivating them, and rendering them more susceptible to further proteolysis by less specific proteases. From such work, a strong case has been made for class-specific initiation of breakdown of the pyridoxal-requiring enzymes in yeast and mammals (*e.g.*, transaminases, serine dehydratase, muscle phosphorylase). Inactivation of these enzymes by an apparently non-lysosomal protease has been documented by KATUNUMA and others (KATUNUMA 1973, KATUNMUA *et al.* 1972, KOMINAMI *et al.* 1972) *. Less complete evidence for initial enzyme-mediated inactivation outside the lysosomes can be marshalled for other cases (see *e.g.*, BUNSIGNORE *et al.* 1968 for glucose-6-phosphate dehydrogenase; BRUSH 1971, KATUNUMA 1973, SCHIMKE 1973). And some curious details of the turnover of a few enzymes might also be explained along such lines. For example, agents that inhibit protein synthesis or RNA synthesis also can inhibit degradation of tyrosine-amino-transferase in liver or hepatoma cells (see STEINBERG *et al.* 1975, KENNEY 1967, McNAMARA and WEBB 1974, REEL and KENNEY 1968); perhaps a rapidly turning over enzyme or other protein is responsible for the key step in beginning the breakdown or in promoting uptake into lysosomes. DE JONG *et al.* (1974)

---

* KATUNUMA *et al.* (1975) and KOMINAMI *et al.* (1975) report the presence of certain of the proteases in mitochondrial fractions, but continue to contend that they are non-lysosomal. One speculative possibility, is that some of the enzymes involved in the events under discussion are situated on the outer (cytoplasmic) surfaces of lysosomes or other organelles (the absence of good evidence for this was noted in Section II.1.4.4).

speculate that the presence of shortened forms of crystallins in bovine lens might be attributable to the action of a proteolytic mechanism that "marks" these molecules for destruction (see also VAN KLEEF *et al.* 1974).

There is ample precedent (*e.g.*, in work on activation of zymogens) for the idea that particular proteins or classes of proteins may be constructed so as to be uniquely subject to specific cleavages by certain proteases.

D. Nucleic acids—a brief note: Surprisingly little is known of the details of nucleic acid breakdown, although enzymes that could participate occur in lysosomes and outside of them. We have already alluded to the unsettled controversies as to whether or not particular mRNA's turn over with age-dependent kinetics (COWAN and MILSTEIN 1974, MATTHEWS 1973, PERRY and KELLY 1973, SINGER and PENMAN 1973). This might foreshadow intensification of the ongoing disputes about such topics as the role of poly-A in stabilizing messengers, or the possible existence in eucaryotes of enzymes with properties resembling those of bacterial restriction enzymes (see *e.g.*, BOYER *et al.* 1973 for references) which can recognize particular features of nucleic acids and selectively initiate degradation. As outlined above, there is some evidence bearing on the degradation of ribosomes but beyond this there is not much on which to base even an informed guess about roles of lysosomes in digesting any of the species of RNA.

III.4.3.3. Aspects of the Turnover of Membrane-Delimited Cytoplasmic Organelles

A. Could autophagy be of major importance? Rough calculations suggest that were a few autophagic vacuoles containing one mitochondrion each, to be formed every hour in hepatocytes, the known rates of mitochondrial turnover could be accounted for (DE DUVE and WATTIAUX 1966). Such intensity of autophagy does not seem unreasonable. And destruction by autophagy provides a simple explanation for a number of observations, such as the findings that mitochondrial (and chloroplast) DNA's seemingly have finite and fairly short life-spans (ALBIN *et al.* 1973, GROSS *et al.* 1969, MANNING and RICHARDS 1972, RABINOWITZ 1973, RICHARDS and RYAN 1974) [16]. From their morphometric studies on rat kidney, PFEIFFER, and SCHELLER (1975) conclude that the known turnover rates for DNA and other components of mitochondria, could result from autophagy if, upon its inclusion in a lysosome, a mitochondrion remained recognizable as such for only a few minutes (roughly 10 minutes). They cite literature, such as reports on organelle degradation in phagocytic vacuoles, suggesting that lysosomes can induce changes with such speed. (The actual degradation of macromolecules probably is more leisurely; what is at issue is the microscopically visible deterioration of structure.)

However, neither for mitochondria, nor for the other membrane-delimited organelles does the body of biochemical data fall into a pattern pointing

---

[16] On the other hand, DURPHY *et al.* (1974) report that mitochondria possess their own DNAses, and too little is known of the duplication, possible recombinations and other features of mitochondrial DNA for confident analysis. See also Eur. J. Bioch. **57**, 213.

clearly toward autophagy as the dominant destructive mechanism (see POOLE 1971 and SIEKEVITZ 1972 a a for especially valuable reviews). For organelles such as mitochondria, endoplasmic reticulum and plasma membrane, the lipids turn over at rates that are heterogeneous and different from those character- izing the proteins, and within the protein population there is additional heterogeneity (Table 2; ARIAS *et al.* 1969, BAILEY and BARTLEY 1967, CUZNER *et al.* 1966, GROSS *et al.* 1969, HOLTZMAN *et al.* 1970, OMURA *et al.* 1967, SIMON *et al.* 1970, TAYLOR *et al.* 1967, 1973). Several groups also report that the same kinds of correlations between subunit size and turnover rates hold for membrane proteins as for soluble ones (DEHLINGER and SCHIMKE 1971, 1972, GURD and EVANS 1973, MELDOLESI 1974, TAYLOR *et al.* 1974).

By and large, membranes seem to grow from preexisting membranes, rather than forming *de novo* (PALADE in: SINGER and ROTHFELD 1973, SIEKEVITZ 1972 a) but it is presently believed by many that a given membrane also can gain and lose individual macromolecules without undergoing extensive destruction and reassembly—a molecule-by-molecule turnover in which the fate of individual macromolecules is dissociated from that of the structure as a whole (SIEKEVITZ 1972). The bases for this belief include the hetero- geneity in steady-state turnover rates of different membrane components coupled with information suggesting that a membrane can undergo changes in composition during organellogenesis (see especially the work on chloroplasts by CHUA *et al.* 1973, DEPETROCELLIS *et al.* 1970, DOUCE 1974, EYTAN and CHAD 1972, EYTAN *et al.* 1974, PALADE 1972 b, PHILIPPOVICH *et al.* 1973) and during such phenomena as the phenobarbital induce hypertrophy of hepatocytic smooth endoplasmic reticulum (ARIAS *et al.* 1969, GIELEN *et al.* 1972, GOLDBERG and CHAD 1970, HIGGINS 1974, JICK and SHUSTER 1966, KURIYAN *et al.* 1969, ORRENIUS and ERICSSON 1966, TAYLOR *et al.* 1973). For some time it has been understood that membrane lipids and lipid subcom- ponents—especially fatty acids—are subject to exchange reactions (DAWSON 1973, PETERSON and RUBIN 1970, SIEKEVITZ 1970), and the fact that mem- brane-associated enzymes of the ER synthesize lipids has formed the basis for proposals of growth and modulation of the reticulum by local accretion of newly made molecules (see *e.g.,* HIGGINS 1974). What is comparatively new is the expansion of experimental approches to the behavior of proteins [*].

There are of course, no necessary inconsistencies between molecule-by- molecule or subunit-by-subunit turnover and turnover in bulk, as by auto- phagy of whole organelles. One might be superimposed upon the other with

---

[*] For lipids, one can dream up schemes in which turnover depends partly upon continual exchanges, in which a given molecule repeatedly enters and leaves membranes until, perhaps by chance, it is trapped in a lysosome. Maybe the decisive event is the entry of the molecule into a portion of the lysosome surface that becomes interiorized within the organelle (Section II.4) during the lipid molecule's sojourn.

Systematic comparisons of the turnover characteristics of proteins that have different associations with membranes (*e.g.,* "peripheral" proteins vs. "integral ones"; SINGER and ROTHFIELD 1973) or of proteins with different arrangements of hydrophobic groups, might prove rewarding. Perhaps there is a relationship between turnover, and the ease with which a membrane protein can enter or leave the membrane.

the relative contributions of each to the behaviour of a given organelle or class of organelles altering under different circumstances. This could tie in nicely, for example, with the fact that organelles such as mitochondria contain proteins in varying association with membranes; some are "free" in the matrix, others are integrated into multienzyme arrays bound to membrane surfaces and still others are probably "buried" deep in the membrane. Furthermore, in work on mitochondriogenesis there is good evidence that some macromolecules of the organelles originate within mitochondria and others outside (see *e.g.*, EBNER *et al.* 1973, HOLTZMAN 1974, RUBIN and TZAGALOFF 1973) and there even is reason to suspect that certain inducible enzymes may be able to enter preexisting mitochondria without requiring other substantial changes in the organelle [17]. RABINOWITZ and co-workers (DRUYAN *et al.* 1969, RABINOWITZ 1973). BEATTIE (1969), and GEAR (1970) among others have published data which seem to demonstrate that the proteins of given morphological compartments in mitochondria (*e.g.*, the inner membrane) show a fair degree of homogeneity in turnover rates even though substantial rate differences exist among the compartments (*e.g.*, the inner membrane components turn over more slowly than do those of the outer membrane). One could make sense of this by conjectural schemes in which the components of different compartments originate in distinctive ways and suffer partly independent fates—for example, portions of the outer membrane might be lost and replaced without perturbing the inner compartments. Unfortunately there are immense gaps in our current knowledge of the details of transport and assembly of mitochondrial components (see *e.g.*, HEREWARD 1974, and KURIYAMA and LUCK 1973 for discussions of the roles of ribosomes in and on the organelles, and the references cited below for continuities of the outer membrane with the ER).

A related perspective emerges from studies on peroxisomes (DE DUVE 1973, POOLE 1971 a) (Fig. 40). In hepatocytes, several peroxisomal proteins show similar half-lives, suggesting that the organelles may be degraded as intact units, presumably through autophagy. After a pulse-label with amino acids, radioactivity is lost from the peroxisome population with simple exponential kinetics and no appreciable lag, so it might be concluded that the degradative mechanisms does not discriminate among "older" and "newer" organelles. However, an alternative possibility is that there are no truly older and newer peroxisomes. This might be the case if individual peroxisomes

---

[17] The evidence that this is what happens when mitochondrial enzymes are induced is not incontrovertible—it consists of observations such as those by SWICK *et al.* (1968) who report that induced aminotransferases are not preferentially localized in small mitochondria, as might be the case if the enzymes entered only newly formed organelles and if newly formed mitochondria were smaller than average. But in any event, since mitochondriogenesis seems to depend upon some sort of growth and division mechanism, it is almost inescapable that the inducible enzymes are integrated into an organelle that existed as an organized structure before the enzymes appeared. (There remains some need for caution since most work on inducible mitochondrial enzymes has relied on determinations of enzyme activities under varying conditions. There is little information about corresponding behavior of radioactive precursors incorporated into the enzyme proteins.)

Summary of Half-Lives Measured
for Catalase

Method	Most probable half-life	95% Confidence interval
	days	
^{3}H-leucine incorporated into catalase	3.74	3.32–4.29
^{3}H-leucine incorporated into whole peroxisomes	3.34	2.81–3.80
Guanidino-^{14}C-arginine incorporated into catalase	2.48	1.60–4.65
^{14}C-δ-amino-levulinic acid incorporated into catalase	1.84	1.59–2.17
Recovery from aminotriazole inhibition	1.73 1.34*	1.37–2.36 1.16–1.57*
Inhibition of catalase synthesis with allylisopro-pylacetamide	1.07*	0.93–1.50*

* Calculated from the data of Price et al.

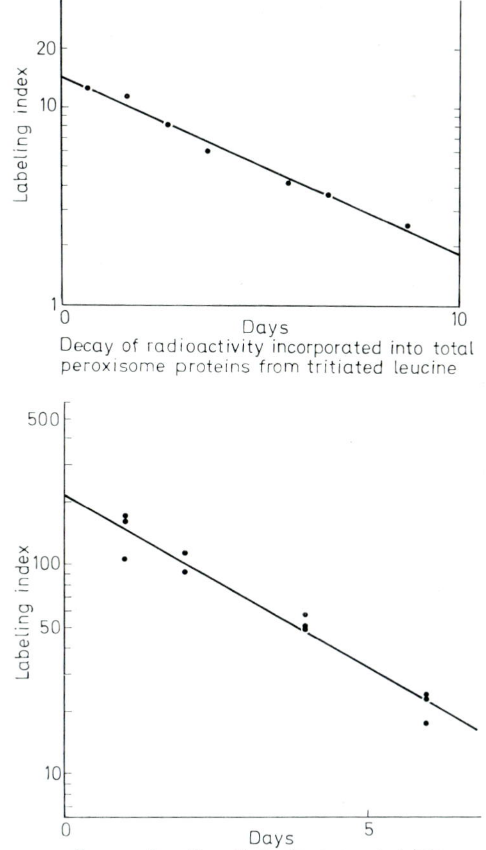

Decay of radioactivity incorporated into total peroxisome proteins from tritiated leucine

Decay of radioactivity incorporated into catalase from δ-aminolevulinic-4-^{14}C acid.

Fig. 40. Turnover of peroxisomal catalase and total peroxisomal proteins of rat liver. The graphs illustrate the time-course of steady-state loss of radioactivity after a short labelling period with the precursor noted. The table compares the half-lives measured with different precursors (leucine, guanidino-labelled-arginine, delta-aminolevulinic acid) and by following the recovery of enzyme activity after a period of inhibition of the activity or of enzyme synthesis. The shorter half-lives almost certainly are the more accurate ones. Note also the exponential fall of radioactivity with time (the data are plotted on logarithmic scales, so straight lines are obtained). (From Poole, B., F. Leighton, and C. de Duve, 1969: J. Cell Biol. **41**, 536—546.)

exchange proteins with one another (Lazarow and de Duve 1973). The organelles seem to arise as dilated buds of the ER and exchanges could occur if the peroxisomes retain continuity with the ER, or perhaps establish intermittent connections with portions of the reticulum or with one another; the requisite continuities among segments of ER and peroxisomes are frequently seen by electron microscopists (Fig. 16; *cf.*, P. M. Novikoff and A. B. Novikoff 1972, Reddy and Svoboda 1973). Autoradiographic studies should eventually help demonstrate whether molecules present initially in a few peroxisomes eventually spread through the population. The analysis could turn out to have broad implications since continuities of other organelles (mitochondria, plastids etc.) with the ER have been noted in several tissues (BRACKER and GROVE 1971, CAROTHERS 1972, CROTTY and LEDBETTER 1973, FRANKE *et al.* 1971, MORRE *et al.* 1971, RUBY *et al.* 1970, see also HOFFMAN and AVERS 1973, 1974, and GRIMES *et al.* 1974 for their discussions of the possibility that what seem to be separate mitochondria in yeast actually are parts of one interconnected network).

Peroxisomes also have been reported to "atrophy" without involvement of lysosomes. Thus, SVOBODA and REDDY (1972) assert that in hepatocytes of some drug-treated rodents, the peroxisomes disintegrate, with their components dispersing in the cytoplasm. Somewhat the same kinds of events may occur in insect fat body (LOCKE and MCMAHON 1971) and in some plant tissues (VIGIL 1973). These observations suggest that there may be alternative routes for peroxisome degradation—autophagy of the organelles is readily observed, but perhaps it coexists with other mechanisms. Complexities of these sorts are not entirely unanticipated since the mobilization, intracellular transport and degradation of peroxisomal enzymes show some unusual and still unexplained features. For instance, the rates of turnover of catalase in mouse liver vary markedly in different genetic strains (RECHCIGL 1971 b; see also GIRANELLO and AXELROD 1973, GANSHOW and SCHIMKE 1969). Also KANAI *et al.* (1974) have reported that when rats are injected with amino-triazole, a catalase inhibitor, hepatic catalase degradation is speeded up. And LAZAROW and DE DUVE (1973; see also LAZAROW 1974) have found that a larger proportion of newly synthesized catalase is present in the soluble fraction of homogenates than might be expected for an enzyme packaged directly into peroxisomes by the ER. (In evaluating microscopic evidence for organelle atrophy, one obviously must be hesitant about dismissing cell death, or poor fixation, as possible factors leading to the appearance of suggestive configurations. The observations just cited raise other specters: Is it conceivable that peroxisome nucleoids lying "free" in the cytoplasm [SVOBODA and REDDY 1972] could form *de novo*, as a result of disarrayed transport, rather than deriving from degenerating peroxisomes?)

B. Turnover of membranes at the cell surface: To illustrate several of the key points that require clarification before adequate assessment of lysosome roles in turnover of membranes will be possible, we will devote the rest of this section to consideration of the membranes and associated materials of the cell surface [see SINGER and ROTHFELD (1973) for a number of useful brief discussions]. The focus will be on several factors whose impact on interpretations of the biochemical data may be major. The sorts of biochemical data available are exemplified by the pioneering studies of WARREN and GLICK (1968) who found little turnover of plasma membranes in rapidly growing tissue culture cells but measured half-lives of less than a day when stationary phase cells were studied with labelled amino acids, sugars or lipid precursors. Turnover could be inhibited by puromycin, dinitrophenol or the omission of essential amino acids. As might be expected, variations from these patterns have been noted in other investigations and reliable generalizations useful for our purposes are very hard to come by. Cell surface turnover often does seem to be affected by growth conditions (see *e.g.*, HUGHES *et al.* 1972), and heterogeneity in rates of replacement among different plasma membrane molecules of a given cell type as well as exchange and acylation reactions affecting the lipids have been noted in many publications (see *e.g.*, GURD and EVANS 1973, OVERTON 1968, PASTERNAK 1973, SCHIMKE 1973, SIMON *et al.* 1970, TAYLOR *et al.* 1973).

1. Movement of membrane and surface coats in bulk between intracellular and cell surface sites can occur by exocytosis and endocytosis (Section II.4.2.2). On the other hand, growing bacteria assemble their plasma membranes and walls without discernible use of exocytosis, and studies on enzyme induction and other phenomena have suggested that components can be added to the bacterial plasma membrane molecule-by-molecule in the non-growing state as well (KENNEDY 1970, MINDICH 1971, 1974, NUNN and CRONAN 1974, SINGER and ROTHFIELD 1973) [18]. In reticulocytes, some of the proteins associated with the plasma membrane, apparently are synthesized on "free" polysomes (polysomes that are not bound to the ER or other membranes; LODISH and SMALL 1975). These proteins may be located at the superficial aspect of the membrane surface that faces the cytoplasm. It is not difficult to envisage molecule-by-molecule addition of such components (see also *e.g.*, ATKINSON 1975, for other data suggesting that eucaryote plasma membranes may be assembled in multiple steps, and LANDRY and MARCEAU 1975, for data on the heterogeneity of turnover rates for plasma membrane proteins). Both in eucaryotes and procaryotes some cell surface materials are lost by sloughing into extracellular spaces (CONE *et al.* 1971, ILLINGWORTH *et al.* 1973, KAPELLER *et al.* 1973). In brief, several mechanisms can accomplish the entry or loss of cell surface molecules. It is often very difficult to disentangle the relative importance of each under given circumstances.

2. If bulk addition mechanisms related to exocytosis are important for maintenance of cell surface membranes, one apparent implication is that the Golgi apparatus is a major intracellular source of plasma membranes, since it is here that packages destined for exocytosis are put together (for morphological and other studies suggesting that the Golgi apparatus contributes membranes and coats to the cell surface see BENNETT *et al.* 1974, BERGERON *et al.* 1973, COOK 1973, DAUWALDER *et al.* 1972, HICKS 1966, FRANKE *et al.* 1971, LEVINE *et al.* 1972, MASUR *et al.* 1972, VAN DER WOUDE *et al.* 1971, WISE and FLICKENGER 1970; this of course does not mean for example that

---

[18] A lively controversy is underway about the details of addition of new molecules to bacterial surfaces (AUTISSIER and KEPES 1971, BEGG and DONACHIE 1973, MINDICH 1974, MINDICH and DALES 1972, OVERRATH *et al.* 1971, SUBBIAH and THOMPSON 1974, WILSON and FOX 1971). Among the key issues being debated are whether new molecules are added at a few localized points or intercalated at many points or at random, and whether proteins enter the membrane alone or in concert with lipids or polysaccharides. At this moment the prevailing views probably are that intercalation can occur at many points, and that new proteins can associate with "old" lipids and thus apparently need not enter as components of preformed lipoprotein complexes. Perhaps ribosomes bound to the plasma membrane are responsible for insertion of new proteins; especially for proteins that span the membrane, such insertion would simplify matters (at least, conceptually). The work on reticulocytes outlined in the next sentence above is, of course, likely to be germane here too.

special portions of the ER such as subsurface cisternae could not also some-times contribute material directly to the surface; *e.g.*, HEREWALD 1974).

As studied in several cell types, the kinetics of radioactive labelling of membranes are consistent with a flow of membrane molecules from ER to Golgi apparatus to cell surface (*e.g.*, CHLAPOWSKI and BAN 1971, EVANS and GURDON 1971, FLEISCHER and FLEISCHER 1971, RAY *et al.* 1967). Further-more, in light of its synthetic capacities, the ER is felt to be the general initial source of membrane lipids (DAWSON 1973, HOKIN 1968, LORD *et al.* 1973, NOZAWA and THOMPSON 1971, PARKES and THOMPSON 1973, SABATINI-SMITH 1971, SIEKEVITZ 1972 a). But while there is some overlapping in enzyme content and other features between Golgi apparatus and plasma mem-brane, on the whole the ER, Golgi membranes and plasma membrane are distinctly different in composition in-so-far as this can be judged from work on cell fractions (*e.g.*, BERGERON *et al.* 1971, FLEISCHER and FLEISCHER 1970, MORRE *et al.* 1971). It is also demonstrable that Golgi-derived packages destined for exocytosis can be delimited by membranes that are dissimilar to those present in microsome fractions containing relatively purified Golgi apparatus or ER (see *e.g.*, the data gathered by MELDOLESI 1974, on pan-creatic zymogen granules). To explain these observations, and still retain the hypothesis that membranes pass from ER to Golgi apparatus and thence to the cell surface primarily by some kind of movement of membrane-de-limited structures (Section II.4.2.2) it has often been suggested that during their travels the membranes undergo extensive modulation of composition by entry and loss of proteins, and through local synthesis, exchanges, glycosyla-tions, acylations and other reactions of the lipids (see *e.g.*, BENES *et al.* 1973, HIGGINS 1974, LEVINE *et al.* 1972, STEIN *et al.* 1968 for pertinent discussion related to the lipids). An alternative possibility is that there are specialized regions of the ER and Golgi apparatus that differ from the rest in com-position and are responsible for the processes under discussion. Efforts to demonstrate marked heterogeneity among the ER membranes of a given cell have thus far met with only a little success (BEAUFAY *et al.* 1974 have observed some variations in composition and enzymes among ER-derived microsome subfractions but in most investigations few such differences have been encountered: ERICSSON and DALLNER 1972, LESKES *et al.* 1971, REMACLE *et al.* 1974. The nuclear envelope, which is a specialized portion of the ER, may turn out to be exceptional in this regard; TATA *et al.* 1972). However, bio-chemical and cytochemical studies have begun to uncover striking divergences in enzyme content and other features among portions of the Golgi apparatus (*e.g.*, FARQUHAR *et al.* 1974; P. M. NOVIKOFF *et al.* 1971) and these may prove highly illuminating for our present considerations. The total area of secretion granule membranes, or cell surface, often is small, compared with the ER (*e.g.*, BOLENDER 1974) and corresponding specialized ER zones might be hard to find, even if they exist.

An important general point to be borne in mind is that in principle, heterogeneity in measured turnover rates among different components of a given class of structures can derive either from entry and loss of individual components or from the existence of subpopulations of structures which vary in composition and kinetics of destruction but each of which is degraded as an integral unit rather than component by component. (That is, if component A in a given type of membrane turns over faster than other components, it could be that the population contains a sub-category of A-rich membranes that are assembled and degraded faster than the average or there could molecule-by-molecule entry and loss.) For none of the membrane-delimited cell organelles is there information adequate to permit sorting out of the relative importance of these alternatives.

3. The models of cellular membranes that are currently gaining favor suggest that at least some of the proteins in a membrane can diffuse laterally within the structure. Thus it may be that the localized exocytic addition of a batch of molecules to a point on the cell surface need not produce a local region that differs permanently from neighboring zones. Through autoradiography, extensive spreading of locally added material has been demonstrated for one case of wall formation in bacteria (RYTER *et al.* 1971; see also footnote 18, p. 145) but fully convincing comparable studies of membranes have yet to be done.

Mobility of membrane molecules may be involved in selective turnover of cell surface components. There is evidence strongly suggesting that the membrane internalized during endocytosis can differ from the "average" plasma membrane of the cell. This has been proposed as the explanation for the finding that mammalian phagocytes can invest much of their plasma membrane in phagocytic vacuoles without removing specific transport enzymes or other molecules from the surface (BERLIN *et al.* 1974, TSAN and BERLIN 1974; also, BODE 1974, DUNHAM 1974, GRIFFIN and SILVERSTEIN 1973). Possibly it also is responsible for the fact that binding of concanavalin A to the surfaces of thymocytes leads to a selective enhancement of turnover of one of the plasma membrane proteins (SCHMITT-ULLRICH *et al.* 1974). Especially for highly polarized cells, where plasma membrane specializations are observable, selectivity in interiorization of membrane proteins could be due simply to the concentration of appropriate receptors and of endocytic activity at localized regions of the surface (DE CAMILLI *et al.* 1974); this may be reflected in data such as those purportedly showing that the proteins of the brush border of kidney turn over with similar half-lives (MCNAMARA *et al.* 1974, QUIRK *et al.* 1973). Less dramatic surface specializations might be present on non-polarized cells as well. But the possibility is now actively being explored that subsequent to binding of material fated for endocytosis, receptors can migrate within or along the membrane and congregate into enriched regions that are then internalized by the cell. In this way an initially homogeneous membrane could lose components selectively. Dramatic clustering of ligands labelled with suitable microscopic tracers can be observed with some cell types—an extreme example is the accumulation at one cell pole ("capping") of antiimmunoglobulin antibodies bound to the surfaces of B-lymphocytes and slated for eventual endocytosis (Section IV.5.2). (One can, of course, always raise the formal objection that movements of ligands along a membrane might not be accompanied by movement of receptors to which they were initially bound.)

From studies with the usual inhibitors, several authors speculate that microtubules, microfilaments or both might control the migrations just discussed (BERLIN *et al.* 1974, also OLIVER *et al.* 1974, RYAN *et al.* 1974). If so, perhaps there is similar intervention of these (or other) "congregating" devices in the turnover of macromolecules of intracellular membranes.

On the basis of his work, using freeze-fracture preparations for electron microscopy, STAEHELIN (1974) asserts that the specialized membranes of tight junctions can sometimes be recognized in lysosomes. He suggests that this reflects the degradative phase of tight-junction turnover. If so, then it would be quite interesting to compare turnover of such membrane regions with that of the rest of the plasma membrane. ALLEN and POTTEN (1975) suggest that phagocytic engulfment of desmosomes, by cells attached at such junctional structures, may be important, or essential, for disrupting the attachments when the associations between the cells alter.

4. Evidence preciously outlined (Section II.4.2.2) indicates that lysosomes can, and do digest membranes that previously resided at the cell surface, including those that are temporarily incorporated in the surface during exocytosis of secretions. In morphological and cytochemical studies, multivesicular bodies seem to be major termini for membrane circulation between intracellular compartments and the plasma membrane. And in direct biochemical work, HUBBARD and COHN (1975) have demonstrated the "destruction" of surface proteins internalized by mouse fibroblasts during endocytosis of indigestible material (latex). But this does not mean that membrane withdrawn from the cell surface is always fated to be immediately degraded. In a few cases it clearly is not—the vesicle-mediated transport across capillary walls depends on the passage of endocytic vesicles from one side of the wall to opposite side, where the vesicles fuse with the plasma membrane and release their contents. For a few other cell types there is circumstantial evidence for repeated cycling of membrane between the interior and the surface without intervening destruction. Thus, actively endocytosing protozoa can interiorize and replace large quantities of plasma membrane in very short times—in a number of cases amounts of membrane equivalent to the entire surface area seem to be replaced in one to a few hours. (BOWERS and OLZEWSKI and BAND 1971, CHRISTIANSEN and MARSHALL 1965, HAUSMANN *et al.* 1972, KLOETZEL 1974, McKANNA 1973 a.) Although we need more adequate information about possible cytoplasmic reservoirs that might contribute to this (ULSAMER *et al.* 1969), various authors have attributed the replacement to reuse of endocytosed membrane through exocytic-like reinsertion in the surface (*e.g.*, CHLAPOWSKI and BAND 1971, GITHENS and KARNOVSKY 1973, McKANNA 1973 a, b). Vesicles budding from larger endocytic structures could participate in such events and there might also sometimes be a contribution from lysosomes that fuse with the surface during egestion of residua (ALLEN 1974). Whether this type of recycling is widespread is not known. According to WERB and COHN (1972 b), when macrophages engulf latex particles they soon reach a point at which new synthesis is needed to restore the depleted surface, but to some degree this could be a special feature of the cell type and there might even be influences of the experimental design

(the presence of large indigestible particles in the vacuoles). However, the notion that subsequent to their internalization, cell-surface components are degraded, also appears to hold for the acetylcholine receptors of developing muscle fibers (DEVREOTES and FAMBROUGH 1975).

Recycling of membrane in secretory cells, and in the synapses of neurons has often been proposed as a sort of conservation device whereby the cell can use the same bit of membrane repeatedly to package components for exocytic release. Initial data of AMSTERDAM et al. (1971) on the parotid gland suggested that this might not be the case since these authors reported that newly synthesized membrane proteins accompany newly synthesized secretory proteins in the secretion granules. But studies on the exocrine pancreas and the adrenal medulla [as well as some more recent work on parotid; CASTLE et al. (1974)] have called this into question: MELDOLESI (1974) and WINKLER et al. (1972) report that the half-lives of secretion granule membranes, as determined by the customary radioactive precursor methods, are substantially longer than those of the secretions—this would be consistent with repeated reuse of the membranes. Unfortunately there are technical ambiguities. One cannot interpret the outcome of these experiments without presently unavailable information about the details of the relations of secretion granules to the ER and Golgi apparatus. If either or both of these systems provide the membranes for secretion granules then measurements of granule membrane turnover based simply upon determination of radioactivity levels in purified granules preparations will give misleading results—the ER and Golgi apparatus could well continue to contribute labelled membrane to newly formed secretory structures long after the labelled low-molecular weight precursors are gone from the intracellular pools. Thus if preexisting membranes give rise to secretion granule membranes the measured half-life for the membranes will be longer than the actual half-life.

Reutilization of synaptic vesicle membranes in repeated rounds of neurotransmission is especially attractive in light of the problems of "resupplying" terminals with macromolecules made in the perikarya, and of the abilities of some neurons to sustain very extensive and rapid transmitter release over prolonged periods. But the relative importance of reuse as opposed to degradation of synaptic vesicle membrane has only begun to be looked into in detail; while repeated recycling of membrane between cell surface and vesicle populations seem likely, few details are clear (Fig. 42 b *; see also CECCARELLI et al. 1972, 1973, DOUGLAS et al. 1974, HEUSER and REESE 1973, HOLTZMAN

---

* From the literature, and from our own studies on synaptic vesicles labelled, through endocytosis, with horseradish peroxidase, we have developed the following working hypotheses: Subsequent to their acquisition of tracers, through endocytic membrane retrieval linked to transmitter release, the vesicles distribute at random within the synaptic vesicle population in a terminal. Some turn up at probable sites of transmitter release, such as along regions of the synaptic ribbons in photoreceptor terminals (SCHACHER, HOLTZMAN, and HOOD, unpublished). Others fuse with multivesicular bodies (Fig. 42 b). Over the course of 12–24 hours following labelling, most of the labelled vesicles are degraded, through transport back to the perikaryon and incorporation in lysosomes. During this interval, however, the vesicles can also participate in exocytosis (TEICHBERG et al. 1975).

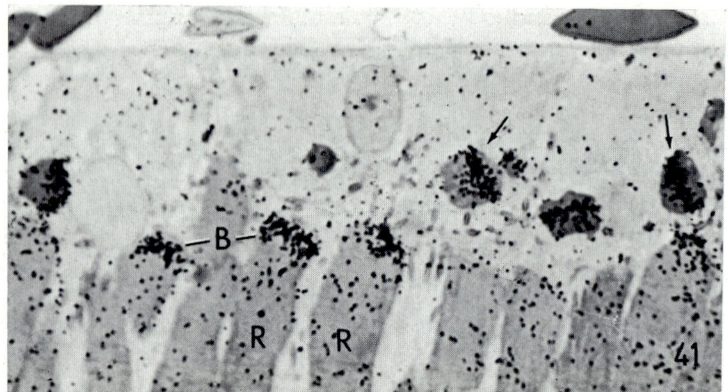

Fig. 41. Autoradiogram showing a portion of a frog retina 61 days after the animal received an injection of ³H-leucine and ³H-phenylalanine. The tips of several retinal rod photoreceptor cells are seen (*R*); the bands of clustered grains (*B*) associated with these cells represent radioactive proteins incorporated in the membraneous discs of the cells (Fig. 42). The groups of discs labelled during the exposure to the radioactive precursors have migrated up the rods and now are about to be shed. Arrows indicate phagocytic vacuoles in pigment epithelial cells; these vacuoles contain recently shed discs and thus show radioactivity (see Fig. 42). ×1,100. (From YOUNG, R. W., and D. BOK, 1969: J. Cell Biol. **42**, 392—402.)

et al. 1971, 1973, TEICHBERG *et al.* 1975). [For another relevant research area in which much interesting data is available, but some perplexing details need explaining see ELSBACH *et al.* (1969), HOKIN (1968), SASTRY and HOKIN (1966) for examples of the many studies showing that endocytosis and exocytosis are accompanied by increases in lipid turnover and exchange reactions. Some of these changes may be more important for processes of membrane fusion and fission than for actual turnover of membranes.]

There is fragmentary morphological information that might bear upon questions of membrane reuse. For example, on rare occasions, endocytosed tracers show up in sacs very closely associated with the Golgi apparatus (see HOLTZMAN *et al.* 1973 for review and PELLETIER 1973). It is possible that these might simply be intermediates in the formation of lysosomes, or "accidents", associated perhaps with abnormal conditions. But one might also expect to see such configurations if internalized membrane is reused for

---

Fig. 42 Membrane turnover in the retina: *a* Portion of monkey retina similar to that shown in the light micrograph in Fig. 41. The arrows indicate phagocytic vacuoles containing discs that have detached from the rod cells and have been taken up by the pigment epithelial cells. The discs in the body at *X* are beginning to show visible signs of degradation. *Lg* = indicates a residual body and *ER* = endoplasmic reticulum. ×16,000. (From YOUNG, R. W., 1971: J. Cell Biol. **49**, 303—318.) *b* This shows the synaptic ends (*S*) of photoreceptor cells from a frog retina maintained in horse-radish peroxidase under low-illumination conditions, and then incubated to demonstrate peroxidase activity. Reaction product is seen in numerous small vesicles that appear to form through endocytic processes compensatory for the exocytic release of neurotransmitters (SCHACHER *et al.* 1974). Peroxidase is also present in multivesicular bodies (*M*) within the synaptic terminals. ×20,000. (Courtesy of S. SCHACHER.)

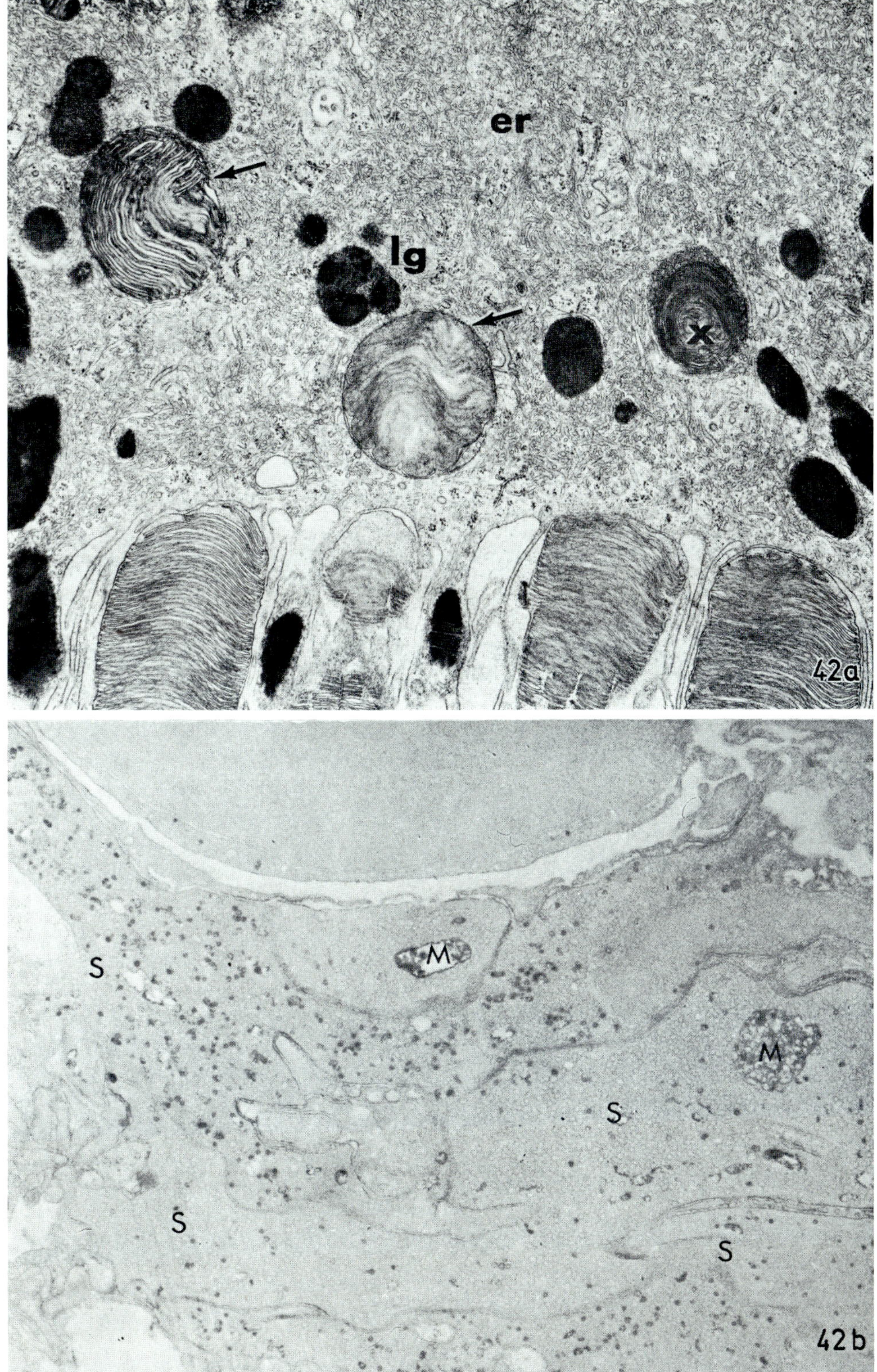

Figs. 42 *a* and *b*.

packaging. Endocytosed tracers seem not to gain access to the rough endoplasmic reticulum (see TEICHBERG *et al.* 1975 for some perspectives on the
problems in deciding whether this also applies to the smooth ER—the grava-
men of the problem lies in the resemblances between smooth ER and tubules
formed in endocytosis). Nevertheless BOLENDER (1974) feels that some of the
configurations be sees in the exocrine pancreas might be vesicles that bud
from the plasma membrane and fuse with the ER.

III.5. Turnover of Photoreceptor Membranes

There are a few situations in which normal turnover of membrane-delimited
structures clearly does depend upon lysosomes. These are the cases in which
portions of cells are sloughed off and digested by special phagocytes. This
occurs, for example, as a regular feature of sperm maturation in vertebrates;
as the spermatid transforms into a spermatozoan much of its cytoplasm
becomes incorporated in a "residual body" which is shed from the maturing
cell and then phagocytosed by Sertoli cells (DIETERT 1966, FOUQUET 1974).
The vertebrate retina provides another particularly interesting case in point
(see YOUNG 1973 for review). The photoreceptor discs of the outer segments
of the retinal rod cells consist of membranes constructed of lipids plus proteins; much of the protein is in the "visual pigment", rhodopsin. The discs
form as invaginations of the plasma membrane at the base of the outer segments via assembly mechanisms that are still imperfectly comprehended;
autoradiography suggests involvement of the Golgi apparatus; HALL *et al.*
1969, YOUNG and DROZ 1968). Subsequently the discs lose their connections
to the surfaces and migrate up the outer segment. After a period of days
to weeks (depending on the species of animal studied) they are shed at the
tip of the outer segment and then degraded through phagocytosis by the
cells of the pigment epithelium (Figs. 41 and 42; HOLLYFIELD and WARDS
1974, ISHIKAWA and YAMADA 1974, O'STEEN and KARSIOGLU 1974, YOUNG
and BOK 1969). The migration can be followed autoradiographically after
a pulse of labelled amino acids (Fig. 41; LAVAIL 1973, YOUNG 1967, 1971).
Since the set of discs labelled during a pulse are still identifiable by their
intense radioactivity at the time of shedding, it seems clear that most of
the proteins incorporated in a disc at the time it forms, remain in that disc
throughout its lifespan, although there may be limited exchanges of the proteins other than opsin (BOK and YOUNG 1972). When a pulse of radioactive
glycerol is used to label the "backbones" of lipid molecules, autoradiography
indicates that some of the label behaves as do the amino acid but some also
disperses through the disc population of a given cell (BIBB and YOUNG
1974 b). Dispersal of label is even more marked when radioactive fatty acids
are used (BIBB and YOUNG 1974 b). Apparently, extensive exchanges of whole
lipid molecules and of the fatty acid portions take place. Thus, in this special
situation lysosome mediated bulk-degradation of membranes coexists with
turnover at the level of individual molecules.

Future studies of the assembly and degradation of retinal photoreceptor
membranes will undoubtedly make use of the fact that the cone cells differ

markedly in their membrane turnover patterns from the rods. In particular the discrete protein labelling patterns observed autoradiographically in the rods are not seen with cones—amino acid labels fairly rapidly become dispersed throughout the entire cone outer segment, suggesting that molecule-by-molecule replacement is extensive (BOK and YOUNG 1972, YOUNG and DROZ 1968, 1968). Since cone discs retain their continuity with the cell surface (and thus, indirectly with one another) whereas rod discs do not, there may be a straight-forward morphological basis for the observed differences. [A preliminary report on human retina suggests that ultimately, cone discs are degraded through phagocytosis by the pigment epithelium (HOGAN and WOOD 1974).]

There are a few retinal disorders in which the turnover of photoreceptor membranes seems grossly abnormal. In one type of hereditary condition, the pigment epithelium appears unable to phagocytose rod discs or test particles such as carbon (BOK and HALL 1972, HERRON et al. 1969, 1971, LA VAIL et al. 1972, SANYAL 1972). This is accompanied by accumulation of excess membrane in the extracellular spaces.

Invertebrate eyes have been little studied with regard to the problems at hand. EGUCHI and WATERMAN (1968) have published findings suggesting that the activity of rhabdomore-type photoreceptor cells may be accompanied by endocytosis of portions of the photoreceptive surface membranes but the underlying molecular events are not known.

IV. Pathology

Lysosomes participate, in one way or another, in a tremendous variety of cellular and organismic disorders. Often, their participation is chiefly through the activities of scavenger cells that clear up cellular debris; corresponding changes in hydrolase levels and numbers and activity of phagocytes have been described for inflammatory responses, infectious diseases, tumors and many other pathological conditions (see e.g., FRANSON and WAITE 1973, GORDON and COHN 1973, VAN FURTH 1970, VERNON-ROBERTS 1972, WILLIAMS and FUDENBERG 1972). For certain disorders there is circumstantial evidence for less indirect lysosomal involvement. But, for very few situations do we have adequate information bearing upon the questions that are most critical from our vantage point—are lysosomes involved as causative agents and if so, through what mechanisms? In other words, it is not difficult to find changes occurring in lysosomes of pathological material but determining the significance of these changes is quite another matter.

We will make no attempt to survey the immense descriptive literature concerning the lysosomes of diseased or injured cells and tissues. Rather, our effort will be to identify and discuss some major issues.

IV.1. Lysosomal Storage Diseases

In these disorders, abnormal amounts or types of macromolecules or related degradation products accumulate within intracellular bodies; sometimes this is accompanied by the presence of high levels of similar materials in the blood and urine. The most widely studied of the storage diseases occur in humans but a number of closely similar abnormalities have been described in animals. The diseases are inherited, and in most, deficiencies in degradative enzymes known or suspected to be lysosomal have been found. The abnormal deposits are usually present in several or many tissues, with the prominence of different sites being characteristic of given diseases; neurons, hepatocytes, fibroblasts, renal tubule cells and cells of the spleen and other portions of the reticulo-endothelial system are commonly involved. The deposits can occupy 25% or more of the cytoplasmic volume of severely affected cells (VAN HOOF 1973).

HERS and VAN HOOF (1973) have recently edited a comprehensive collection of papers on the storage diseases; see also BERNSOHN and GROSSMAN (1971), VOLK and ARONSON (1971) and Ann. Rev. Biochem. **44**, 357.

IV.1.1. Lipidoses and Polysaccharidoses

Severe hereditary deficiencies in protease activity or in nuclease activity have yet to be detected. Defects in protein or nucleic acid metabolism might be expected to manifest themselves early in development. Therefore, if lysosomes are important in relevant aspects of normal cellular biochemistry, defective lysosomes might lead to very early death and consequent non-detection of the abnormality.

The best known of the storage diseases involve lipids or polysaccharides (see pp. 156–161 in: HERS 1973 for an excellent summary chart). Thus, for example, in Tay-Sachs' disease, hexosaminidase A activity is deficient (O'BRIEN 1973) and the ganglioside GM_2 (Fig. 43) accumulates in neurons and elsewhere (Fig. 44; TERRY 1971). In metachromatic leukodystrophy, aryl sulfatase A activity is lacking and bodies rich in cerebroside sulfate are prominent in glial cells (AUSTIN 1973, ETO et al. 1974, JATZKEWITZ and MEHL 1969). In the Hurler syndrome, α-L-iduronidase activity is missing and greatly elevated levels of the mucopolysaccharides, dermatan sulfate and heparan sulfate and related degradation products are found in the urine and within several cell types (BACH et al. 1972). In Pompe's disease (glycogenosis type II), "acid maltase" (α-glucosidase) is absent and abnormal glycogen deposits are found in hepatocytes and muscle cells, among other sites (HERS and DE BARSY 1973). (See also Fig. 43 for some additional disorders.) In these and other lipidoses and polysaccharidoses much of the abnormal intracellular accumulation is within membrane-delimited bodies that often have a characteristic morphology (Figs. 44 and 45) related to the nature of the predominant "stored" materials. In several cases the presumption that these bodies are lysosomes has been verified through cytochemical demonstrations of hydrolase activity, and for Tay-Sachs' disease among others, biochemical studies have provided additional substantiation (TALLMAN et al. 1971).

The enzymatic defects and other characteristics of an increasing number of the lipidoses and polysaccharidoses are well enough known that confident prenatal diagnosis can be made through study of cultured cells obtained by amniocentesis (e.g., DESNICK et al. 1973, KABACK and HOWELL 1973, O'BRIEN

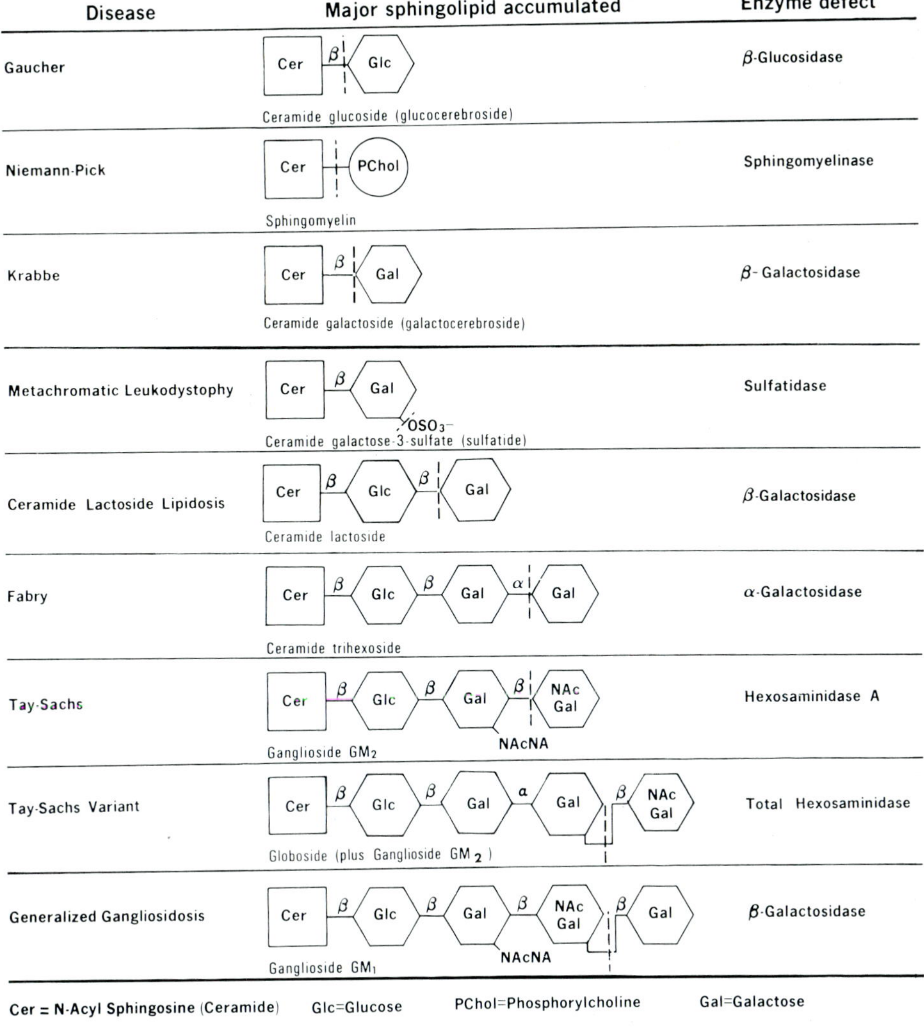

Fig. 43. Summary of enzymatic deficiencies in storage diseases characterized by inabilities to degrade sphingolipids. The dotted lines indicate the bonds that cannot be hydrolyzed due in each case to the lack of the enzyme noted at the right. *Cer* = ceramide (N-acyl-sphingosine); *NAcNA* = N-acetylneuraminic acid; *Gal* = galactose; *Glc* = glucose; *PChol* = phosphorylcholine; *NAcGal* = N-acetylgalactosamine. (From Brady, R. O., 1972: In: Basic neurochemistry. [Albers, R. W., G. J. Siegal, R. Katzman, and B. W. Agranoff, eds.], pp. 485—495. Boston: Little Brown and Co.)

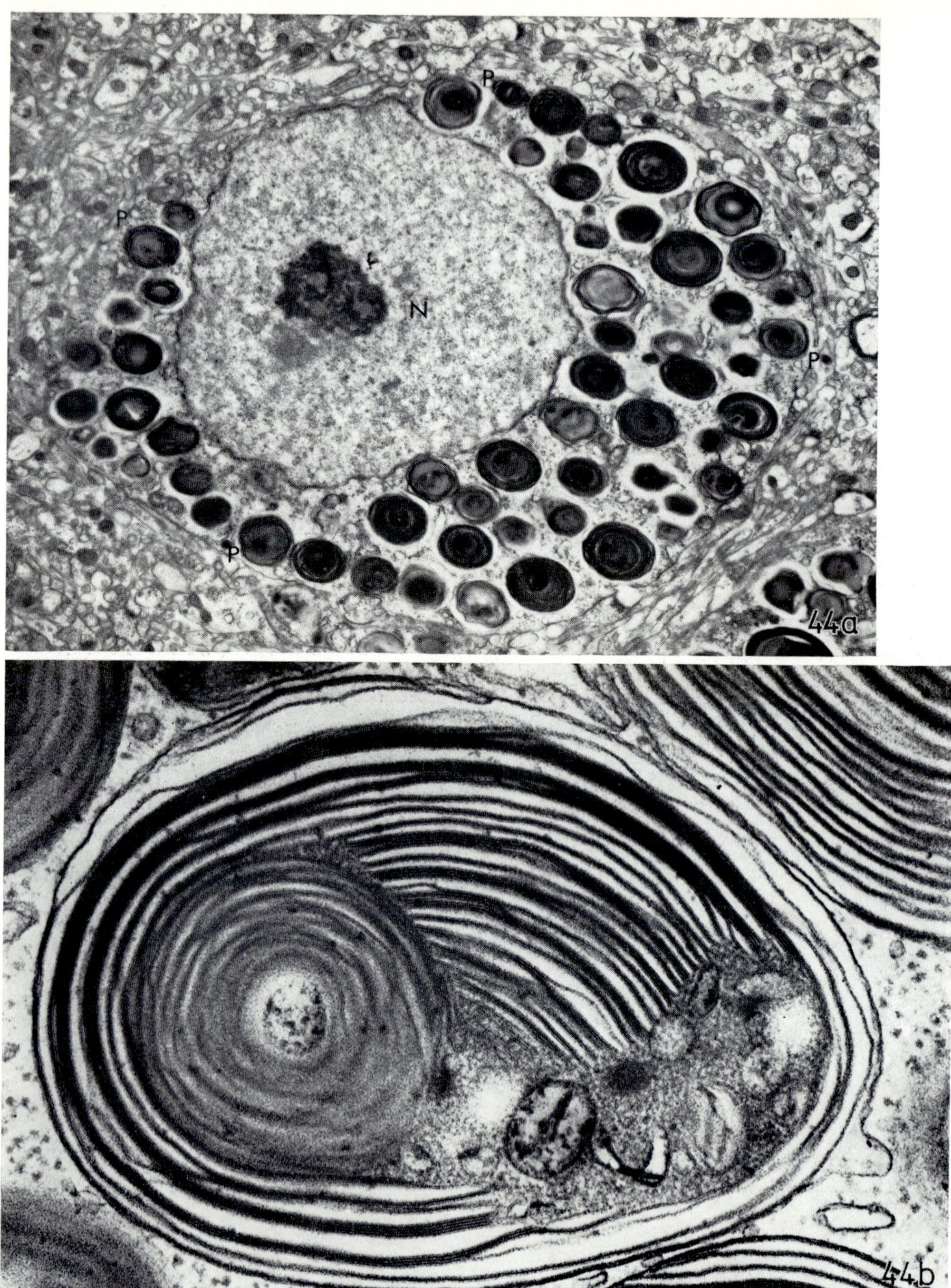

Fig. 44. Portions of neurons from the cerebrum of a Tay-Sachs patient. *a* is a low-power view (×8,000); much of the perikaryal cytoplasm is occupied by the prominent "membraneous cytoplasmic bodies" that characterize this disorder. The nucleus is indicated by *N* and the plasma membrane by *P. b* shows one of the cytoplasmic bodies at higher magnification (×70,000). (From TERRY, R. D., and M. WEISS, 1963: J. Neuropath. exp. Neurol. **22**, 18—55.)

1973). Further, since exogenous proteins can gain access to lysosomes via endocytosis, efforts to replace missing hydrolases can be made. Cells from patients with several naturally-occurring storage diseases have been "cured" in tissue culture by inclusion of appropriate hydrolases in the medium (*e.g.,* LAGUNOFF *et al.* 1973, O'BRIEN *et al.* 1973). For several mucopolysaccharidoses, cocultivation of normal and abnormal fibroblasts results in greatly

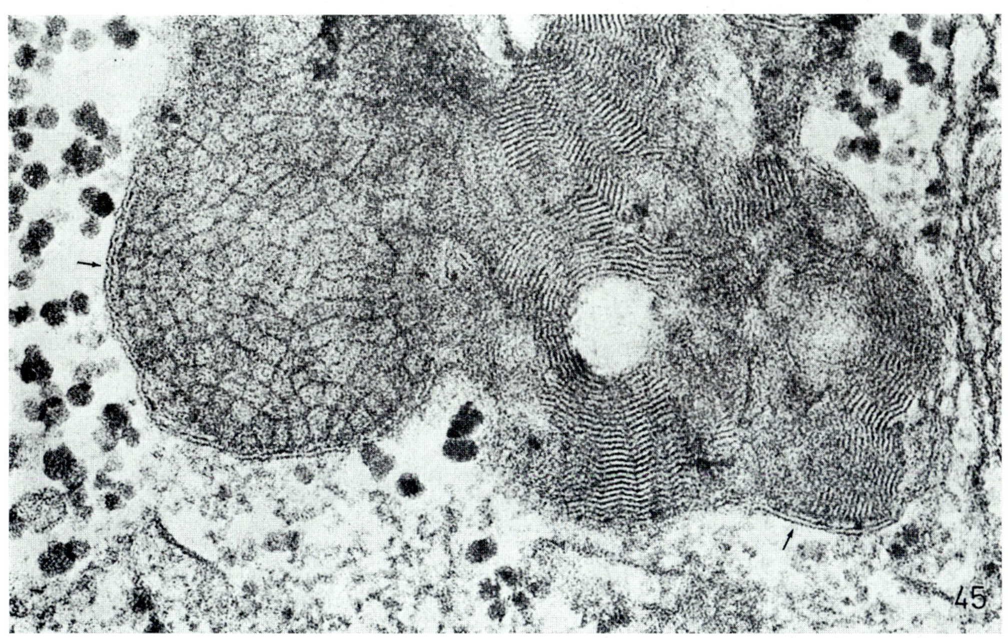

Fig. 45. Intracellular storage body from a sulfatase-deficient (metachromatic leukodystrophy) fibroblast maintained in culture in the presence of exogenous sulfatide (sulfated glycolipid). The body shows a distinctive ultrastructure, characteristically different from other types of storage bodies (*e.g.,* compare with Fig. 44). Arrows indicate the delimiting membrane. ×120,000. (From RUTSAERT, J., R. MENU, and A. RESIBOIS, 1973: Lab. Invest. **29**, 527—535.)

diminished abnormal deposits, presumably as a result of the release of enzymes from one group of cells (by unknown mechanisms) and the incorporation of the enzymes by the other cells (NEUFELD and CANTZ 1973). Studies utilizing this approach have revealed the interesting fact that cells from patients with "I cell" disorders produce several enzymes that substitute poorly for missing enzymes in other cells with which they are cocultured; [the "I cell" disorder itself results in characteristic multiple hydrolase deficiencies in several tissues of affected patients (DACREMONT *et al.* 1974)]. To explain these findings, HICKMAN and NEUFELD (1972) advance the speculative proposal that release of hydrolases to the extracellular space and their subsequent efficient "capture" by endocytosis is normally important for lysosome formation or function and is significantly disrupted in certain disorders; the "I-cell" hydrolases might enter key cells of "I-cell" victims in inadequate amounts. Or, perhaps the defects in the hydrolases that affect their uptake by cultured cells, *in vivo* primarily affect their intracellular transport or yield abnormal leakage of lysosomal enzymes from the cells (WIESMANN *et al.* 1971). But regardless of the validity of the release-recapture hypothesis it has helped focus attention on the growing body of data showing that endocytic uptake of hydrolases can occur at high rates and with much specificity (BROT *et al.* 1974, LAGUNOFF *et al.* 1973, NICOL *et al.* 1974, O'BRIEN *et al.* 1973, VON FIGURA *et al.* 1974 a, b). In addition, there are a few non-pathological situations for which investigators have proposed that enzymes reach lysosomes from sources outside the cell. PRICE (1973 b) suggests that insect salivary glands produce proteases that eventually are utilized within cells of the fat body, and we have already alluded to KLOCKAR and OSSERMAN's (1974) idea that lysozyme in kidney lysosomes is acquired through endocytosis, and to the probability that alkaline phosphatase in the lysosomes of intestinal mucosal cells is taken up from the brush border (WILLIAMS and BECK 1969).

Efforts to cure patients with storage diseases by inclusion of missing enzymes in their circulation have been stymied, probably in part by the low level of endocytosis characterizing certain key tissues like cardiac muscle, plus the presence of circulatory barriers such as the systems that preclude entry of much protein from the blood stream into the cerebrospinal fluid. Some encouragement may be derived from the fact that cells whose lysosomes are readily accessible to circulating proteins can show decreased "stores" *in vivo* experiments; this was noted for example for the hepatocytes of a Pompe's disease patient given *Aspergillis* maltase intravenously (see HERS and DE BARSY, 1973 for a review of their work and that of BAUDHUIN, HUG, SCHUBERT, and others). BRADY *et al.* (1975) optimistically review evidence for beneficial effects of enzyme infusion in diseases such as Fabry's or Gaucher's disorders, which affect endocytically active cell types in the spleen and other sites, but have little major impact on the nervous system. REITRA *et al.* (1974 b) outline the more pessimistic perspectives for diseases in which the nervous system, cardiac muscle etc. are heavily involved.

Perhaps treatment regimens will eventually be designed that can overcome "blood-brain barrier" problems and other difficulties. Experiments currently underway with hydrolases encapsulated in "liposomes" also are of interest (GREGORIADIS and BUCKLAND 1973, GREGORIADIS and NEERJUN 1974, GREGORIADIS *et al.* 1974); one could expect with such preparations to minimize hydrolase damage to extracellular materials, and to deflect immunological responses, while freeing active enzyme within lysosomes through degradation of the encapsulating material by the target cells' own hydrolases.

Another therapeutic approach being explored is the enzymatic treatment of patients' blood to clear it of the circulating abnormal components; this has been tried by simply introducing enzymes into the circulation (BRADY 1973, SWEELEY *et al.* 1971) but more sophisticated attempts involving percolation of blood past enzymes immobilized on columns, and so forth, are underway or anticipated (BRADY 1971, 1973). The hope here is to clear extracellular stores and to upset equilibria between circulating and cell-stored materials; one might thereby induce cells to "unload" their abnormal contents, and also decrease uptake of the abnormal components by actively endocytic tissues (kidney, liver, spleen and so forth). The longterm prospects for such treatments depend upon presently unknown features of the disorders and of cellular egestion. In particular, the approach might deal adequately with materials stored outside of cells, those in the circulation or urine and those released by dead cells (ARONSON and DAVIDSON 1968, VAN HOOF et al. 1971). But there is much more uncertainty as to whether viable cells can be stimulated to dump their stores, or whether when extracellular materials are removed, the cells' defecation mechanisms will be able to catch up with the processes leading to intracellular storage.

IV.1.2. Etiological Aspects

It is an article of faith for modern biology that genetic approaches can help resolve many difficult problems. Such approaches have been used only sparingly for work on lysosomes. The investigations by PAIGEN and others on β-glucuronidase (Section II.2.2; FELTON et al. 1974, MEISLER and PAIGEN 1972), initial efforts to map locations of genes for hydrolases through somatic hybrid techniques (BRUNS and GERALD 1974) and a few studies on hydrolases of lower forms (e.g., those of DIAMOND et al. 1973 on slime molds) are among the handful of examples that come to mind. Study of mutation-engendered storage diseases should eventually provide much information about normal lysosome functions. But at present there is only fragmentary data available concerning the relations of lysosome defects to the havoc wreaked upon cells and organisms and for none of the storage disorders do we have more than a very dim perception of the disease as a whole. Given the widespread cellular abnormalities seen with most of the disorders, the very high rates of early fatality are not surprising. Cardiac muscle involvement resulting in heart failure, gross mental retardation, circulatory abnormalities, decreased resistance to infection, renal disorders and so forth are commonly present. Although detailed information is largely lacking, one can make reasonable guesses (and occasionally, more firmly-based deductions) as to the relations of the cellular disorders to the grosser symptioms. Abnormalities in the leukocyte and macrophage lysosomes might be expected to result in decreased bactericidal effectiveness and modifications in "scavenging" or turnover. Similarly, remodelling of bone and cartilage during development or involution of organs such as the thymus could be distorted by lysosome malfunction, and kidney functions dependent upon endocytosis might also be impaired. The presence of numerous very large inclusions within cells might affect their shape and mechanical properties with various potential ramifications ranging from weakening of vascular walls (KINT et al. 1973) and interference with normal muscle functions to abnormal neuronal growth and synapse formation. It might even be that very large lysosomes are susceptible to rupture in cells such as muscle fibers where vigorous mechanical agitation might occur (e.g., HERS 1973). Multiple secondary consequences could result from extensive release of hydrolases to extracellular spaces that might take place as cells die or through other events (Section IV.3, will consider such possibilities in more detail).

It may also prove important that a number of the lysosomal hydrolases that split complex lipids, also mediate transfer reactions of the types responsible for lipid formation; this has been demonstrated *in vitro* (DISTLER and JOURDAIN 1973, LEDEEN and YU 1973, RAGHAVAN *et al.* 1974) but its *in vivo* significance is still unclear.

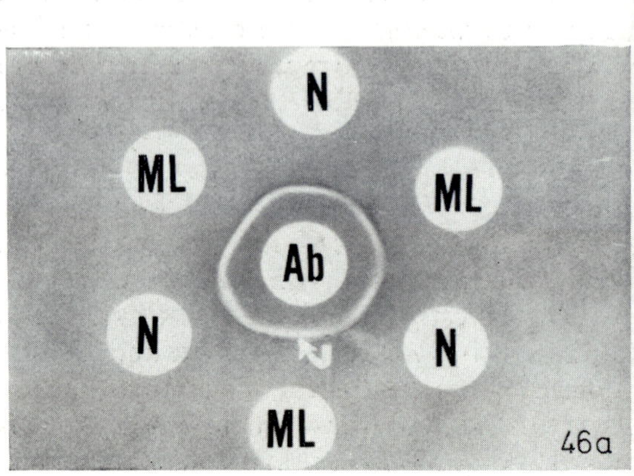

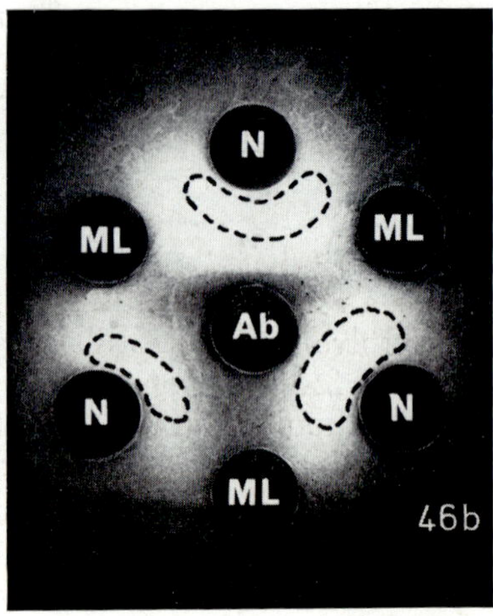

Fig. 46. Immunodiffusion patterns obtained when rabbit antibody to human aryl sulfatase *A* (*Ab*) is tested against human liver enzyme preparations from normal (*N*) and *ML* (metachromatic leukodystrophy) sources. In *a* a precipitin line (arrow) is seen between the central well and all of the peripheral wells, indicating that both the normal and the *ML* tissues contain protein with the immunological characteristics of sulfatase *A*. In *b* the same preparation has been stained to demonstrate locations of sulfatase enzyme activity and the resulting pattern has been printed in reverse photographic contrast. The bright areas are sites of reaction product. From a comparison of *a* and *b* it is clear that the immunologically-identifiable sulfatase of the *ML* tissue is not enzymatically active. (From STUMPF, D., E. NEUWELT, J. AUSTIN, and P. KOHLER, 1971: Arch. Neurol. **25**, 427—431. Copyright, 1971, Amer. med. Assoc.)

Most of the storage disorders are thought to be manifestations of simple recessive mutations (see *e.g.*, KOLODNY 1973). In partial confirmation of this, immunological methods in several cases have demonstrated the presence of the expected enzymatically inactive proteins (Fig. 46; AUSTIN 1973) [19]. However, apparent simplicity may sometimes mask intriguing complications. For instance, in Sandhoff's disease (SANDHOFF and HARZER 1973), both

---

[19] RIETRA *et al.* 1974 report that in Fabry's disease the lack of α-galactosidase activity is paralleled by a lack of immunologically detectable enzyme protein. Perhaps the mutant protein is abnormally sensitive to the intralysosomal environment. But the possibility that one is dealing with a mutation in a regulatory gene should not be overlooked. Phenomena resembling induction of certain hydrolases have been encountered (*e.g.*, phosphate deprivation leads to synthesis of specific phosphatases in *Euglena* and there may be some similar effects for sulfatases in *Aspergillis*; APTE *et al.* 1974, SOMER and BLUM 1965; see also footnote 9 on p. 99), but what this signifies about the mechanisms that regulate lysosomal hydrolase synthesis has yet to be studied, and other approaches to such mechanisms are badly needed. Ho's (1975) concept that hydrolases acting on lipids may be regulated by special "activator" proteins also merits mention.

hexosaminidase A and hexosaminidase B activities are lacking. The disorder reflects an autosomal recessive mutation and resembles Tay-Sachs' disease (hexosaminidase A deficiency) in many respects, including the accumulation of GM$_2$ ganglioside. Why should two enzyme activities, characteristic of two distinguishable proteins be missing when only one gene seems to have mutated and when one of the enzymes can be affected separately by a different mutation? From careful immunological comparisons plus the beginnings of detailed genetic studies such as the analysis of complementation in somatic hybrids of Sandhoff and Tay-Sachs cells (GALJAARD *et al.* 1974, LALLEY *et al.* 1974) it has been proposed that the A and B enzymes share a subunit or subunits in common but also have distinctive polypeptide components (see also CARROLL and ROBINSON 1974, SRIVASTAVA and BEUTLER 1974). Other multiple hydrolase deficiencies (*e.g.*, those involving sulfatases; MURPHY *et al.* 1971) are known and may have similar explanations (VAN HOOF 1973) but they have not yet been investigated to the point where one could argue with much conviction. Even for the hexosaminidases, the issue is not settled. For example, GILBERT *et al.* (1975) find that the genes controlling hexosaminidases A and B, segregate entirely independently of one another, in somatic hybridization experiments with cultured human and mouse cells. This would not be expected from the simpler schemes for subunit sharing. GILBERT *et al.* speculate that their findings may imply that the mutation leading to Sandhoff's disease is in a regulatory gene of some kind, or perhaps in a gene controlling hexosaminidase packaging in the lysosomes (see also footnote 19 on p. 160 and Ann. Rev. Biochem. **44**, 357).

Commonly, secondary alterations in levels of a number of hydrolase activities accompany the primary absence of a given enzyme in storage disease cells (see HERS 1973 for an extensive compilation and, *e.g.*, the reports by KINT *et al.* 1973, 1974 and by HERS 1973 that iduronidase deficient cells of Hurler's syndrome show partial deficiencies in β-galactosidase and "hyperactivity" of some other hydrolases). To an extent this might grow out of the operation of currently unknown control mechanisms governing enzyme synthesis. However, the presence of abnormal amounts or types of material in the lysosomes may also have a substantial influence for example through binding of normal hydrolases and consequent interference with their activities (KINT 1974, KINT *et al.* 1973). Inhibition of lysosomal sialidase by the Tay-Sachs GM$_2$ ganglioside (or by a complex of the lipid and inactive hexosaminidase) has been invoked by BRADY (1973) to explain why cells of Tay-Sachs' patients do not utilize a potential alternate degradative route, whose enzymes, such as sialidase, are apparently normal and could, in theory, degrade the stored lipids. On the other hand, stabilization of hydrolases against denaturation or degradation, through formation of enzyme-substrate complexes might account for apparent "hyperactivity" (HERS 1973). Very likely the overall operative enzymic patterns of storage disease tissue can deviate increasingly from the norm through a kind of chain reaction that spreads the impact of an initially simple lesion.

That tissues normally vary in their content of hydrolases and in the substrates the enzymes encounter also would be expected to influence the patterns of effects of storage diseases. For instance, it is claimed for tissue culture preparations that normal fibroblasts show little hyaluronidase activity (THOMPSON *et al.* 1973) (although the experience with other connective tissue hydrolases such as collagenase warns against too rapidly drawing the conclusion that the enzyme is actually absent, as opposed to its being difficult to demonstrate, due perhaps to binding with substrates or inhibitors). This normal lack of hyaluronidase supposedly could explain the fact that in mucopolysaccharidoses attributable to other hydrolase deficiencies, the fibroblasts sometimes seem to store higher molecular weight components than are present in the tissue fluids and in other cells where endopolysaccharidases appear to have acted.

Investigation of the storage diseases has already aided in clarifying the likely specificities *in vivo* of enzymes detected initially with artificial substrates, and in establishing probable sequences for breakdown of several components that are degraded in multiple steps, such as the sphingolipids, and the complex interlinked networks of proteins and polysaccharides present in connective tissue matrices. In several cases it is known that the failure of one step in a sequence can markedly inhibit the subsequent steps even if the appropriate enzymes are present. This is the situation, for example, with several of the sulfated polysaccharides; if sulfates are not removed due to lack of sulfatases (Hunter's syndrome; SJOBERG *et al.* 1973) or sulfamidase (San Fillipo A disease; KRESSE 1973), exoglycosidatic attack is prevented even in cells or tissue extracts containing normal exoglycosidases (DORFMAN *et al.* 1971, MUIR 1973, NISHIDA-FUKUDA and EGAMI 1970). Such considerations, coupled with the facts that a variety of lipids and polysaccharides are built from the same building blocks and linkages, that lipids and carbohydrates are linked in various associations with proteins, and that large lipid-rich structures can pose difficulties even for normal lysosomes, help explain the widespread and heterogeneous biochemical results of a given hydrolase defect. It also is not surprising that there are numerous cases in which unexpected linkages persist despite the presence of normal enzymes that can attack them under favorable experimental conditions. And, in some storage disorders, various components are degraded despite the absence of key enzyme activities thought to be necessary; these cases can lead to discoveries of new pathways.

Detailed elucidation of the genesis of the cellular abnormalities seen in storage disorders awaits clarification of important features of intracellular metabolism and turnover of cells and connective tissue matrices. Some progress is being made. For instance, it is increasingly evident that membrane components contribute to the abnormal deposits in many disorders (LEDEEN and YU 1973). Derivatives of lipids and other macromolecules characteristic of red and white blood cell plasma membranes often are prominent in the deposits in the spleen or liver (*e.g.*, in Gaucher's disease; BRADY and KING 1973, KALTLOVE *et al.* 1969, WOLFE *et al.* 1974). This is consistent with the suspected importance of spleen and liver lysosomes in turnover of blood cells. DAWSON and STEIN (1970) make the point that a disorder is known corresponding to each step in the catabolism of red cell globoside. But, especially for cells that do not engage in extensive endocytosis, the situation is less clear (SPICER *et al.* 1974). And for example, the genesis of the intra-cellular stores in fibroblasts in the mucopolysaccharidoses is not understood (NEUFELD and CANTZ 1973). Is a pathway for "crinophagic" degradation of secretory materials affected (VAN HOOF 1973)? Are abnormalities in the handling of cell surface or extracellular materials responsible? Is some key clue to be found in the observation that cytochalasin B treatment engenders increased hydrolase release, decreased intracellular hydrolase levels, decreased endocytosis and increased intracellular storage of glycosaminoglycans in cultured fibroblasts (VON FIGURA and KRESSE 1974 b; these authors interpret their findings as supporting the hypothesis about release and endocytosis of hydrolases mentioned in the previous section).

Why are nervous-system cells so frequently among the most markedly affected by storage diseases—how do lysosomes fit into the patterns of cell death, cell growth and migration, myelination and so forth that characterize the perinatal and post-natal periods when neurological abnormalities begin to become strikingly evident[20]? Does the prominence of sulfatides similar to those of myelin, within the glial cell lysosomes of metachromatic leuko-dystrophy signify an indirect role of lysosomes in myelin formation or remodelling, such as participation in turnover of the glial cell plasma membranes that give rise to myelin? (AUSTIN 1973; see also footnote 25, p. 183.) Do the neuronal inclusions, which often are very rich in derivatives of membrane lipids arise through autophagic phenomena, possibly related to the membrane circulation processes discussed in the last chapter (Section III.4.3.3)? We have proposed that perikaryal lysosomes are agents for the ultimate degradation of the membranes of the synaptic vesicles that release transmitters (HOLTZMAN *et al.* 1973, TEICHBERG *et al.* 1974, 1975). One might think that this lysosomal function, or perhaps participation in membrane turnover linked to the maintenance of the axon surface is tied in with neuronal susceptibility to the effects of storage disease. Synaptic activities probably involve intensive mobilization and circulation of membranes and neurons are characterized by an unusually extensive surface area. More thorough studies of various cells that secrete by exocytosis might help test these ideas—there have been relatively few reports on the status of gland cells in storage diseases, although in a few cases some have been seen to contain prominent inclusions (AUSTIN 1973).

On the other hand, when excess lipids are delivered by endocytosis to the lysosomes of otherwise normal tissue-cultured neurons, the lysosomes transform into bodies strongly resembling those of storage diseases and these persist for extended periods (STERN 1972, STERN *et al.* 1971). Perhaps this is to

---

[20] For examples of the kind of data available on the timing of appearances and changes of hydrolases and potential substrates in developing brain see GILBERT and JOHNSON 1972, and VERITY and BROWN 1968. Descriptions of the disorders are provided in BERNSON and GROSSMAN (1971), VOLK and ARONSON (1971), and HERS and VAN HOOF (1973)—see *e.g.*, SUZUKI and SUZUKI's (1973) account of Krabbe's leukodystrophy in which a galactocerebro-side-β-galactosidase activity is deficient and apparently as a consequence, myelination is grossly abnormal. DECKER (1974, 1974 b) has done correlated biochemical, cytochemical, and electron microscopic studies on developing amphibian brains.

be interpreted as indicating that neurons have unusually limited capactites to expel their lysosomal contents. The case could be made that neurons are particularly prone to the effects of storage disorders merely because of this and of the cessation of division by mature nerve cells.

There are those who maintain that in at least one lipidosis, Tay-Sachs' disease, the abnormal lipid accumulations in neurons turn up first in the endoplasmic reticulum (ADACHI *et al.* 1971, 1974) where the molecules probably are synthesized. While the biopsy material available has not yet provided pictures that unambiguously demonstrate this, there is no present basis for ruling out any reasonable possibility for the mechanisms through which neuronal storage disease inclusions bodies are formed. Perhaps as excessive stores of lipids build up they interfere with the normal transport of materials from the ER. Bodies with the characteristic "Tay-Sachs" morphology form in the test tube when appropriate macromolecules are present (TERRY 1972, SAMUELS *et al.* 1974) and thus they might be expected to appear in the cell wherever the Tay-Sachs ganglioside is present in sufficient concentration. The prior discussions of ER-lysosomes relations may also be germane here (Sections II.2.2 and II.5.2).

The abnormal glycogen deposits of Pompe's disease merit some comment. Autophagy of glycogen in normal muscle is pronounced during the perinatal period (KOTOULAS and PHILLIPS 1971, KOTOULAS *et al.* 1971, SCHIFFINO and HANZLIKOVA 1972 a) and during some types of induced glycogenolysis (see *e.g.*, BECKER and CORNWALL's 1971, report on phlorizin effects), and glycogen is found in autophagic vacuoles in various other tissues. But, conventional schemes of glycogenolysis do not assign a major role to the lysosomal α-glucosidase missing in the disease, or to other lysosomal enzymes and it is not demonstrated that the cells in tissue affected by the disorder cannot adequately utilize glycogen (HERS and DE BARSY 1973). Thus, the key effect of the lysosomal dysfunction on the cells could be a secondary consequence of the accumulation of increasing numbers of large intracellular inclusions rather than a problem in mobilizing low molecular weight carbohydrates.

The presence within cells, of storage-disease inclusions might have a variety of effects in addition to those directly linked to the absence of an enzyme activity or to mechanical problems, such as interference with orderly cellular movement, or alterations in diffusion pathways. The finding by DAVIES (1973) that Triton WR-1339-loaded lysosomes are not accessible to endocytosed proteins suggests that engorged lysosomes may cease to participate in normal fusions. This types of effect could intensify the consequences of a hydrolase deficiency for turnover, the maintenance of pools and so forth. The piling up of large quantities of given metabolites could lead to secondary or indirect biochemical changes of varying types, such as feedback inhibition of synthetic pathways and abnormalities in metabolism and transport of macromolecules not directly involved in the disorder. And, the presence of excessive amounts of charged macromolecules in the lysosomes could lead to changed patterns of accumulations of small molecules of various types by the mechanisms discussed in Sections I.3.2.3 and II.1.4.5; this might have injurious effects on cellular ionic balances, intralysosomal pH, etc.

IV.1.3. Some Interesting Disorders of Uncertain Status

In many of the circumstances in which abnormal lysosomes abound in cells there are no signs of intrinsic defects in the hydrolase complement. Rather, the lysosome alterations may reflect "efforts" by the cell to deal with metabolic abnormalities originating elsewhere; obviously once the changes in lysosomes have begun there are many possibilities for pyramiding of effects.

We mentioned in the last section that when neurons are cultured in the presence of excess gangliosides (STERN 1972) their lysosomes acquire considerable amounts of the exogenous lipids, which persist for many days even after the culture medium is changed back to normal. This type of finding is one element in an important, though circumstantial argument which can be made for the proposition that in a number of the pathological situations in which cells accumulate lipid-rich lysosomes, the primary defects relate to the amounts or types of lipids that the lysosomes encounter or to processes that affect the "digestibility" of the lipids. For example, it has been speculated (ZEMAN and SIATOKOS 1973; see Section II.5.2) that vitamin E deficiencies could conceivably result in a build-up of lipofuscin-like bodies as a consequence of "disinhibition" of oxidation or peroxidation processes that give rise to indigestible lipid derivatives. DE LA IGLESIA and co-workers (1974) similarly propose that secondary phospholipidoses can result from the presence of abnormal lipids, or complexes of lipids with other macromolecules, in humans treated with certain drugs. And several investigators (e.g., FERRANS et al. 1971, 1973, LANDING et al. 1971) believe that hyperlipoproteinemias can lead to extensive lysosomal lipid deposits in abnormal sites, due to saturation of the abilities of the reticuloendothelial system to metabolize the lipids.

Probably the most important disease to which this line of reasoning may apply is atherosclerosis. Biochemical and cytochemical studies indicate that the smooth muscle cells and macrophages of the aortae of animals maintained on cholesterol-rich atherogenic diets accumulate much esterified cholesterol within their lysosomes (PETERS et al. 1972, SHIO et al. 1974); apparently the limits to the abilities of the lysosomes to deal with lipids are surpassed [21].

---

[21] Reports bearing on these capacities include the finding by BLACK et al. 1972 that smooth muscle lysosomal cholesterol esterase is notably less effective with some cholesterol esters involving saturated fatty acids than with the corresponding unsaturated acids. WERB and COHN (1971, 1972) have demonstrated that cholesterol can exchange between extracellular media and the membranes of macrophage lysosomes and they have suggested that the capacities of the exchange system to clear cholesterol from the lysosomes may be overwhelmed when much cholesterol enters the cell as in xanthomas or atherosclerosis. JOHNSON et al. (1975) have published preliminary findings that macrophages have trouble digesting liposomes that contain high proportions of cholesterol.

PETERS et al. (1972, 1974) have isolated a cholesterol-rich sub-population of lysosomes from atheromatous tissue, and have found these particles to be relatively poor in cholesterol esterase and in some other hydrolase activities. WOLINSKY et al. (1975) review work indicating that hypertension induces changes in the lysosomes of arteries, similar to alterations evoked by an atherogenic diet; as do others, these authors propose that one factor in atherosclerosis may be a sequence of events in which lysosomal overloading contributes to cell death, with the dying cells releasing their hydrolases (and lipids) to the extracellular spaces in the artery wall.

The numerous abnormal lipid-containing inclusions in the smooth muscle cells give them their characteristic "foamy" appearance. From the proposals of several laboratories (*e.g.*, PETERS *et al.* 1972, WOLINSKY *et al.* 1975) one can envisage processes wherein factors such as initial alterations in the vascular endothelium, or excessive hyperlipemia, might lead to heightened uptake of lipids by smooth muscle and other cells. The abnormal lysosomes generated in this way could then participate in mediating additional effects that increase the severity of the lesions. (ROSS and GLOMSET 1973, review the development of atherosclerotic lesions, and Sections IV.3 and IV.4 will outline ways in which lysosomes may contribute to cell and tissue damage.)

In β-thalassmeia, normoblasts, erythroblasts and other stages in erythroid maturation show prominent inclusions, some of which reportedly contain cytochemically demonstrable acid hydrolases (ALLISON 1972). The inclusions are rich in hemoglobin α-chains (FESSAS *et al.* 1966) which are present within the cells in excess due to a genetically based defect. Further substantiation and detailed study of the association of the hemoglobin with lysosomes could lay the basis for valuable insights into the mechanisms by which cells "select" material to be autophagocytosed; ALLISON has begun to construct hypotheses wherein the absence of stabilizing β-chains leads to changes in the α-chains including denaturation or precipitation and consequent enhanced affinities of the α-chains for membranes.

Wilson's disease is a hereditary disorder in which copper accumulates in hepatocyte dense bodies and other lysosomes (see GOLDFISCHER and MOSKAL 1966 for background and references). The current prevailing opinion seems to be that the lysosomal accumulations are a secondary reflection of imbalances elsewhere. Both *in vivo* and in culture, various cells exposed to excess exogenous metals show much storage of metals within lysosomes (BRUN and BRUNK 1973, BRUNK and BRUN 1972, EVANS 1973, GOLDFISCHER 1967). It also appears that intracellular molecules such as ferritin and heme-proteins and circulating molecules such as the copper-containing protein, ceruloplasmin, can contribute metals to lysosomes during normal turnover. GOLDFISCHER and colleagues (1970) were struck by the observations that in toad liver and in newborn human liver, there are present very high levels of lysosomal copper without any evident abnormal consequences. In hepatocytes from patients with marked Wilson's-disease symptoms, the lysosomal copper is accompanied by unusually heavy nonlysosomal cytoplasmic deposits; it is as if lysosomal mechanisms have proved quantitatively inadequate to "protect" the cell (GOLDFISCHER and STERNLIEB 1968). Copper homeostasis is still poorly understood (see EVANS 1973 for a recent review) but there are reports that Wilson's disease material show abnormally tight binding of copper by cytoplasmic metallothionin (EVANS *et al.* 1973) and there also are claims of deficiencies in the amounts or activities of other copper-containing proteins such as ceruloplasmin, cytochrome oxidase components, or perhaps even superoxide dismutase (see *e.g.*, SHOKHEIT *et al.* 1969). In Evans' view, the inability of hepatocytes to discharge copper into the bile at a rate fast enough to maintain intracellular copper levels in balance, is an essential factor in the disease.

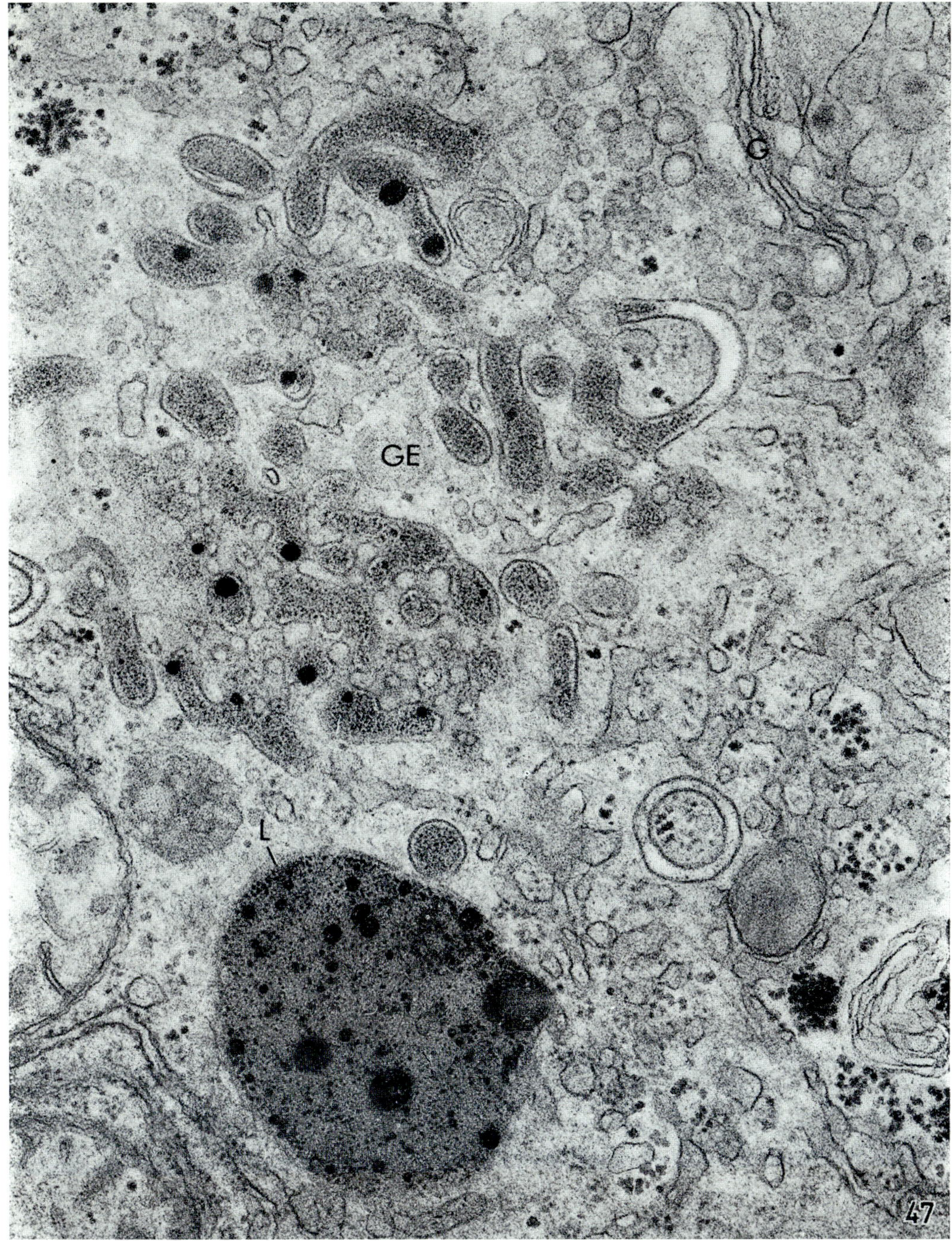

Fig. 47. Portion of an hepatocyte from a beige mouse (a homologue of Chediak-Higashi disease). Numerous large lysosomes containing lipid droplets and fine granular material are present in the cells (*L*). The hepatocytes also show an elaborate network of interconnected tubules or sacs with a finely granular content (*GE*); this network shows continuities with lysosomes and apparently corresponds to hypertrophied GERL. *G* = indicates the Golgi apparatus. ×40,000. (From Essner, E., and C. Oliver, 1974: Lab. Invest. **30**, 596—607).

This might bring Wilson's disease into line with some of the lipid-storage phenomena discussed earlier.

In Chediak-Higashi disease and its animal counterparts (beige mouse, Aleutian mink, etc.) some cell types show striking enlargement of certain cytoplasmic organelles such as the pigment granules of melanocytes and of

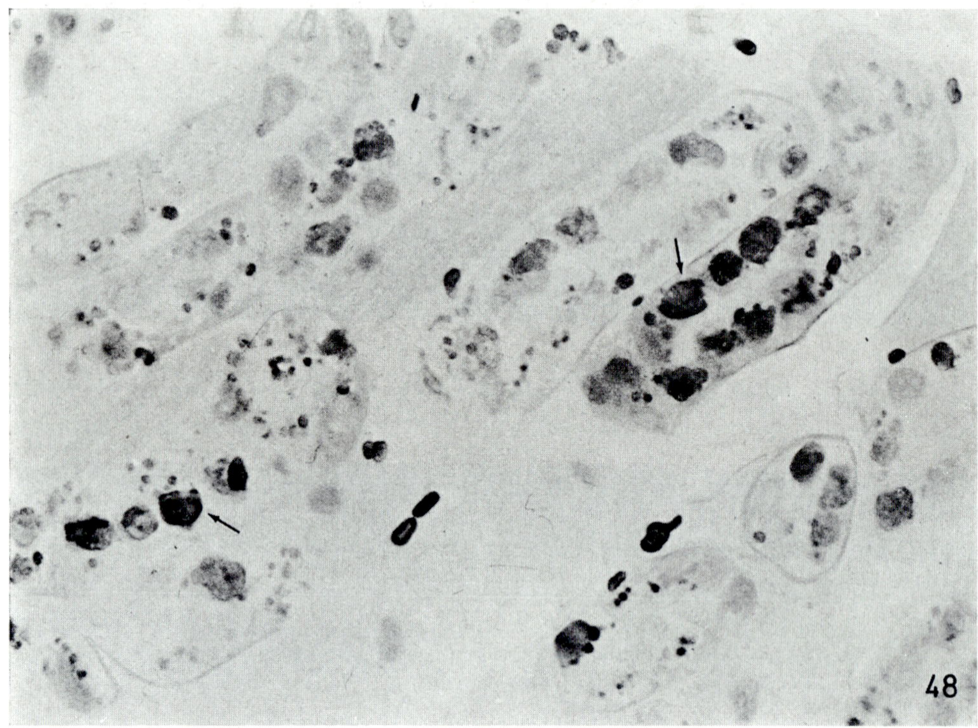

Fig. 48. Portion of kidney from a beige mouse (*cf.*, Fig. 47) after administration of horse-radish peroxidase and incubation to demonstrate peroxidase activity. Arrows indicate two of the peroxidase-containing protein absorption droplets (heteophagic lysosomes) present in the renal tubule cells. Such bodies are quite large even in normal mice, but they are substantially enlarged in the disorder. ×700 (approx.). (From ESSNER, E., and C. OLIVER, 1973: Amer. J. Path. **73**, 217—232.)

the retinal pigment epithelium (LUTZNER and LOWRIE 1969, ZELICKSON *et al.* 1967), leukocyte granules (DAVIS *et al.* 1971, WHITE 1966) and renal protein absorption droplets (Fig. 48; ESSNER and OLIVER 1973; SUNG *et al.* 1969). Functional problems are noted such as low resistance to infection (DAVIS and DOUGLAS 1972; GALLIN *et al.* 1974) and slow digestion of proteins in renal lysosomes (ESSNER and OLIVER 1973, PRIEUR *et al.* 1972) that probably reflect abnormalities of the lysosomes. The disease is hereditary but no thoroughly persuasive case has yet been made to support a specific model for its genesis. OLIVER and ESSNER's (1975) report that PMN granules of normal size form initially but then fuse to give rise to the large ones typifies one of the patterns observed but other authors attribute more importance to hypertrophy of individual organelles into giant ones and there is no general agreement as to the relative significance of fusion vs. hypertrophy in different cell types. From our viewpoint it is tantalizing that so many of the inclusions that are prominently affected are lysosomes and also that the disease apparently can result in hypertrophy of the packaging systems that generate lysosomes (Fig. 47; ESSNER and OLIVER 1974). The disorder may prove to affect directly transport or packaging of hydrolases (*cf.*, LUTZNER *et al.* 1966) or, for example, there may be abnormalities in membranes that influence patterns of fusion among lysosomes or between lysosomes and endocytic or autophagic bodies (ALLISON 1972, BRANDT *et al.* 1974, DAVIS and DOUGLAS, ESSNER, and OLIVER 1974, ROOT *et al.* 1972). If either of these possibilities pans out, Chediak-Higashi tissues may turn out to be of exceptional utility for analysis of lysosome function.

J. OLIVER *et al.* (1975) have published a brief report indicating that the surface membranes of PMN leukocytes may be different in beige mice, as compared with normal mice. The differences affect the behavior of surface bound ligands, such as concanavalin A. This may provide the beginnings of a direct approach to the molecular alterations underlying the Chediak-Higashi syndrome.

IV.2. Lysosomes and Infection

Lysosomal hydrolases can degrade most components of bacteria and of viruses, and lysosomes of phagocytes (macrophages, PMN leukocytes, etc.) and other cells obviously are intimately involved in defense against infection. However, particularly in light of the variety of infectious organisms it is not surprising that the bactericidal and virucidal systems of organisms are complex, and involve many factors in addition to the acid hydrolases (see *e.g.*, VERNON-ROBERTS 1972). For example, while phagocytes can kill bacteria in the absence of serum (*e.g.*, ELSBACH *et al.* 1973), circulating antibodies can strongly influence the rates at which foreign organisms are phagocytosed and they may have direct effects upon the viability of some bacteria and viruses. Cytophilic antibodies bound to phagocyte surfaces also are important in recognition and uptake phenomena. At the intracellular level, the lysosomes and their hydrolases may be aided by other organelles and proteins that take part in killing and digestion (the low intralysosomal pH probably also is unfavorable for some bacteria). The specific granules of PMN leukocytes contribute lysozyme to phagocytic vacuoles and this enzyme is important in the degradation of many bacterial cell walls [*]. These granules also provide lactoferrin, which seems to have antimicrobial activities (HIRSCH 1972); perhaps, by chelating metals it helps maintain an environment favorable for lysozyme's activities (BAINTON 1972). Other PMN granule proteins may bind to bacteria and aid in their destruction, possibly through modifying their

---

[*] As mentioned earlier, the lysosomal (azurophilic) granules apparently also contribute lysozyme. Interestingly, a strain of rabbits that lacks detectable leukocyte lysozyme, shows little evident effect of the deficiency (PRIEUR *et al.* 1974).

surfaces non-enzymatically; some cationic proteins of the azurophilic granules
are under investigation in this regard (BAINTON and FARQUHAR 1968, HIRSCH
1972, KLEBANOFF 1971, SPITZNAGL 1972, ZEYA and SPITZNAGL 1966, see also
GLEICH *et al.* 1974 for basic proteins of eosinophils and MACRAE and SPITZ-
NAGEL 1975, for cytochemical evidence that some of the cationic proteins are
in the azurophilic granules of PMN leukocytes.)

KLEBANOFF (1971) and his co-workers stress the importance of reactions
catalyzed by the myeloperoxidase of PMN leukocytes in the killing of bacteria and perhaps of viruses as well. *In vitro,* when provided with peroxides
and halogen ions, the enzyme can catalyze reactions with bactericidal outcomes. KLEBANOFF's experiments suggest that iodinations are particularly
effective. *In vivo,* chlorinations might be more important, given the ionic
milieu of a cell (although there has been interesting speculation about thyroid
hormones as potential sources of iodine and about the possible attendant
interactions between hormone levels and responses to infections; KLEBANOFF
and GREEN 1973; WOEBAR *et al.* 1972). Several authors have begun to look
into other reactions of peroxides or comparable agents that might be related
to antimicrobial activities—superoxides are now receiving a flurry of attention
(DE CHATELET *et al.* 1974, CURNUTTE and BABIOR 1974, KLEBANOFF 1974).
That the enzymes and reactions being studied are important ones is suggested
by the existence of the disorder, chronic granulomatous disease, in which the
iodination abilities of leukocytes are markedly reduced, due perhaps to abnormalities in the metabolic generation of hydrogen peroxide (KARNOVSKY 1973,
KLEBANOFF 1971, SIMMONS and KARNOVSKY 1973). Despite the seeming
normalcy of their lysosomal hydrolase complement, children with this disorder
are highly susceptible to infection with *Staphylococcus* strains and other
bacteria. In addition, LEHRER and KLEIN (1969) report that leukocytes from
patients deficient in myeloperoxidase are markedly less able to kill *Candida*
and *Staphylococcus* despite their apparently normal capacities for phagocytosis.

On the other hand, as already indicated (Section II.1.3.2) some categories
of macrophages may lack lysosomal peroxidases but nonetheless can kill bacteria (PAUL *et al.* 1973 assert that rabbit alveolar macrophages do carry out
iodinations, but this is in contrast to the mouse peritoneal populations and
other types discussed by SIMMONS and KARNOVSKY 1973 and by others, such
as HIRSCH and FEDORKO 1970). In some cases, lysozyme plus the acid hydrolases might be adequate for inflicting fatal damage (BRAUNSTEIN and
SCHMALZL 1970), especially on Gram-positive organisms. On the whole,
the key "take-home" lessons from the considerations we are discussing probably are that the hydrolases require the "aid" they obtain from other systems
to cope rapidly with many of the organisms with which they are confronted,
and that there still is much to learn about the mechanisms by which phagocytes kill microorganisms.

The last point is given emphasis by the present lack of understanding of
the roles of the peroxidases in eosinophil lysosomes (BUJAK and ROOT 1974)
or platelets; for platelets, whose peroxidase is contained in a system of intracellular tubules separate from the lysosomes (BRETON-GORIUS and GUICHARD

1972, WHITE 1972 b) spadework has begun on evaluation of the possible involvement of the enzyme in prostaglandin metabolism (HAMBERG and SAMUELSSON 1974).

It is not even certain how the peroxides, involved in the microbicidal events discussed above, are generated *in vivo*. BRIGGS *et al.* (1975) find that the myeloperoxidase in phagocytic vacuoles of living PMN leukocytes, will oxidize the cytochemical peroxidase substrate, diaminobenzidine (Section I.3.2.2), when this compound enters the vacuoles during suitably designed phagocytosis experiments. This provides direct confirmation of the probable presence of peroxide in PMN-leukocyte vacuoles. As do many others, BRIGGS *et al.* believe that nucleotide oxidases produce the peroxide. HOHN and LEHRER (1975) hold that an NADPH-oxidase is the best candidate, but there is no definitive evidence as to how this enzyme, or other possible contributors, brings about the accumulation of peroxide in phagocytic vacuoles. (In HOHN and LEHRER's view, chronic granulomatous disease involves a deficiency of the NADPH oxidase, but this is still a matter of dispute.)

It is far beyond the scope of the present endeavor to attempt a review of the manifold aspects of the interactions of infectious organisms and defensive systems. We will confine ourselves to certain topics directly germane to lysosome function. However, mention should be made of the fact that as is true for inflammatory responses in general, the presence of foreign organisms can engender mobilization of specific populations of defensive cells and also call forth the production of some important macromolecules in addition to antibodies (*e.g.*, in responses to viral infections, interferons that aid in protection of cells against viruses are synthesized by a number of cell types, including macrophages [SMITH and WAGNER 1967, VAN FURTH 1970, VERNON-ROBERTS 1972, WILLIAMS and FUDENBERG 1972]). Also, when phagocytes take up bacteria or other organisms they undergo a variety of interesting metabolic changes, including several directly related to their capacities for phagocytosis, and alterations in oxidative metabolism that may be involved in the generation of peroxide (KARNOVSKY *et al.* 1970, 1973). And finally, bacteristatic roles of extracellular hydrolases have been little explored but it should be kept in mind that hydrolases may be released from lysosomes of viable cells at sites of inflammation (Section IV.3) and from lysosomes of dead cells.

IV.2.1. The Entry of Structures with Macromolecular Dimensions into Cells: Viruses and Toxic Proteins

The successful infection of a cell by a virus involves the uncoating of the virus and the penetration of at least one cellular membrane; the latter process is essential if the nucleic acid of the viral core is to gain direct access to the cytoplasmic synthetic machinery and other components of the cell interior. For bacteriophages, the well-known "injection—like" introduction of DNA into a potential host accomplishes both uncoating and penetration. For animal cells the situation is far more controversial. When cells are exposed to viruses, the most readily observable initial intracellular populations are those within endocytic vesicles and vacuoles, and sometimes within the corresponding lysosomes. For one virus, reovirus, the lysosomal populations probably are the major successful invaders (DALES 1969, SILVERSTEIN *et al.* 1973). The core of reovirus, composed of double-stranded RNA plus some proteins, resists the lysosomal hydrolases. SILVERSTEIN and his co-workers (SILVERSTEIN and DALES 1968, SILVERSTEIN *et al.* 1973) have developed strong evidence that lysosomal proteases uncoat the reovirus core, and that subsequently, the beginnings of viral replication can occur within the lysosomes (Fig. 49). Eventually, by unknown processes, reovirus nucleic acids appear outside the lysosomes and the infection spreads within the cell *.

For other viruses, this route seems precluded as an effective path for infection since the lysosomal enzymes are probably able to degrade most viral constituents. Thus, mechanisms by which the lysosomes are evaded must exist; heat-inactivating viruses or coating them with antibodies makes these mechanisms fail (DALES 1969, SILVERSTEIN 1970). Perhaps as a reflection of differences among viruses, there is no present general agreement as to the nature of viral entry into the cytoplasm of animal cells. Some investigators hold that endocytosis is a usual step in infection with the viruses escaping from endocytic structures (DALES 1969, 1973) while others maintain that a small population of viruses avoids endocytosis and crosses the plasma membrane directly [see DALES (1973) for review and discussion and MORGAN and ROSE (1968), MORGAN *et al.* (1969), NEWTON (1970), for examples]. Only one or a few viruses need succeed in entering the cytoplasm in an effective state for an infection to result and the problems of analyzing these against the background of "unsuccessful" viruses are formidable. (Such considerations even lead to residual doubts regarding the reovirus case, although the experimental backing for the picture presented above is increasingly convincing.)

By what sorts of mechanisms might viruses penetrate membranes? They are far too large to cross by conventional permeability pathways. On the basis of evidence that is still quite scanty, and by analogy with bacteriophages, many assume that viral coats can somehow interact with membranes so as to permit at least the viral cores to cross. Viral surface enzymes such as neuraminidases might aid in these events. For a number of viruses it is thought that the membranes of vacuoles surrounding endocytosed particles dissolve or disintegrate. For some of these cases (*e.g.*, that of *Vaccinia virus*) the presence of a lipoprotein membrane, belonging to the virus itself, might be important; *e.g.*, if this membrane fused with a vacuole membrane, the viral core could be released into the cytoplasm (DALES 1973, NEWTON 1970, SAROV and SOCKLICK 1972) leaving behind a residual membraneous structure that might be quite difficult to distinguish from other cellular membranes. Direct fusion of membrane-enveloped viruses with the cell surface may also take place.

---

* According to MARVIN and WACHTEL (1975), filamentous viruses exit from bacteria, without cellular lysis, through a sequence of events in which viral proteins become inserted in the cell's plasma membrane, and then the viral DNA interacts with these proteins. Does reovirus use a similar mechanism to get its RNA out of the lysosomes?

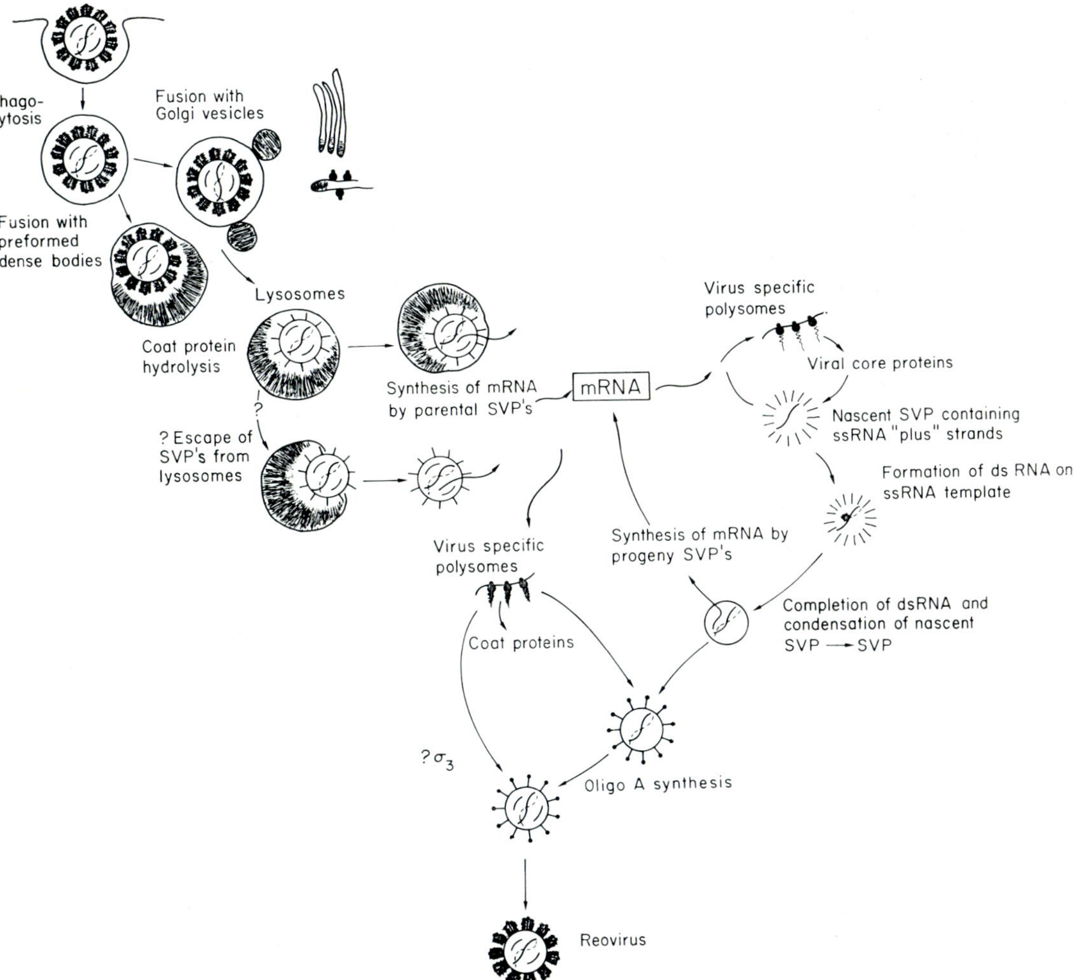

Fig. 49. Diagram of the reovirus replicative cycle in *L*-strain fibroblasts. Virus-containing phagocytic vacuoles fuse with lysosomes and the viruses appear to undergo uncoating in the resulting digestion vacuoles. *SVP* = subviral particles; *ssRNA* = single-stranded RNA; *dsRNA* = double stranded RNA. (Courtesy of S. SILVERSTEIN: modified from a diagram in SILVERSTEIN, S., 1973: Adv. Biosci. **11**, 1—27.)

Under proper experimental conditions, certain macromolecules other than those responsible for conventional viral infections also seem able to penetrate membranes, although with extremely low efficiency (and there is much additional pertinent speculation about the abilities of cells to transfer active macromolecules to one another; see *e.g.,* KOLODNY 1973). In general, when cells are exposed to macromolecular tracers, the vast majority of tracer molecules that enter the cells seem to do so in membrane-delimited endocytic structures and to remain within membrane-delimited compartments. However when biological tests sensitive enough to detect one or a few molecules are available, evidence can be obtained for penetration of very small numbers of certain molecules into the cytoplasm outside of the compartments. This is true for nucleic acids in bacterial transformation and for other infectious nucleic acids in procaryote systems and eucaryotes; with these, entry into the cells results in genetic alterations or proliferation of viruses (BISHOP and KOCH 1967, PAGANO *et al.* 1970). Infectivity is enhanced by use of hypertonic or hypotonic media or by adding charged macromolecules such as DEAE-dextrans, which has led to suggestions that some small-scale surface "damage" along with charge-masking or electrostatic binding phenomena are important for entry of the nucleic acids (*e.g.*, KATCHALSKY 1964, RYSER 1967, SELJELID *et al.* 1973, WEIGERS and KOCH 1972).

The penetration of certain proteins into the cytoplasm is suspected on the basis of evidence that the proteins act on cells through direct enzymatic attack on other cytoplasmic macromolecules. Thus, both colicins such as E_3 which affect bacteria (BOON 1971, REEVES 1971, TAI and DAVIS 1974) and diphtheria toxin, which affects eucaryotic cells (GILL *et al.* 1973) are believed to alter components of the protein synthesizing system. In these cases, the proportion of molecules that penetrate is thought to be so small, that usual analytical methods cannot detect them against the background of molecules adsorbed to cell surfaces or taken up by endocytosis. Hence, little is known of the relevant mechanisms. It cannot even be unequivocally demonstrated that the molecules penetrate all the way in and come to lie free inside the cell. Conceivably, portions of the proteins could remain attached to membranes while biologically active parts are exposed to the cytoplasm although given the rapidity and extent of the effects this seems unlikely. Experiments with red blood cells have shown that enzymatically active exogenous proteins can be loaded into the cytoplasm through brief hemolysis (IHLER *et al.* 1973) so localized and limited "damage" to the cell membrane may operate for entry of proteins as well as for nucleic acids.

Interestingly, diphtheria toxin protein is composed of two parts. One (the B fragment) is relatively hydrophobic, and therefore might interact effectively with membranes. The other (the A fragment) is responsible for the enzymatic action of the toxin (GILL *et al.* 1973, PAPPENHEIMER and GILL 1973, UCHIDA *et al.* 1973). In the test tube the A fragment becomes active when proteolytically cleaved from the remainder of the molecule and it is quite stable to extremes of pH and to proteolysis, unlike the B fragment. Thus the possibility is open that by virtue of proteolytic attack by cellular (lysosomal?) hydrolases only the A fragment actually enters the cytoplasm, with the B fragment, having served as a "carrier", being left behind adhering to a membrane (or even within a lysosome). The toxic plant proteins abrin and ricin also seem to be of two parts; in these proteins the parts are separable by disulfide reduction (REFSNES *et al.* 1974). NICHOLSON (1974) has attempted to follow ferritin-labelled ricin as it enters 3T3 tissue-culture cells. The proteins are found first in endocytic vacuoles and then free in the cytoplasm, but whether their liberation precedes cell death is not clear.

The matters discussed in this section provide tentative evidence that lysosomes are not particularly effective in "detecting" foreign molecules present in the cytoplasm. This is suggested also by the fact that viable heterokaryons, somatic hybrids and other mixed systems can be made between cells of quite different sorts and by the observations that RNA's injected directly into the cytoplasm of large cells such as oocytes, survive for prolonged periods (GURDON *et al.* 1973). However, ALLENDE *et al.* 1974, point out that while mRNA's, tRNA's, poly-A and ribosomes are stable when injected into oocytes, certain synthetic polynucleotides, and protein-free RNA are rapidly degraded (which could provide some nice possible openings for experimental inroads into the *terra incognito* of nucleic acid degradation mechanisms). In addition, oocyte lysosomes may not be typical in some of their features (*cf.*, the discussion of yolk formation in Section V.3) and it also is worth noting that some proteins injected into oocyte cytoplasm are hydrolyzed fairly rapidly (DEHN and WALLACE 1973). These questions are of more than academic interest since from experiments with model systems it appears feasible to introduce specific molecules into the cytoplasm of cells by encapsulating the molecules within some phospholipid-rich liposomes that can fuse with the cells surface (*cf.*, the article by S. BATZRI and E. D. KORN, 1975: Jour. Cell biology **66**, 621—634); one might hope, in this way, to accomplish enzyme replacement, or other therapeutic intervention, for the cytoplasm outside the lysosomes.

IV.2.2. Potentially Instructive Failures of Defenses

Microorganisms and viruses have evolved a number of mechanisms by which they evade destruction, ranging from the synthesis of toxins that attack potential "defending" cells (ALLEN 1969) to the "invasion" of lysosome-poor cells, as in the case of the malarial parasites' entry into erythrocytes. TRAGER (1974) claims that certain intracellular parasites such as the Microsporidia avoid phagocytosis and are "injected" directly into the cytoplasm but this needs further study. Especially malevolent are the mechanisms that permit some organisms to survive and poliferate within lysosomes. This can, for example, result in the use of the relatively long-lived migratory phagocytes such as macrophages, as vectors for spread of the infection within tissues of the host (SMITH 1972) and perhaps also in restricting the contact of the organisms with the immune system or drugs (MANDEL 1973). Perhaps the organisms also depend on the hosts' lysosomal contents as sources of nutrients.

We have already mentioned reovirus which utilizes lysosomal proteases for uncoating. There also are bacteria such as strains responsible for leprosy, and protozoa such as *Leishmania* (MAUEL *et al.* 1973, TRAGER 1974) that thrive within phagocytic vacuoles (see ALLEN 1969 for review); some of these organisms survive the bactericidal and hydrolytic agents within the vacuoles partly as a consequence of the resistance of their cell coats to the lysosomal enzymes (for leprosy bacteria see ALLEN *et al.* 1965, D'ARCY-HART *et al.* 1972; for additional discussion see DRAPER and REES 1970, DRUTZ *et al.* 1974). Particularly important for our concerns are the several cases in which phagocytosed microorganisms seem able to elude exposure to hydrolases. This seems to be the situation with the protozoan *Toxoplasma gondii* (ANDERSON and REMING-

TON 1974, JONES and HIRSCH 1972, JONES *et al.* 1972) and with bacteria such as *Chlamydia psittaci* (FRIIS 1972), which are implicated in diseases of vertebrates. Previously killed organisms of these species are taken up by cultured cells in normal fashion and are degraded within phagocytic digestive vacuoles. But with exposure of the same mammalian cells to viable organisms, a substantial proportion of the bacteria or protozoa come to reside in phagocytic vacuoles that do not acquire hydrolases (see also Fig. 50 for an experiment in which the lysosomes were prelabelled with Thorotrast to demonstrate the absence of lysosomal fusion with vacuoles containing living *Toxoplasma*). Comprehension of the mechanisms by which the microorganisms exert their influences on the cell would be quite valuable for understanding processes governing the normal movement and fusions of lysosomes. But very little is known beyond such observations as the presence of layers of endoplasmic reticulum and mitochondria around the vacuoles that contain living *Toxoplasma*.

Recently, EDELSON and COHN (1974) and GOLDMAN (1974 a) have observed that binding of lectins such as concanavalin A to the plasma membrane inhibits subsequent fusion of lysosomes with phagocytic vacuoles formed from the lectin-charged membrane. As with *Toxoplasma*, the implication is that events occurring on one side of a membrane bounding a phagocytic vacuole, can affect the properties of the other side of the membrane or of the adjacent cytoplasm. LOWRIE (1975) has found macrophages that have phagocytosed living *Mycobacteria*, to contain elevated levels of cyclic-AMP. He speculates that this nucleotide may be responsible for the observed lack of fusion of lysosomes with vacuoles containing the phagocytosed organisms. If so, it will be particularly important to follow up his initial findings suggesting that it is the bacterium itself that generates the cyclic-AMP.

In recent years there has been growing interest in symbiotic relationships in which one cell type lives within another. It is speculated that such intracellular symbiosis may have been responsible for the evolutionary origin of membrane-delimited organelles such as mitochondria and plastids. But only a little experimental work has been done on lysosome behaviour in cases of intracellular symbiosis. For example, KARAKASHIAN and KARAKASHIAN (1973) have described the survival of certain phagocytosed algae within *Paramecia*, but the underlying mechanisms are still to be elucidated. [For other descriptive studies of symbiosis and of survival of phagocytosed cells or organelles, see

---

Fig. 50. Portion of a mouse peritoneal macrophage that was grown for several hours in the presence of the electron-dense colloidal tracer, Thorotrast and then exposed for one hour to *Toxoplasma gondii*, a parasatic protozoan. The tracer is seen in numerous bodies that probably are secondary lysosomes (*S*). One of these seemingly has just fused with a vacuole containing a *Toxoplasma* (arrow): the protozoan (T_1) is abnormally electron dense, and is dead or dying. In contrast, the *Toxoplasma* at T_2 seems healthy, and no Thorotrast is present in the vacuole surrounding it, indicating that no secondary lysosomes have fused with this vacuole. One frequently finds living and dead protozoa within the same cell; apparently whatever it is that inhibits fusion, operates at a local level rather than a cell-wide one. $\times 36,000$. (From JONES, T. C., and J. G. HIRSCH, 1972: J. exp. Med. **136**, 1173—1194.)

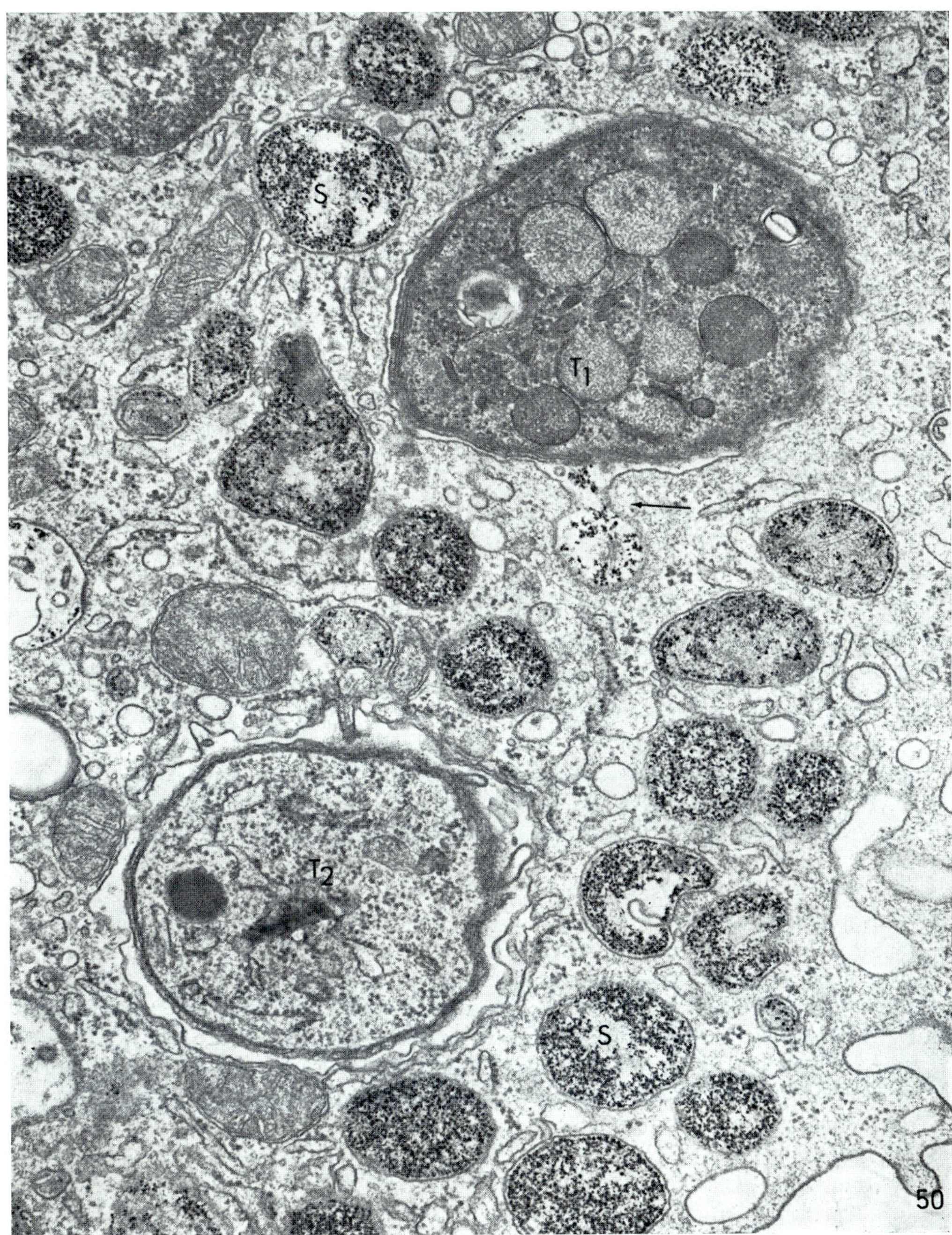

Fig. 50.

e.g., TRAGER 1974, MARGULIS 1970, MUSCATINE 1970, and LOUIS and GIANOTTI 1974. Occasionally (*cf.,* LOUIS and GIANOTTI) the claim is made that certain microorganisms within other cells are surrounded only by their own membranes and walls, as if they had escaped from host cell vacuoles. This is, of course, a plausible possibility, given the discussions above. But, the evidence must be scrutinized carefully; in some of the cases, the intracellular organisms, like their extracellular counterparts, are surrounded by fairly complex arrays (*e.g.,* multiple membranes), and it is difficult to rule out a host cell contribution.]

IV.2.3. Malaria

The lysosomes of the infectious parasitic protozoa (*Plasmodia*) responsible for malaria, are of special interest. The parasites enter red blood cells and thrive within the erythrocyte equivalent of phagocytic vacuoles (TRAGER 1974). They digest hemoglobin in heterophagic vacuoles that are generally thought to be lysosomes (WARHURST 1973) although histochemical studies have not given consistent results (*cf.,* LEVY and CHOU 1973, SCORZA *et al.* 1972, WARHURST 1973) due perhaps to problems in achieving adequate fixation. Antimalarial drugs such as chloroquine have striking effects on the plasmodial lysosomes—the intracellular pigment bodies resulting from hemoglobin digestion clump and are incorporated into a large structure, seemingly a sort of autophagic vacuole that also includes other portions of the plasmodial cytoplasm (ALLISON 1968, ALLISON and YOUNG 1969, WARHURST and HOCKLEY 1967, MACOMBER *et al.* 1967). Degradation of ribosomal RNA ensues; this effect, and the "starvation" of the parasite, deprived of adequate functioning lysosomes, could contribute to the demise of the organism. According to autoradiographic studies by AIKAWA (1972), chloroquine is concentrated within lysosomes of plasmodia within erythrocytes. HOMEWOOD *et al.* (1972) have advanced the interesting hypothesis that the drug, being a moderately strong base, can raise the pH within the digestive vacuoles in which it accumulates and thus affect hydrolase activity. WARHURST, however, believes that the labelling observed by AIKAWA was over autophagic structures, rather than heterophagic vacuoles (WARHURST 1973); that is, the accumulation of chloroquine within lysosomes might occur at a late stage of the effect of the drug. WARHURST also points out (personal communication and 1973) that there are inadequacies in the explanation of primary chloroquine effects on the basis of alterations in lysosomal pH; for example, when the antimalarial agent quinacrine, which is related to chloroquine, is administered in dilute solutions (10^{-7} M) it can be seen by fluorescence microscopy to accumulate at the limiting membrane of the parasite and not in the heterophagic vacuoles. DE DUVE *et al.* (1974) suggest that chloroquine may directly inhibit lysosomal hydrolases.

SCHNITZER *et al.* (1972) report that some of the plasmodia infecting monkey red blood cells can be removed by "pitting" in the spleen (*cf.,* Section III.1.1.2).

IV.3. The "Pathological" Release of Enzymes to Extracellular Spaces; Arthritis, Inflammation, and Related Phenomena

In various forms of arthritis, cartilage or bone matrices are extensively eroded. Synovial fluids from arthritic joints show elevated levels of extracellular hydrolases. These types of observations coupled with the findings that acid hydrolases accumulate extracellularly during autolysis of explanted cartilage or bone rudiments (Section III.2.1.1) have led to the concept that degradative aspects of the damage to connective tissues in arthritis are due in large measure to abnormal release of lysosomal enzymes from cells of the perichondrium or synovium, from chondrocytes and from phagocytes that enter the damaged tissue (reviewed by DINGLE 1973, LACK 1969, PAGE-THOMAS 1969, WEISSMANN 1972, ZIFF 1973, see also ALI and EVANS 1973, ARSENIUS et al. 1971, BARTHOLOMEW and PERCY 1973, HAMERMAN et al. 1969, HUFFER 1973, ORANSKY et al. 1973, PEARSE 1974). Particularly from the work on explants it is argued that much of this release is from viable cells.

In inflammatory responses to tissue injury (ROSS and ODLAND 1968) or to the presence of irritants, and in related phenomena, such as the Arthus reaction accompanying the formation of antibody-antigen complexes (GRANT et al. 1967, SOROKIN et al. 1970) a sequence of cellular mobilizations and migrations to the affected sites in the organism takes place. Often PMN leukocytes and other granulocytes accumulate first and then monocytic precursors of macrophages appear in increasing numbers[22]. The details and the outcome of inflammatory events depend upon the nature of the insult, the status of the immune system and many other factors (see e.g., SPECTOR and RYAN 1970, VAN FURTH 1970, VAN FURTH et al. 1973, VERNON-ROBERTS 1972, ZWEIFACH et al. 1974, for descriptions of various types of inflammations such as certain chronic categories in which indigestible irritants may be stored for prolonged periods within macrophages that remain at the site of inflammation, released when the cells eventually die and then phagocytosed anew). The processes of sequestration of irritants and clearing of debris can result in release of lytic enzymes into extracellular spaces. To some extent this is almost inevitable, particularly in acute inflammations or when infectious organisms with cytotoxic capacities are present and extensive death occurs in the phagocyte population (see e.g., ROJAS-ESPINOSA et al. 1974, for a description of cathepsin release from dying macrophages in tuberculosis lesions). And, of course, the damaged tissues themselves can release enzymes as their cells die. But recent work has provided clear indications that hydrolases can also be "secreted" from viable phagocytes. As alluded to earlier, PMN leukocytes and macrophages treated with cytochalasin B and then exposed to stimuli that normally induce phagocytosis (e.g., antibody-coated red blood cells or certain components of the immune "complement" system of

---

[22] The granulocytes are much shorter lived than the macrophages and the latter cells predominate in the later stages of inflammation; as the influx of macrophages increases one often finds them phagocytosing dead PMN leukocytes; ROSS and ODLAND 1968, WISSE 1974.

the serum) release hydrolases through exocytic-like fusion of lysosomes with the cell surface (Fig. 51). A comparable "massive" release from PMN leukocytes not exposed to agents like cytochalasin B can occur under some conditions of contact with immune complexes, a fact that is of obvious potential interest both for responses to exogenous inflammatory stimuli and for auto-immunity disorders. Normally, such complexes can be phagocytosed and digested in the cell, but when PMN leukocytes encounter Millipore filters covered with large aggregates of antigen-antibody complexes (comparable to surfaces of glomeruli or blood vessels in some forms of nephritis and other human diseases; HAWKINS and PEETERS 1971, HENSON 1972) they spread over the surface as if trying to surround it. Then their cytoplasmic granules fuse with the plasma membrane adjacent to the filter as though they have been "fooled into believing" that a phagocytic vacuole had formed [23]. Macrophages also engage in extensive hydrolase release when confronted with materials such as components of group A-*Streptococcal* walls (ACKERMAN and BEEBE 1974, CARDELLA *et al.* 1974, DAVIES *et al.* 1974). Less dramatic, but perhaps of more general significance is the limited "leakage" of enzymes that can accompany normal phagocytosis of many materials (including immune complexes in suitable forms) as a result of the fusion of lysosomes or other granules with forming phagocytic vacuoles that are not yet separated completely from the cell surface. [For reviews and correlations of hydrolase release and human disorders see Ann. N.Y. Acad. Sci. 256, 1–450 (1975).]

Some major features of enzyme release from phagocytes remain to be worked out. For example, cultured macrophages can secrete a good deal of lysozyme (GORDON *et al.* 1974) but the intracellular source of the enzyme, and the mechanism of secretion are not known (GORDON *et al.* 1974, LEAKE and MYRVIK 1964, STERNBERGER *et al.* 1970); release is not dependent upon phagocytosis. Clearly, the lysosomes are not the only potentially significant source of extracellular enzymes—the specific granules of PMN leukocytes are one obvious alternative, and the prior discussion of the release of proteases of non-lysosomal origin (Sections III.2.1.1 and III.2.1.3) should not be forgotten. For macrophages, ALLISON (1970, 1974) and co-workers have shown that gold taken up during a prior round of endocytosis is not released when cytochalasin B-treated cells are induced to secrete acid hydrolases; this suggests that the secretion may involve some special subpopulation of macrophage lysosomes. (As we have previously noted, that macrophages retain their secondary lysosomes "makes sense" in terms of sequestration of injurious materials.) (For macrophage hydrolase secretion see J. Exp. Med. **142**, 346; 361.)

Little is known in detail of relevant contributions of extracellular enzymes made by eosinophils, basophils or other pertinent cell types (although mast

---

[23] There are other types of situations in which phagocytes have to deal with bodies too large for uptake. When they encounter such bodies, groups of phagocytes may surround them and release hydrolyses onto their surfaces. See *e.g.*, KALINA *et al.* (1971) for a description of such handling by macrophages and PMN cells, of yeast encapsulated in a large wall. HIBBS *et al.* 1974 have also asserted that under some conditions macrophages can kill other cells by somehow releasing hydrolases into them. The existence of this unusual mechanism must be verified by more detailed study.

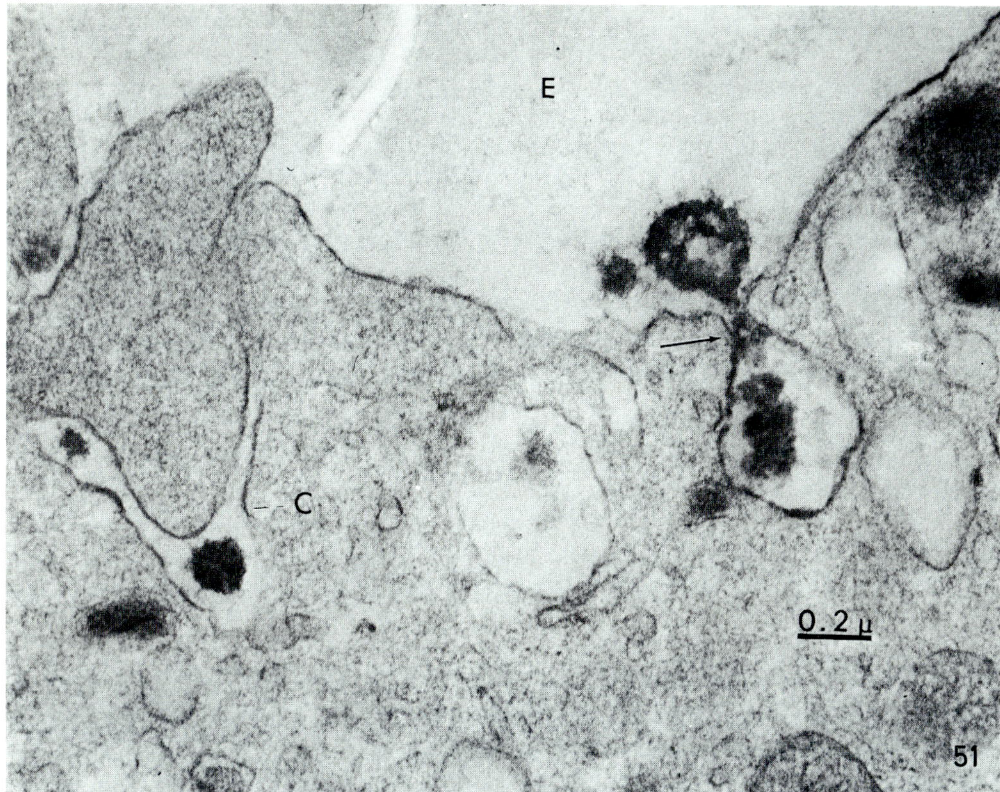

Fig. 51. Surface of a human PMN leukocyte treated with cytochalasin B and then with zymosan to induce release of lysosomal contents to the extracellular space (*E*). The preparation was incubated to demonstrate the endogenous myeloperoxidase activity of the azurophil granules (lysosomes). The arrow indicates a configuration interpretable as a peroxidase-positive granule that has fused with the cell surface and is releasing its contents. The irregular channel at *C* almost certainly is continuous with the cell surface; it contains what seems to be the content of another peroxidase-positive granule. ×40,000. (Courtesy of S. Hoffstein; from Zurier, R. B., S. Hoffstein, and G. Weissmann, 1973: J. Cell Biol. **58**, 27—41.)

cells seem capable of releasing proteases; Park *et al.* 1973, Vensel *et al.* 1971; see also Zweifach *et. al.* 1974, for reviews and background). Platelets are suspected to participate in various ways in conditioning the extracellular environment at inflammatory sites. They do contain an appreciable population of lysosomes and can release hydrolases (Bentfield and Bainton 1974, Chesney *et al.* 1974, Droller *et al.* 1974, Hawkins 1972, Henson 1970, Holmsen *et al.* 1969, Legrand *et al.* 1973, Nachman and Weksler 1972, White and Estensen 1974 b). Under a number of experimental conditions, platelets engage in activities resembling phagocytosis but in at least some cases this amounts more to trapping of particles in tubular systems continuous with the cell surface than to true uptake into intracellular vacuoles (Mant and Firkin 1972, Movat *et al.* 1965, White 1972 a). Perhaps sometimes this process eventuates in leakage of hydrolases.

These various uncertainties are marginal to the central conclusion that extracellular enzymes must be considered in analysis of a variety of disorders. It is noteworthy that injection of disrupted lysosomes can produce sustained inflammations (WEISSMANN *et al.* 1969) and that some of the other materials known to engender inflammatory responses, such as bacterial cell wall components or dental plaque (DAVIES *et al.* 1974 b) can be shown also to evoke extensive release of acid hydrolases from *in vitro* populations of phagocytes. The direct damage that could be done by extracellular lysosomal hydrolases or other enzymes would depend on several incompletely understood factors such as the pH levels in extracellular fluids and the effects of circulating protease inhibitors. POOLE (1973), for example, suggests that invasive tumors, in which necrosis produces high levels of extracellular enzymes, may maintain an acid environment as a result of glycolytic production of organic acids which pass out of the cells. The enzymes could facilitate disruption of tissues as the tumor invades (FISER-SZAFARZ and SZAFARZ 1973). According to ACKERMAN and BEEBE (1974) and JANOFF (1972), individuals deficient in a circulating antitrypsin inhibitor often come down with panlobular emphysema; circumstantial evidence suggests that in this disorder, elastin in the lung is degraded by hydrolases released from alveolar macrophages or invading PMN leukocytes [24]. One can even conceive of circumstances in which extracellular myeloperoxidase derived from leukocytes could catalyze cytotoxic halogenation reactions (EDELSON and COHN 1973). And VAN SNICK *et al.* (1974) propose that iron deficiencies in acute inflammations may be due in part, to the "massive" release of lactoferrin from leukocytes; hyposideremia could ensue as the lactoferrin acquires iron from transferrin and is subsequently degraded by the reticuloendothelial system.

The presence of extracellular hydrolases might also have some beneficial or useful aspects. The enzymes could, for example, aid in combatting infections at sites of inflammation, or normally, at locales such as the lining of the respiratory tract. And in the views of PARR *et al.* (1967), the contraceptive effects of intrauterine devices depend in large measure on enzymes released from PMN leukocytes that accumulate in response to the foreign body.

As inflammatory lesions develop, various "factors" appear at the site or in the circulation. These include, among others, pyrogens, vasoactive substances, chemotactic agents and materials that evoke histamine release from mast cells (see *e.g.*, MICKENBERG *et al.* 1972, WARD 1969, WEISSMANN 1967). Many of the factors almost certainly are proteins or peptides (although few have been fully characterized and this whole facet of inflammation is still quite murky, particularly in terms of details of mechanisms). The contributions of lymphocytes and other components of the immune system, such as complement, are often prominent. But, the phagocytes we have been discuss-

---

[24] The elastolytic enzyme thought to cause damage to tissue elastins in arteritis, emphysema and other disorders (JANOFF and BLONDIF 1973, JANOFF and ZELIGS 1968) can also hydrolyze certain cross-links involving alanine in bacterial cell walls. This hints at some interesting evolutionary balances between advantageous and disadvantageous aspects of the possession of such an enzyme.

ing seem also to be important. For example, PMN leukocytes *in vitro,* exert chemotactic and other influences on the movement of other PMN's and of macrophages (VAN FURTH 1970, ZIGMOND and HIRSCH 1972). In addition, leukocytes and platelets reportedly release cationic proteins or other factors that stimulate histamine release (HENSEN 1972, NACHMAN and WEKSLER 1972). The intracellular sources of the chemotactic or releasing factors are not known; the possibility that some are released along with the lysosomal hydrolases and other enzymes is certainly open. Further, the enzymes themselves may be important in these respects (*e.g.,* ALI and EVANS 1968, MOVAT *et al.* 1973). Certain circulatory factors require proteolytic activation before becoming biologically active (*e.g.,* Mayer 1973). Thus, the complement component C'-3 is cleaved to leukotactic fragments when incubated with proteolytically active extracts of rat tissues (HILL and WARD 1969). And macrophages and other cells can release proteases that activate serum plasminogen, converting it to the fibrinolytic enzyme, plasmin (BARNHART 1965, GORDON *et al.* 1974 b, UNKELESS *et al.* 1974 a, b); the latter enzyme not only can dissove clots but it also can generate vasoactive peptides and other active factors, raising the possibility of an elaborate cascade of effects. GREENBAUM (1972) summarizes evidence that leukocytes contain enzymes that can produce active components ("kinins") from plasma proteins and also enzymes that can degrade them.

Since the lysosomal hydrolases can degrade hormones and similar mole-cules, one might expect the local environment at an inflammatory site to differ from adjacent normal tissues in the concentrations of such potentially signifi-cant agents as the hypertensive factor, angiotensin (MCDONALD *et al.* 1974). There could be far-reaching consequences for controls of blood vessel perme-ability, local circulatory patterns, aspects of cellular metabolism and so forth (see also YANG *et al.* 1973 for findings on the degradation of hormones by extracellular enzymes in insects; one can envisage a sort of feedback system wherein cell death triggered by hormones releases hydrolases that degrade the hormones).

Finally there are some fairly profound consequences of hydrolase release that are not beyond the realm of possibility, although they are still little more than conjectures. Proteins or other macromolecules altered by exposure to the enzymes might have new antigenic features that could set off autoimmune phenomena (WEISSMANN 1965); direct evidence concerning this is hard to come by, but there is some speculation concerning several disorders that can be stretched to accomodate this concept (see *e.g.,* BROSTOFF *et al.* 1974 for allergic encephalomyelitis). If some of the propositions about turnover described earlier prove correct, then increases in levels of extracellular enzymes might entrain altered rates of turnover of circulating molecules or erythrocytes [25]. In addition, it is observed that when cells are grown in

---

[25] BRADY and QUARLES (1973) have put forth the imaginative suggestion that in normal myelinogenesis, extracellular hydrolases are responsible for altering Schwann cell or glial cell plasma membranes to give rise to membranes of the myelin sheath. If normal membrane modulations actually occur in this way, long-term changes in extracellular hydrolase levels, as in some chronic conditions, could lead to all sorts of mischief.

media containing hydrolases, interesting changes in their properties take place, ranging from dramatic ones such as modified shape and mobilities (OSSERMAN *et al.* 1973) or stimulation of proliferation (SCHEBLI and BURGER 1972, POSTE 1971, 1972) to such "subtle" changes as the alteration in the type of collagen synthesized by cartilage cells when they are exposed to acid hydrolases (DESHMUKH and NIMNI 1973). It is assumed that some of the enzymes' effects follow from their attacking specific cell-surface constituents. In extending this analysis, a number of authors have proposed that secreted (or cell-surface) proteases and other enzymes released from altered or sublethally injured cultured cells (BOSMAN 1974, POSTE 1971) can be important in the responses of the populations to various perturbations, including for example, oncogenic transformation (BOSMAN *et al.* 1974). Correlations have been noted between the tumorigenicity or other features of transformed cells, and the levels of release of proteases that can activate plasminogen (see *e.g.*, CHRISTMAN *et al.* 1974, RIFKIN *et al.* 1974). Experimental efforts to explore relevant influences of endogenous proteases have been based upon addition to cell cultures of what are hoped to be agents that specifically inhibit classes of proteases (*cf.*, Section III.4; BOREK *et al.* 1973). This can, for example, reduce growth rates of transformed cells (COLLARD and SMETS 1974, SCHEBLI and BURGER 1972) or alter cellular adhesiveness (WHUR *et al.* 1974). But one is given pause by observations such as those of DARZYNKIEWICZ and ARONSON (1974) who found that some supposed effects of proteases at cell surfaces are not seen with exogenous enzymes bound to inert beads. Overall, while the work on supposed protease-induced, cell surface-mediated effects on cells has turned up some fascinating phenomena, there still are too many complications and possible nonspecificities and too many problems in distinguishing cell surface events from intracellular changes for it to provide definitive detailed explanations for sequelae of hydrolase release *in vivo*.

IV.4. The Intracellular Release of Hydrolases: Lysosome Fragility, Labilizers, and Stabilizers

IV.4.1. Methodological Problems

Effects of a variety of agents and influences on cells have been tentatively attributed to intracellular release of hydrolases as a primary event—proposals of this type have been made for phytohemagglutinin mitogenesis (WEISMANN *et al.* 1972, 1973), chemical carcinogenesis (ALLISON 1969), killing of cells by photoreactive dyes or silica (ALLISON 1970, ERICSSON *et al.* 1974), the cytotoxic actions of bacterial endotoxins and viruses (WEISMANN 1967), and other situations (*e.g.*, SLOR *et al.* 1973, SZEGO and SEELER 1973). For some of these cases it is argued that lysosomal rupture or perforation is the fatal lesion. For others, it is hypothesized that limited enzyme release into the cytoplasm or nucleus produces sublethal injuries or other alterations. The literature is a frustrating one because so many of the schemes proposed are attractive (occasionally they are ingenious) but definitive evidence for primary involvement of lysosomes is exceedingly hard to acquire.

In most studies, changes in the distribution of hydrolases in cell homogenates are adduced to support the investigator's viewpoint. In particular, the ratios of soluble to particle-bound enzymes are given close attention (see *e.g.*, SLATER 1969 for a review). But by its nature, such data cannot provide a critical test. First, when cells are dying, one would expect hydrolases to be released, irrespective of what had killed them and so, causal relations cannot be determined. Second, homogenization of tissues, even by the best designed of methods inevitably ruptures at least a few organelles. For example, when one observes an increase in lysosomal enzymes in the supernatant of homogenates of cells responding to some non-fatal perturbation, it is rarely possible to distinguish even tentatively among alternatives: have hydrolases been released within the cells? ... have a few cells died? ... is there a change in the frequencies of different categories of lysosomes or in the sizes of lysosomes that might affect their resistance to rupture during homogenization?

Cytochemistry is of relatively little help here. There are notorious artifacts that entail diffusion of enzymes or of reaction product within cells, leading to "false" localizations (BRUNK and ERICSSON 1972, GOLDFISCHER *et al.* 1963, HOLT 1959). It is not at all uncommon for incorrectly prepared tissues to show reaction product in their nuclei or diffuse in the cytoplasm when there is no reason to suspect the presence in these locales, of the enzyme being studied. And even when artifact is unlikely, cytochemistry cannot resolve the cause-effect ambiguities.

Clever efforts to surmount some of these difficulties and to obtain cytochemical information about "subtle" pathological alterations in the properties of lysosomes have been made. For example, following leads developed by BITENSKY (1963), several laboratories have attempted to quantitate the formation of reaction product in cytochemical preparations by carefully monitoring the time at which product first becomes visible (a test of cytochemical "latency") and more recently through optical determinations of rates of product deposition. When such approaches are used on cells and tissues subjected to different regimens *in vivo* but then put through cytochemical procedures kept as comparable as possible, reproducible differences are obtained, at least in some hands (BITENSKY 1973, REED and WENZEL 1975, GAHAN 1967, 1971). Sometimes the changes in rates of reaction product deposition are referred to as reflecting alterations in lysosome "fragility" or membrane stability since it was proposed by BITENSKY that lysosome permeability is a decisive element, determining the rates of delivery of cytochemical substrates to the enzymes. The finding that after various stresses, or in diseases, tissues may show faster staining with no evident change in hydrolase quantity is interesting (see *e.g.*, GAHAN 1965 for a study of hepatic cells after ether anaesthesia and YATES *et al.* 1970 for work on retinal pigment epithelia in retinosis pigmentosa). But it is not really clear how observations of this type are to be interpreted. There are some major technical points requiring clarification. The processes by which cytochemical reaction product forms a visible precipitate are incompletely understood. And, the influences of the permeability properties of the plasma membrane on the rates of substrate penetration to the lysosomes require exploration (SILVERSTEIN 1970); there is no conclusive evidence in the work under discussion that it is the lysosome membrane that constitutes the rate-limiting barrier. However, even if matters of this sort are left aside, the implications of changed lysosome "fragility" for cell function are hardly self-evident. One can speculate that were the lysosome membrane to become more permeable, osmotic or pH changes in the organelles might affect their internal environment and small molecules of various types might pass in or out more readily, but there is very little to go on [26]. We will return to this question in a slightly different guise when we take up "labilizers" and "stabilizers".

ERICSSON and his co-workers (HAWKINS *et al.* 1974) are trying to look at suspected intracellular lysosomal enzyme leakage by use of macromolecular tracers, such as ferritin or Thorotrast that are introduced in the lysosomes through heterophagy prior to exposure of the cells to injurious treatments. Appreciable escape of the tracers into the cytoplasm presumably would occur only as a consequence of some (catastrophic?) change in the lysosome membrane (see also NADLER and GOLDFISCHER 1972). Some of the results are perplexing: for example, dying tissue culture cells or cells fixed in a solution of low osmolality may show reaction product for acid phosphatase diffuse throughout the cytoplasm relatively early, at a time when the tracers remain confined within the lysosomes (BRUNK and ERICSSON 1972 b, ERICSSON *et al.* 1974). One interpretation put forth is that the hydrolases have escaped through "morphologically intact" membranes but since little is established about the possibilities for binding of the tracer molecules within lysosomes, definitive attribution would be premature. In following up proposals by ALLISON and other (ALLISON and YOUNG 1969, SLATER and RILEY 1966) on light-induced killing of cells previously "loaded" with "photosensitizing" dyes, such as acridine orange, ERICSSON, BRUNK, and ARBORGH (1974) have argued from their cytochemical work that release of hydrolases is an early event, and probably is responsible for killing the cells. An essential consideration for the latter inference is that with appropriate concentrations and timing, the dye becomes sequestered within lysosomes without conspicuous accumulations elsewhere; thus light-induced, dye mediated photoreactions leading to damage to membranes would be likely to affect primarily the lysosomes. But the caveat is often advanced that inconspicuous accumulations of dye at the plasma membrane or nucleic acids might go undetected and nonetheless represent the site of the initial fatal lesion in the cell. To help counter this argument, ERICSSON (personal communication) points out that the obvious manifestations of the destructive effects of acridine orange on cells do not occur at low temperature, even in cells that were initially exposed

---

[26] A curious effect noted by GOLDMAN (1973) may be of interest in this connection. She found that when isolated lysosomes are exposed to low concentrations of dipeptides or methyl esters of amino acids, they often rupture. Apparently the esters and dipeptides enter the lysosomes and there they are hydrolyzed. The amino acids produced by the hydrolysis are zwitterions, and GOLDMAN feels that they penetrate lysosome membranes more poorly than do the progenitor dipeptides and esters; the result is the progressive accumulation of amino acids in the lysosomes, leading to an osmotic influx of water. Effects of this type could amplify the influences of changes in lysosome permeability.

to the dye at normal temperatures. This is not intended as a definitive clue to initial events, but it does suggest that the actual disruption of the cell depends on enzymatic processes rather than physicochemical ones such as osmotic influx or efflux of water and ions through a damaged cell surface. Another bit of evidence that may eventually prove useful is the fact that light can induce hydrolase release from isolated dye-loaded lysosomes (WILLIAMS and SLATE 1973). [Light induced cell-death may be of practical interest since there are metabolic disorders in which photoreactive molecules seem to play a causal role; see *e.g.*, WILLIAMS and SLATER's (1973) discussion of a disease in cattle related to a by-product of chlorophyll metabolism, and ALLISON 1968.]

Acquiring evidence for lysosome roles in injuries that are not lethal to the cell is especially problematic since the supposed injuries are themselves usually conjectural and the presumed sublethal intracellular release of hydrolases might involve only a few molecules. Even if such release does occur, there are factors yet to be assessed such as pH's (POOLE and DE DUVE 1973) and the presence of inhibitors that could prevent the action of the hydrolases (some of these inhibitors are found in the cytoplasm outside the lysosomes; DAVIES *et al.* 1971, JANOFF 1972, LAZARUS *et al.* 1972, LENNEY *et al.* 1974).

IV.4.2. Silica and Uric Acid

Probably the most convincing case for killing of cells by rupture of their lysosomes has been made for the cytotoxic effects of certain types of silicon dioxide particles (ALLISON 1970, ALLISON *et al.* 1970); that some of the evidence is circumstantial is perhaps unavoidable given the limitations of current techniques.

When silica particles of phagocytosable dimensions are introduced into the medium, cultured macrophages or other cells are killed. Microscopic observations indicate that the particles are phagocytosed, suggesting that the cells are viable, at least up to this stage. But as the cells die the particles appear to escape from the phagocytic vacuoles into the cytoplasm. That the lysosomes rupture is also indicated by the experiments of NADLER and GOLDFISCHER (1970) who showed that horseradish peroxidase previously endocytosed by cultured macrophages becomes diffuse in the cytoplasm when the cells are exposed to silica (NOVIKOFF *et al.* 1972, have pointed out that even with peroxidase, diffuse cytochemical reaction product must be regarded with some skepticism but for the present case, artifact seems only an outside possibility). Particulate silica in proper form can be membrane active (*e.g.*, hemolytic effects have been reported; *e.g.*, SCHNITZER *et al.* 1971) apparently through hydrogen bonding and other surface chemical reactions. As expected from this, when the particle surfaces are covered with non-digestible polymers, the particles lose their cytotoxic capacities. Probably when it is introduced into culture media (or into organisms) particulate silica acquires a coating of denatured serum proteins which inhibits the effects on membranes until the coating is digested off by the lysosomal hydrolases.

Killing of alveolar macrophages by inhaled fibers and subsequent repeated release of the particles from dead cells and re-uptake could well be a major factor in silicosis. It could contribute heavily to the inflammatory responses that eventually stimulate collagen deposition by fibroblasts; this deposition and other processes generate respiratory symptoms and related features of the disease.

Asbestos fibers may produce lesions through a different train of events. DAVIES *et al.* (1974 a) report that they induce release of hydrolases to extracellular spaces.

Evidence analogous to that outlined for silica has been marshalled to support the contention that monosodium urate crystals also kill cells through "perforation" of the lysosomes (WALLINGFORD and McCARTHY 1971, WEISSMANN and RITA 1972). This could be significant for gouty inflammations.

IV.4.3. *Labilizers, Stabilizers, Drugs, and Inhibitors*

Lipid-soluble forms of vitamin A, such as retinol, have a marked stimulatory effect upon certain of the processes in which lysosomes participate (ROELS 1969, VAES 1969, WEISSMANN 1969). A good example is the increase in degradation of matrix seen in cartilage rudiment explants (Section III.2.1.1) and in rabbit ears exposed to excess retinol. Other compounds, notably some of the corticosteroids seem to have inhibitory effects on lysosome-related processes (DINGLE 1953, VAES 1969, WEISSMANN 1969). Cortisones are used as anti-inflammatory agents, and they often antagonize vitamin A effects. Lysosomes isolated from cells exposed to high levels of cortisones are less susceptible to disruption by a variety of treatments (less "fragile"); apparently they have been "stabilized" in some fashion. In a similar sense, vitamin A derivatives seem to "labilize" lysosomes (DINGLE 1963). Liver lysosomes are also affected oppositely by vitamin A and by cortisones under a number of experimental conditions in which the particles are exposed to the agents subsequent to isolation (DINGLE 1963, WEISSMANN 1969).

From the information just outlined, from work on artificial lipid systems and from evaluations of hemolytic capacities of the compounds in question, it seems likely that vitamin A, corticosteroids and other "labilizers" and "stabilizers" can affect lysosome membranes (*e.g.*, WEISSMANN 1969 for discussion of a number of steroids and non-steroidal drugs). Still to be determined however are the relations between these effects and the effects on cells and organisms. Clearly, if intracellular leakage of hydrolases turns out to be widely important for cell pathology, then the properties of the lysosome membrane that could influence leakage or intracellular rupture would be quite significant. Similarly, if lysosome permeability to small molecules is important for controlling interactions with the cytoplasm or in influencing, for example, the susceptibility of lysosomes to osmotic rupture, one can think of many ways in which known effects of labilizers and stabilizers could be mediated. For instance, BADENOCH-JONES and BAUM (1973) have reported data suggesting that progesterone can affect the permeability of isolated lysosomes to the extent that osmotic rupture occurs in solutions of sucrose, which normally cannot penetrate the membrane. Or, perhaps as a consequence of modified membrane fluidity, it could be the fusion of lysosomes with other membrane-delimited structures, that is chiefly altered by cortisone or vitamin A; this might influence phagocytic events, the "secretion" of hydrolases etc. (DAEMS *et al.* 1969, DINGLE 1968, LOCKHARD *et al.* 1973, MERTOW *et al.* 1968, WEISSMANN 1969). Unfortunately, the area is muddy since agents like cortisones have diverse actions on cells and organisms. For example, in appropriate experimental circumstances, corticosteroids can alter rates of enzyme synthesis in various tissues, change rates of pinocytosis, or influence the mobilization of phagocytes from the bone marrow (ALLISON and DAVIES 1972, BEISEL and RAPOPORT 1969, FAUVE 1972, RECHCIGL 1971 b, THOMPSON and VAN FURTH 1970; see also LOCKHARD *et al.* 1973 for supposed *in vivo* roles of corticosteroids in mediating stress-related effects on phagocyte capabilities). Sorting out the primary from the secondary and determining lines of causation is difficult. This is particularly true since "labilizers" and "stabilizers" can alter other classes of cellular membranes in addition to those of the lysosomes. It might even be argued that sometimes the changes in lysosome membranes are secondary consequences of modifications in other membranes that give rise to lysosome surfaces, such as the plasma membrane or endoplasmic reticulum. For example, LAURENT *et al.* (1975) propose that the diazafluoranthen derivative, AC 3579, accumulates in membranes of smooth ER, and then secondarily in lysosomes (through autophagy?). The drug is a largely non-polar compound and its effects seem to include the rendering of membrane phospholipids less susceptible to phospholipases. Overall, despite observations of a host of tantalizing phenomena, we still lack a clear picture of where and how the crucial actions of labilizers and stabilizers take place.

Many drugs and other compounds of interest to cell pathologists accumulate in lysosomes (*e.g.*, hydrocarbon carcinogens; ALLISON 1969, ALLISON and MALLUCCI 1964). Some, such as chloroquine, call forth enhanced formation of autophagic structures (ABRAHAM *et al.* 1968, COHN 1970, FEDORKO *et al.* 1969). It is always tempting to ascribe special importance to sites of the heaviest accumulation of agents being studied. Whether or not this is fully justified in particular cases, it does seem reasonable to hope that lysosomal accessibility to exogenous materials will prove a useful feature for therapeutic efforts and to worry about how the intralysosomal environment might affect drugs (DE DUVE 1972). Work on enzyme replacement was outlined in Section IV.1.1.[27]. Pioneering successful attempts have also been made to inhibit lysosomal enzymes through endocytic incorporation of anti-lysosome antibodies (DINGLE *et al.* 1973, TULKENS *et al.* 1970, WESTON and POOLE 1973), compounds such as Trypan Blue (WILLIAMS *et al.* 1973), or inhibitory lipid analogues (DAWSON *et al.* 1974). At present, such work is chiefly directed to

---

[27] An effort to clear lysosomal stores of toxic plutonium compounds by administration of chelating agents within liposomes has met with some success (RAHMAN *et al.* 1975). Precisely how the chelators and the lysosome contents interact, and how the plutonium then leaves the cells are still to be determined. But the experiments support the idea that administration of non-enzymatic materials, as well as enzymes may be a useful approach for efforts to clear out some intralysosomal stores. For example, maybe surface-active agents could render some lipid deposits more digestible.

analysis of lysosome functioning and to study of such topics as the teratogenic effects of Trypan Blue [LLOYD and BECK (1968) think these effects may have something to do with alterations in yolk-sac lysosomes]. But the approach might also open avenues for treatment of some disorders in which lysosomes are important. For example, WEISSMANN (1969) is interested in the fact that gold salts, useful in treating arthritis, also inhibit acid hydrolases. And, Triton WR-1339 apparently can "aid" phagocytes in dealing with tuberculosis bacilli, probably by intralysosomal effects of the detergent (ALLISON 1968, DE DUVE 1968).

Finally, the groundwork is being laid for design of "lysosomo-tropic drugs" (DE DUVE 1972) and treatment regimens that will make specific use of the tendancy for exogenous materials to be sequestered in particularly high concentrations in lysosomes of endocytically active cells. Experiments with liposomes as "endocytosable" carriers for drugs (GREGORIADIS 1973) may foreshadow work in which carriers will be designed to link with specific cell surface receptors in an organism, and thus to deliver agents to one or a few cell types. TROUET and co-workers (1972) have found that DNA-daunorubicon complexes are taken up rapidly by cultured tumor cells into heterophagic lysosomes. The chemotherapeutic agent (daunorubicon) is then released within the lysosome, apparently as a consequence of digestion of the DNA, and since when not complexed with macromolecules, daunorubicon can cross membranes, it now diffuses out of the lysosomes. One thus has the elements of a way to administer a drug that by itself is quite toxic, with decreased damage to the organism. Malignant cells often are quite active in endocytosis and therefore are good targets. DE DUVE *et al.* (1974) have reviewed the therapeutic uses of various types of agents that affect, or accumulate in lysosomes.

IV.5. Lysosomes in Immune Responses

There is little information pertaining directly to roles of lysosomes in immunology. However, phagocytes participate in diverse phenomena linked to immune responses (COOMBS and FELL 1969, VAN FURTH 1970, VERNON-ROBERTS 1972). Thus, in many types of responses [*e.g.*, in delayed hypersensitivity reactions (DAVID 1970, MACKANESS 1970)] factors such as the migration inhibitory factor released by lymphocytes aid in bringing large populations of leukocytes and macrophages to sites of foreign materials. The macrophages and granulocytes engulf antibody-antigen complexes and immunologically foreign cells (Fig. 52). In various circumstances, circulating antibodies opsonize materials for phagocytosis and "cytophilic" antibodies (TIZARD 1970) bind to the surfaces of phagocytes, "equipping" them to discriminate among the molecules and particles with which the cells come in contact (RABINOVITCH 1970). The fact should be borne in mind, that cytotoxic effects of the immune system probably do not necessarily require initial participation of phagocytes; direct complement-mediated mechanisms that seem to affect the cell surface are known to exist (ILES *et al.* 1973, KOLB and MULLER-EBERHARD 1973). On the other hand, macrophages are very much involved, at least as scavengers, in many circumstances of cell death engendered by immune

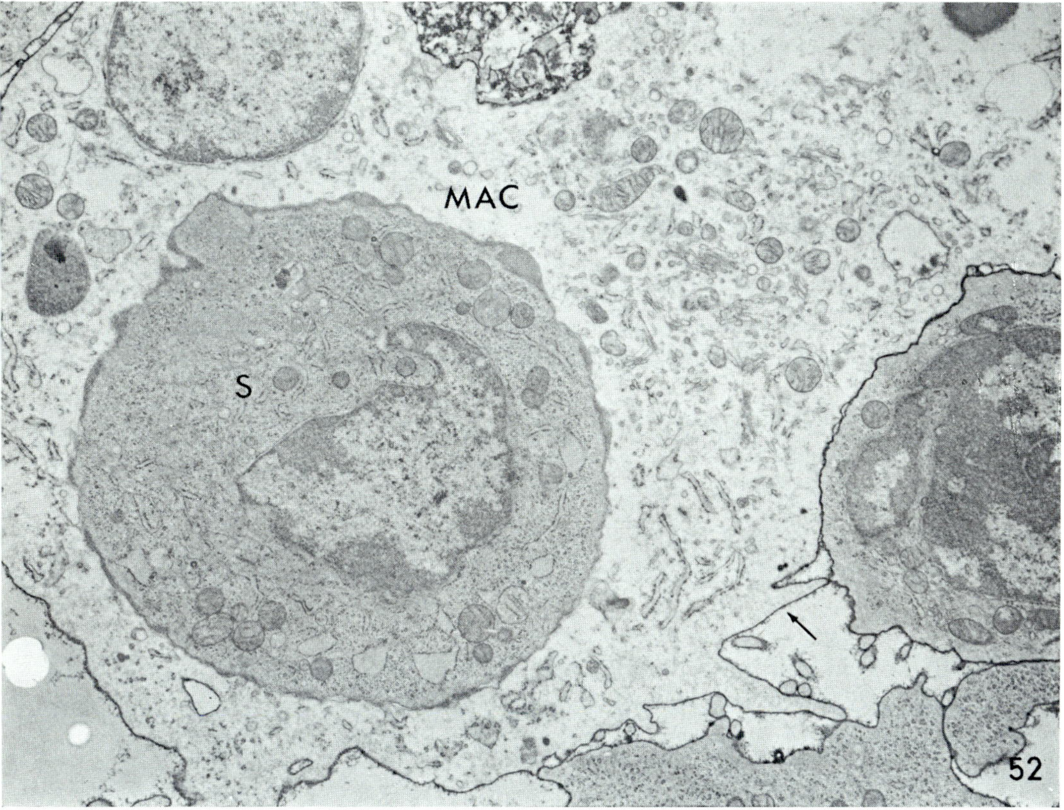

Fig. 52. Macrophages from the peritoneal cavity of a mouse that had previously received an injection of sarcoma I cells were plated in a culture dish and then exposed to sarcoma I cells in the presence of cell-free mouse immune ascitic fluid. The "alloimmune" macrophages (*MAC*) phagocytose the sarcoma cells (*S*). The fact that the engulfed cell in this figure is within a sealed-off phagocytic vacuole, rather than a deep infolding of the macrophage surface, is indicated by the observation that the electron dense tracer, ruthenium red, added with the fixative has stained the macrophage surface (arrow) but has not reached the enclosed sarcoma cell. ×7,500. (From CHAMBERS, V. C., 1973: J. Cell Biol. **57**, 874—878.)

responses, as well as in the steady-state turnover of lymphocytes in sites such as the thymus, where there is constant extensive death in the lymphocyte population (BOWERS 1969).

Are there important roles of phagocytes in addition to their functioning as "effectors" or scavengers? One possibility relates to the fact that antibody synthesis in an organism confronted with a given antigen can be critically dependent upon the quantity of antigen presented, and on the timing, especially in terms of the developmental status of the responding organism (DRESSER and MITCHISON 1968). Thus, neonatal mammals exposed to high levels of foreign proteins often turn out to be immunologically tolerant of these proteins later in life although naive adults challenged with the same proteins will respond by producing antibodies. Such tolerance phenomena may be linked to the ability of organisms to distinguish "self" from "non-self". NOSSAL and ourselves, among others, have shown that the barriers to penetration of antigens into potentially important sites, such as the parenchyma of the thymus, are markedly more permissive in neonatal animals than in older ones (GERVIN and HOLTZMAN 1972, NOSSAL *et al.* 1968 a, b). Thus, for instance, a far greater proportion of thymic lymphocytes is exposed to circulating exogenous horseradish peroxidase in newborn mice than in young adults (GERVIN and HOLTZMAN 1972). Actively endocytosing macrophages are a conspicuous element in the "barrier" system that appears to restrict entry of foreign proteins into the adult thymic parenchyma (Fig. 53; GERVIN and HOLTZMAN 1972, RAVIOLA and KARNOVSKY 1972). Perhaps the exposure of thymocytes to potential antigens in the neonatal animal contributes to subsequent tolerance. Macrophages are also important in the "filtering" functions of lymph nodes. [Since macrophage-lymphocyte interactions can be involved in inducing immune responses (Section IV.5.1) conclusions about the significance of apparent barrier activities of macrophages must be tentative; the cells may actually be playing subtles roles.]

Related to this discussion is the consideration that organisms may not be immunologically tolerant of some of their own intracellular components—when severe and massive cellular necrosis occurs, auto-immune phenomena can follow (DE HEER *et al.* 1974). It may be that the normal turnover and modulation mechanisms discussed in the last chapter and the scavenging activities of phagocytes in pathological conditions serve usually to prevent dangerous contacts of cell contents with the immune system. We have already mentioned the idea that components modified by lysosomal hydrolases might have abnormal immunological properties.

IV.5.1. Macrophages and Some Other Cells in the "Processing" or "Presentation" of Antigens

It has been known for many years that the efficiency of induction of immune responses by potential antigens frequently is enhanced by the simultaneous administration of antigen and "adjuvant" materials that evoke inflammatory responses (*e.g.*, the very widely employed Freund's adjuvant contains killed mycobacteria and emulsified oils). Perhaps this is partly a question of mobilization of potentially immunocompetent cells and of other "mundane" matters, such as the subdivision by phagocytes, of large aggregates of foreign substances into "manageable" portions. But it also may hint at some important functions of macrophages as "helper" cells: they might participate in the "processing" of antigens into special forms suitable for "reception" by lymphocytes, or in facilitating interaction of the B and T lymphocytes, or conceivably in transporting foreign materials to sites of potential antibody production (see VAN FURTH 1970 for reviews of various possibilities). In favor of such concepts one can cite observations that certain antigens are far more "immunogenic" when they are permitted to interact with macrophages and then the macrophages are injected into a suitable organism (ASKONAS and JARASKOVA 1970, KOLSCH 1970, MITCHISON 1969, PERKINS

1970). Related work with tissue culture systems, suggests that in some circumstances, macrophages must be present if lymphocytes are to respond "correctly" to antigens (EDELMAN and MOLLER 1973, FELDMAN and NOSSAL 1972, SHORTMAN *et al.* 1970).

The interpretation of information of this type is still being debated. For a time, it was felt that intracellular processing of antigens by macrophages might be responsible—such processing, perhaps involving the lysosomes,

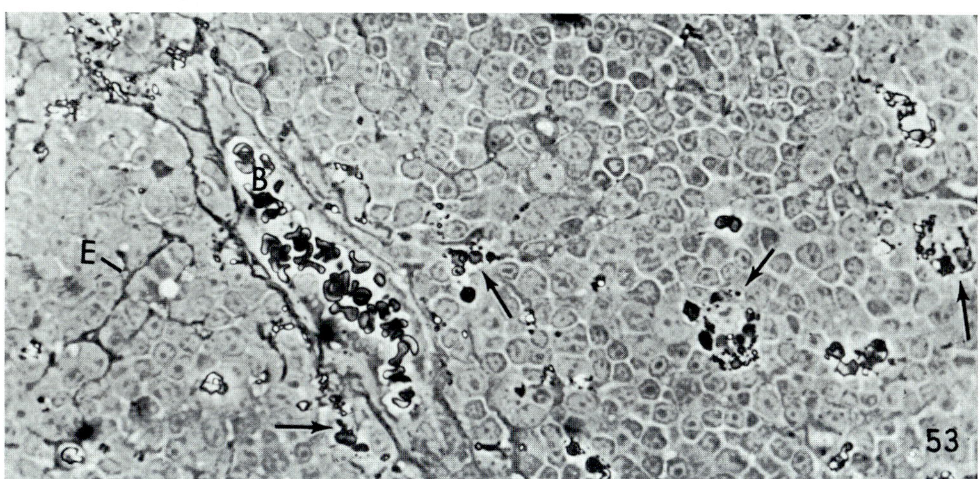

Fig. 53. Phase-contrast micrograph of a thick section prepared from an Epon-embedded preparation of the thymus of a young adult mouse; the animal had received horse-radish peroxidase intraperitoneally 6 hours before sacrifice and the preparation has been incubated to demonstrate peroxidase activity. Reaction product is seen in the extracellular spaces (*E*) surrounding a few thymocytes near a blood vessel (*B*; probably a venule), but most of the extracellular space is free of detectable tracer. Numerous macrophages containing exogenous peroxidase in intracellular globules are scattered throughout the parenchyma (arrows). ×600. (From GERVIN, S. W., and E. HOLTZMAN, 1972: J. Histochem. Cytochem. **20**, 445—462.)

could generate a crucial fragment of the antigen (or an informational nucleic acid) for transfer to a lymphocyte. But, the evidence for this sort of phenomenon is still very shaky. (For various viewpoints see the discussion following BOWERS 1970, and FISHMAN and ADLER 1970, HUMPHREY 1970, ROELENTS and GOODMAN 1969, VERNON-ROBERTS 1972.) An alternative possibility is that a proportion of the antigen molecules with which a macrophage comes in contact, survives undergraded either at sites at the cell surface (UNANUE and CERETTINI 1970) or in some intracellular compartment that can slowly return the antigen to the surface (*e.g.*, an endocytic vesicle that fails to fuse with lysosomes (KOLSCH 1970, KOLSCH and MITCHISON 1969). Antigen molecules bound to the macrophage surface might interact with lymphocyte surfaces in some special fashion (*e.g.*, the achievement of a critical molecular orientation or concentration might be facilitated, or the endocytosis of the antigen by the lymphocyte might be retarded). In this connection, it is interesting that especially for secondary immune responses

many investigators maintain that the biologically important antigen is that
which binds to the surfaces of "reticular" cells in the spleen or lymph nodes
(see *e.g.*, HANNA and SZAKAL 1966); these cells are relatively inactive in endocytosis, and lymphocytes seem to migrate across their surfaces, as if making
contact with the antigens that appear to reside there for prolonged periods
(NOSSAL *et al.* 1968 a, b, WHITE *et al.* 1970). There is no unequivocal evidence
relating to these possibilities. Some investigators (*e.g.*, RYSER 1968, 1971)
are convinced that endocytosing cells do not degrade a small proportion of
the molecules they take up. But, for example, STEINMAN and COHN (1972)
failed to detect either a persistent intracellular store or a cell-surface store
of horseradish peroxidase when macrophages were exposed to this protein.

IV.5.2. Lysosomes in Lymphocyte Activation

It is not yet known how interaction with antigen triggers the subsequent
entry of lymphocytes into the mitotic cycle, or the synthesis of immunoglobulins, although there seems to be a growing consensus that the initial event is
the binding of the antigen to a lymphocyte surface receptor that probably
is an immunoglobulin (EDELMAN and MOLLER 1973, VITELLA and UHR 1974).
When B-lymphocytes are exposed to soluble antigens or antibodies that bind
to their surfaces they often undergo the characteristic "capping" reaction we
have mentioned before (ANTOINE *et al.* 1974, DE PETRIS and RAFF 1973,
EDELMAN and MOLLER 1973, KARNOVSKY *et al.* 1972, ROBINEAUX and PINET
1959, STACKPOLE *et al.* 1974, YOSHINAGA *et al.* 1972). The bound molecules
move to one pole of the cell, presumably along with the membrane-molecules
to which they are bound. The capped material is then endocytosed (some
may be shed). Lymphocytes do contain lysosomes [28] and exogenous molecules accumulate within these; most if not all of the components internalized
after capping presumably suffer this fate. How this sequence of events is
related to the subsequent development of the cell is uncertain. Among other
possibilities the turnover of surface intrinsic in capping might sometimes serve
to clear the surface of "inappropriate" or surplus bound molecules.

Interestingly, when T-lymphocytes are stimulated by a number of agents,
including the mitogenic lectin, phytohemagglutinin (PHA), they synthesize
acid hydrolases and form additional lysosomes (COHNEN *et al.* 1973, NADLER
et al. 1965, PARKER *et al.* 1965). Populations of stimulated cells release into
surrounding media some hydrolases, as well as a variety of cytotoxic, chemotactic and other biologically active factors (*e.g.*, YOSHINAGA *et al.* 1972). In
addition, in comparison to normal lymphocytes, homogenized PHA-treated
cells show increased ratios of soluble to particulate hydrolases, and when

---

[28] The atypical properties of rat lymphocyte lysosomes were alluded to in Section II.6:
Bowers has found a cathepsin D-rich class of lysosomes that seem to be present in lymphocytes of spleen and thymus. He also has noted that thoracic duct lymphocytes contain an
unusual form of cathepsin D that is either not particle bound, or is present in a structure
that does not survive homogenization. This enzyme differs from the typical cathepsin D
in sensitivity to inhibitors and other properties (BOWERS 1969, 1970, 1973; species other than
rat seem not to have the unusual cathepsin D; BOWERS, personal communication).

lysosomes are isolated, they are found to be more readily disrupted by membrane-active agents (HIRSCHHORN *et al.* 1968). As hypotheses for future evaluation it has been proposed that the lysosomal changes may be aspects of the "remodelling" of the lymphocyte as it enters a new stage of its life cycle (HIRSCHHORN *et al.* 1965) or that limited hydrolase release into the cytoplasm results in passage of neutral proteases or some other enzymes into the nucleus where they act to alter the chromatin and thereby activate cell division or other processes (WEISSMANN *et al.* 1972, 1973). Either of these proposals is consistent with the findings that PHA effects are inhibited by leupeptin and by epsilon-amino-caproic acid, agents that can inhibit proteases (HIRSCHHORN *et al.* 1971, SAITO *et al.* 1973). DARZYNKIEWICZ and AARONSON (1974) feel that changes in RNA metabolism may be important in these regards since in their hands various protease inhibitors suppress mitogen-induced increases in lymphocyte incorporation of uridine.

PHA stimulation of lymphocytes is one of several situations in which altered rates of cell division have been attributed to lysosome mediated effects, or in which intracellular hydrolase release is supposed to engender changes in the chromosomes. BARKA (1974) has reviewed the circumstantial evidence applicable to a number of these cases (*e.g.*, "stabilizers" sometimes can decrease and "labilizers" increase rates of cell division). But results of COHEN and HIRSCHHORN (1971) indicate that dangerous pitfalls may abound. In trying to determine whether chromosome breakage induced in lymphocytes by mitomycin or streptonigrin could be ascribed to release of lysosomal hydrolases, they observed promising inhibitory effects of cortisone. But they went on to make the disquieting observations that the actions of cortisone could be mimicked by heat or cold shock and that cortisone was more effective if added hours after the supposed releasing agents, than it was if added before the agents.

IV.5.3. The Transfer of Maternal Antibodies

Both in mammals and in birds, recently born (or hatched) animals show the presence in their circulation, of antibodies derived from the mother (WILD 1973). "Passive" acquisition of immunity in this sense can occur prenatally or postnatally; this varies from species to species. Much of the transfer of antibodies from mother to offspring apparently depends upon endocytic routes, wherein vesicles acquire proteins at one surface of an epithelium and transport them to the other surface, releasing the proteins there by exocytic processes (KRANEBUHL and CAMPICHE 1969, LEV and ORLIC 1973). Such transfer is thought to take place in yolk sac or intestine.

From his electron microscope tracer studies of rat, in which antibodies are transferred postnatally via the intestine (from milk or colostrum) RODE-WALD (1973) has reported that there is some specificity as to what can pass from the postnatal intestinal lumen into the circulation. In recently born rats, species-homologous immunoglobulins can cross the epithelium of the jejunal regions of the small intestine by vesicular transport with only limited delivery of the proteins to epithelial cell lysosomes. Some non-homologous antibodies (*e.g.*, those from chicken) are not endocytosed extensively by cells in this region of the intestine. Bovine immunoglobulins are endocytosed; some molecules traverse the cells and are released but many seem to end up in multivesicular bodies. After a few weeks, vesicle-mediated transepithelial transport ceases and endocytosis switches to its "normal" fairly non-specific and lysosome-terminated path; this path predominates from the outset in the distal reaches of the intestine (see also Orlic and Lev 1973, and Parsons and Boyd 1972 and Phil. Trans. Roy. Soc. Lond. **B 271**, 402, 408).

The membrane features responsible for this behavior are not known. Presumably both selectivity in endocytosis and "avoidance" of fusion with lysosomes are implied. The latter possibility warrants some additional comment. One could argue that in cases where endocytosed proteins are transported across epithelia and released, the cells in question have very limited heterophagic capacities and that these are "overwhelmed" by the rapid influx of exogenous molecules. A variant of this probably applies in some sense, to capillary endothelia; their geometry is such that it is quite unlikely that an endocytic vesicle will encounter one of the cell's lysosomes before it encounters a cell surface (see *e.g.*, Green and Casley-Smith 1972, and Shea *et al.* 1969 for estimate of the rates of movement and the distances involved in transcapillary transport) *. But for antibody transport an additional possibility has been suggested (Wild 1973, Waldmann and Jones 1973). Perhaps the binding of certain immunoglobulins within the endocytic vesicles that carry them, affects the fate of the vesicles or the immunoglobulins. Fusion with lysosomes might be avoided (*cf.*, Section IV.2.2). Or if fusion does occur, the immunoglobulins might be bound in a protective fashion that enables them to survive; some presently undetected escape mechanism would then be needed whereby the proteins could reach the intercellular spaces. Rodewald (1974) suggests that the pH of the surrounding medium may be of decisive importance in promoting the binding of immunoglobulins to the membrane at the luminal surface of the intestinal cells and the dissociation of the proteins when endocytic vesicles subsequently fuse with the other cell surfaces. His results with electron-microscopic tracer proteins suggest that the immunoglobins bind more tightly to the surface membranes at pH's below 7 than they do at higher pH. How this would apply to endocytic vesicles fused with lysosomes is an open question.

A priori, one might suspect that the ability of certain tissues in fetal or neonatal animals, to transport immunoglobulins intact, might be linked to paucity, or inactivity of lysosomal hydrolases. This seems not to be the case, at least for such tissues as the yolk sac splanchnopleur of fetal rabbits; Jones and Hemmings (1971) found considerable acid protease activity in this tissue, and in consequence, suggested that even for homologous antibodies, there might be a "competition" between degradation and trans-epithelial transport by the yolk sac.

---

* Simionescu *et al.* (1975) have presented evidence that the endothelial-cell vesicles may function, to some extent, by fusing with one another in such patterns as to open patent channels from one cell surface to the other. How this relates to the matters under consideration is not yet clear.

V. Some Special Topics and Some Loose Ends

V.1. Lysosomes in Plant Cells

Structures that can reasonably thought of as lysosomes have been identified in many different types of plants ranging from yeast and fungi to various higher plants (for reviews see MATILE 1969, 1974 and GAHAN 1967, 1973, WILSON 1973). The criteria for their identification are substantially the same as those used for animal cells—the presence of characteristic acid hydrolases within membrane-delimited bodies. The hydrolases have been found with biochemical procedures (*e.g.*, HOFFMANN 1971, MATILE 1974, PITT and GILPIN 1973) as well as with cytochemical methods for demonstration of several enzymes (*e.g.*, BERJACK 1968, COULOMB 1971, GAHAN 1973, PITT 1968, REISS 1974, WILSON 1970). PITT and COOMBS (1968) have also successfully employed several of the vital stains found to accumulate in animal cell lysosomes (*e.g.*, neutral red, euchrysin).

In many cases the structures referred to by plant cytologists as the *vacuoles* are prominent among the sites of acid hydrolases. These are a category of osmotically-sensitive membrane-delimited compartments which sometimes occupy a very large proportion of the cell volume (see LEDBETTER and PORTER 1970 for a concise description). (See also page 212.)

In a few plant cell types, cytochemically demonstrable acid phosphatase has been found in Golgi-associated sacs and vesicles (COULOMB and VICENTE 1973, COULOMB and COULON 1971, COULOMB *et al.* 1972, DAUWALDER *et al.* 1969, PICKETT-HEAPS 1967, POUX 1963 a, b), and it seems appropriate to adopt the working hypothesis that hydrolase transport in plant cells can occur via routes similar to those prevailing in animal cells. Acid phosphatase, peroxidase or aryl sulfatase activities have also been reported in the endoplasmic reticulum of some plant cells (CATESSON and CZANINSKI 1968, POUX 1969) although for the most part the significance of this for lysosome formation is not established. According to MARTY (1971 b) peroxidase occasionally is detectable in plant cell lysosomes; it is not certain that this has functional importance. (Some plant peroxidases are held responsible for inactivation of growth-controlling hormones, such as auxin.)

Despite strong elements of comparability to animal cells and despite the likelihood that lysosomes are very widespread among plant species and cell types, ideas about the details of roles of lysosomes in plant cells' economies are still somewhat shadowy, with a few important exceptions. In part this may reflect the fact that heterophagy is very difficult to demonstrate in most plant cells and may in actuality play little role in these cells. (Heterophagic roles of animal cell lysosomes are the most readily studied.) In some special cases, such as the ameboid cells of slime molds, phagocytosis-like uptake of particles can be demonstrated (ASHWORTH and WEINER 1973, GITHENS and KARNOVSKY 1973). And a few very suggestive observations of what seems to be the formation of large intracellular vesicles from the plasma membrane have been made on intact living cells of some higher plants, such as the hair cells of *Tradescantia* (MAHLBERG 1972). But for most plants, cell walls preclude clear viewing of living cells, and impede the use of macromolecular tracers; presumably they also prevent access of ordinary macromolecules to the cell surface. There have been efforts to work with protoplasts from which walls have been stripped, using a variety of tracers and microorganisms to follow endocytosis. Occasional results have been suggestive; thus, COCKING (1966, 1970) contends that protoplasts can take up ferritin, latex and viruses, and CARLSON (1973) has reported that tobacco cell protoplasts can incorporate chloroplasts, which survive within the cell. But the studies have been plagued with numerous difficulties: in some investigations, tissue preservation has been inadequate to completely quell skepticism; in others, there has been no clear demonstration that tracers are actually within sealed-off vesicles inside the cell (as opposed to being included in infoldings that retain continuity with the cell surface). BRADFUTE, CHAPMAN-ANDRESEN and JENSEN (1964) also point out that some reports of pinocytosis by higher plants cells may reflect artifictitous uptake of tracers attendant upon severe cell injury; their own efforts to demonstrate uptake of fluorescent proteins in a favorable cell system (liquid endosperm) failed. In other words, there is much reason to hesitate before drawing general conclusions. Nevertheless, it can be argued for example that the multivesicular bodies found in plant cells (COULOMB 1973, MATILE 1969) might participate in a cycle of exocytosis and endocytosis involved in the maintenance and growth of cell wall and plasma membrane—the exocytic addition of material to the cell surface is commonly observed (*e.g.,* VIAN and ROLAND 1972, DAUWALDER *et al.* 1969) and configurations consistent with endocytic membrane retrieval mechanisms are encountered.

TRUCHET and COULOMB (1971, 1973) have reported that bacteria can become sequestered within membrane-delimited vacuoles in plant cells. This seems to be the case in the interactions of *Rhizobium* with pea roots—such interactions are fundamental to nitrogen fixation. The details of bacterial entry into the cells and the events that ensue are incompletely understood. However, DEWEY and COCKING (1972) report that during the removal of cell walls to produce protoplasts, pea cells can incorporate bacteria into vacuoles, through processes involving invagination of the plasma membrane, attendant upon osmotic plasmolysis. TRUCHET and COULOMB (1973) have studied pea root nodules cytochemically, and have developed an appealing set of hypotheses: They maintain that the intracellular bacteria are treated by the "host" plant cells in a manner reminiscent of heterophagy in animals; through fusion of host cell lysosomes, the vacuoles in which the bacteria reside, acquire acid hydrolases that ultimately destroy the microorganisms. Subsequently the host cell itself dies. The studies are complicated by the fact that the bacteria themselves elaborate a cytochemically demonstrable acid phosphatase, perhaps indicative of the release of hydrolases involved in bacterial penetration into the host. TRUCHET and COULOMB's views contradict more traditional notions of a fairly stable symbiosis, and the matter still is under dispute.

Roles of plant cell lysosomes in processes other than heterophagy are gradually becoming clarified. The next few sections will be concerned with autophagy, cell death and enzyme secretion, phenomena for which there is increasing relevant evidence. Far more speculative but worth keeping in mind are some other possibilities. For example, the guard cells of stomata possess organelles known as spherosomes, some of which contain cytochemically demonstrable hydrolases. SOROKIN and SOROKIN (1968; see also GAHAN 1973) propose that these bodies might strongly influence the osmotic balances of the cells; osmotic phenomena control the opening and closing of the stomata. The spherosomes might act by contributing osmotically active small molecules to the cytoplasm, and by destroying or producing molecules which affect the enzymes that determine the rates at which low molecular weight carbohydrates are polymerized into starch. Osmotic phenomena are thought to be important for various other types of plant cell shape-change or elongation. Perhaps in many of these cases the lysosomes exert an influence by hydrolyzing (special?) stored macromolecules and thereby liberating osmotically active hydrolysis products.

V.1.1. Senescence and Cell Death

Increases in acid hydrolase activity are often observed in plant organs (leaves, flowers, etc.) undergoing senescent changes that eventuate in their death and often in absicission from the plant. The enzymes appear to be responsible for the degradation of macromolecules that is extensive in these tissues and that may lead to the conservation of metabolites, since degradation products can pass into surviving tissues via the circulation. Cycloheximide can inhibit the enzyme increases (MATILE and WINKENBACH 1971) suggesting that new synthesis is involved. Cell death and dissolution is also central to the development of vascular tissues in higher plants.

Several investigators report that processes they believe to be autophagic (see below), are prominent during senescence. More general agreement obtains for the view that as the cells die, hydrolases are released from membrane-delimited bodies into the cytoplasm, as in animal cell autolysis (BERJACK and VILLIERS 1970, BUTLER and SIMON 1971, GAHAN and MAPLE 1966, MATILE 1974, SEXTON and SUTCLIFFE 1969, WARDROP 1968). For at least some plant tissues, a major source of the released enzymes seems to be the *vacuoles:* the *vacuole* "tonoplast" (delimiting membrane) is often seen to be ruptured in dead cells (BARTON 1966, BERJACK and VILLIERS 1970, GAHAN 1973, MATILE 1974). BERJACK and VILLIERS (1970) speculate that the *vacuoles* burst due to osmotic swelling engendered by the hydrolysis of polysaccharides, which they suggest might be secreted into the *vacuoles* at the appropriate time. But as usual, discriminating among causes and effects in this sort of phenomenon is not easy.

V.1.2. Autophagy

Structures quite similar to animal cell autophagic vacuoles and their precursors have been noted in some plant cell types (*e.g.,* COULOMB and COULOMB 1973, MARTY 1971 b, 1972, POUX 1963 c). Such structures increase in frequency under conditions reminiscent of those discussed for animals. This is true, for starved *Euglena,* for embryos within seeds subjected to prolonged abnormal conditions of storage (BERJACK and VILLIERS 1972, VILLIERS 1967,

1972) and for cellular remodelling during spore formation (BRACKER and WILLIAMS 1966). THORTON (1968) proposes that autophagic vacuoles are important in controlling the peroxisome population of some developing cells (see also VIGIL 1970, 1973). And apparent disposal of secretory materials through autophagy has also been observed (MARTY 1971 a). Marked changes in hydrolase levels, in extracellular release of hydrolase and in degradation of cellular proteins and nucleic acids are also found to characterize stages in the development of slime molds, but how (or if) this relates to autophagy has yet to be studied (ASHWORTH and WEINER 1973, HAMES and ASHWORTH 1974, SUSSMAN and SUSSMAN 1969).

A special type of autophagy has been claimed by numerous investigators to be prominent both in normal plant cells and during senescence or other changes such as the acridine orange induced decline of protein content of yeast cells (MATILE 1970). Cytoplasmic organelles are sometimes seen within the *vacuoles* and from observations such as the one represented by Fig. 54 it has been proposed that cytoplasmic blebs or bulges can protrude into a *vacuole* and eventually pinch off (see *e.g.*, FINERAN 1972, MATILE 1970, VILLIERS 1972 and many other reports). Perhaps this is one aspect of the supposed role of the *vacuole* as a "dumping grounds" or disposal depot (the presence of a cell wall may "require" the use of an internal "sink" for some materials, including acid hydrolases and other contents of primary or secondary lysosomes). Or, if the *vacuole* itself derives from the ER or Golgi apparatus (a matter still being debated *cf.*, LEDBETTER and PORTER 1970), it might be that this process is a geometrical variant of the autophagic sequestration processes mediated by membranes of the ER or Golgi apparatus. MARTY (1970) feels that the very large central vacuole seen in such cells as laticifer cells actually grows at the expense of the rest of the cytoplasm by autophagy of the type being considered. Residual doubts about the existence or significance *in vivo* of configurations like the one in Fig. 54 have still to be dealt with; there are the usual difficulties in extrapolating from static electron micrographs to dynamic processes, and with regard to possible artifacts. The argument against the configurations been artifactitious is somewhat strengthened by the observation of comparable structures in freeze-etch preparations; MATILE 1970, FINERAN 1972.

Relevant data on macromolecular turnover within plant cells is too fragmentary to support a discussion of the involvement of lysosomes (Ann. Rev. Plant Physiol. 25, 363, 1974). DICE *et al.* (1970) have data on pea seedlings suggesting that the same relations of turnover to subunit size seen with animal cells may also apply to plants. Yeast should be very useful and convenient material for future studies of problems related to turnover. For example, it is found that increases in hydrolase levels accompany the enhanced turnover that is evident during such metabolic shifts as the adaptation of anaerobically grown yeast cells to aerobic environments (MATILE 1974, MATILE *et al.* 1971) and there also are variations of enzyme levels at different stages of the cell growth and division cycle. Proteolytic enzymes and other yeast hydrolases are located within the *vacuoles* (HASILIK *et al.* 1974, LENNEY *et al.* 1974) and these structures also contain appreciable pools of amino acids (WIEMKEN

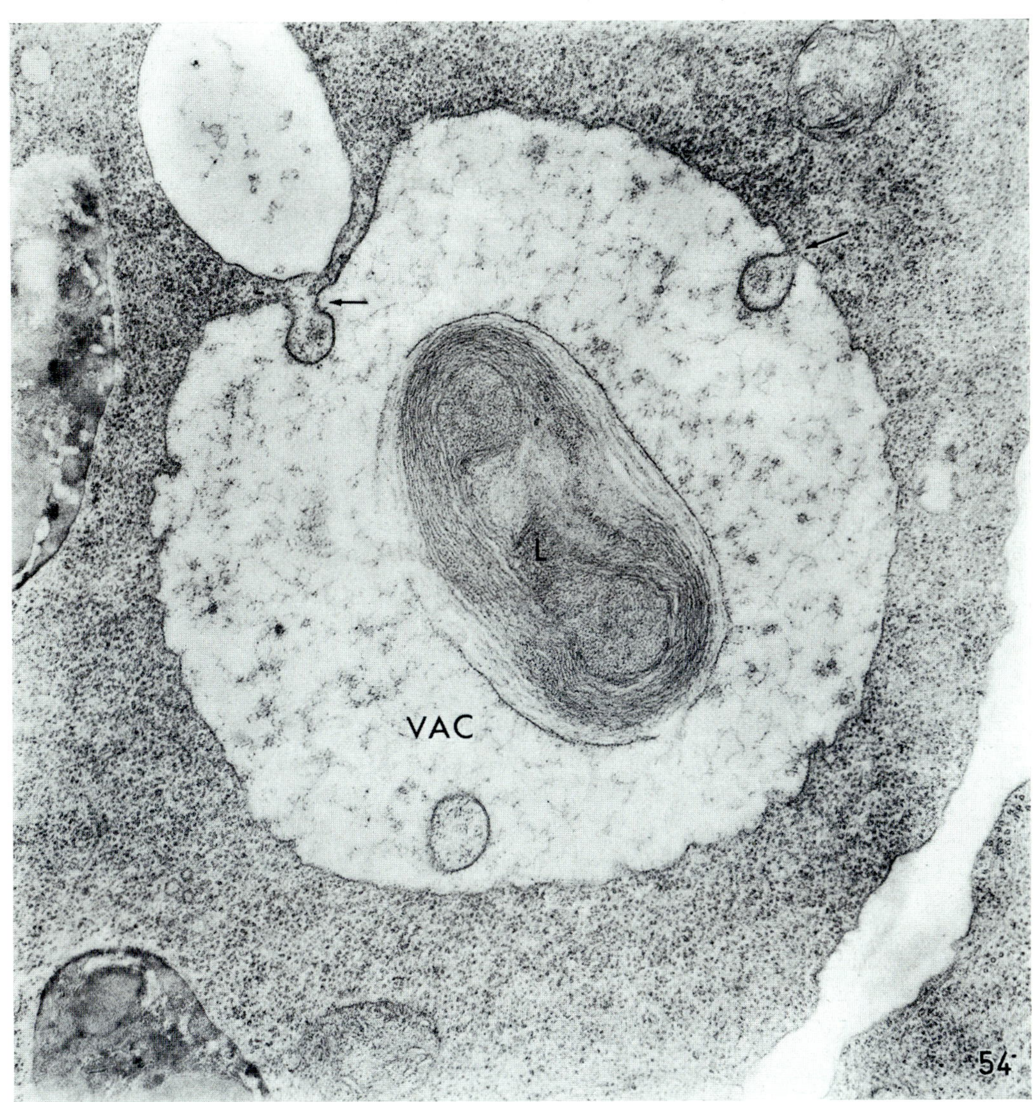

Fig. 54. Portion of a cell from a young root of the plant, *Scoronera hispanica*. The arrows indicate points at which cytoplasm seems to be budding into the *vacuole (VAC)*. The lamellated material at the center of the *vacuole (L)* may have arisen through such budding. ×28,000. (From COULOMB, C., 1973: C. R. Acad. Sci. [Paris] **276**, 1161—1164; obtained courtesy of PH. COULOMB and MME. C. COULOMB.)

and Nurse 1973; see also Weiss 1973 for a report that some amino acid pools in *Neurospora* are maintained within membrane-delimited structures). Protease inhibitors in yeast cells might function to control the enzymes or somehow protect the cell. Inhibitors have been detected which probably are located in the cytoplasm outside the vacuole (see Betz *et al.* 1974 and Lenney *et al.* 1974). Among other roles, the inhibitors and certain yeast proteases may interact in the adjustment of the cells to varying metabolic conditions since some of the proteases are capable of activating or inactivating other metabolically significant enzymes such as tryptophan synthase (Cabib *et al.* 1973, Hasilika and Holtzer 1974).

As mentioned in Section III.4.3.1, protein degradation in other unicellular plant forms also varies in different growth media, and at different stages in the life cycle. A few comparable reports have been made for higher, multi-cellular, plants (*e.g.*, Humphrey and Davies 1975, claim that transfer of *Lemna* to a nutrient-poor medium, results in enhanced turnover of soluble proteins). The possible participation of lysosomes in these phenomenon has yet to be evaluated.

Protein storage or "aleurone" grains and some other storage structures in a number of plant cells are reported to contain acid hydrolases. This raises the possibility of lysosomal involvement in intracellular digestion of food stores (*cf.*, Section V.3; see *e.g.*, Plant Physiol. **56**, 292 [1975]).

V.1.3. Extracellular Hydrolases

Appreciable levels of extracellular hydrolases are observed in various plant systems. Many of these enzymes have acid pH optima and other features in common with lysosomal hydrolases, but in general their relations to intra-cellular structures are not known *. Some probably are lysosomal but some, such as the protease released by *Neurospora* are secreted without simulta-neous release of other enzymes and show additional properties suggesting they may not come from lysosomes (Heiniger and Matile 1974, Matile 1969). Extracellular enzymes are thought to play a variety of roles in different situations. For example, they can contribute to local or general cell wall softening or dismantling during ripening of fruit (Pressey and Avanta 1973), growth and division of yeast and other cells (Forkes *et al.* 1973, Mullins and Ellis 1974), germination or release of spores and related structures ("microspores", microcysts; Iten and Matile 1970, Stieglitz and Stern 1973, Wilson *et al.* 1970), invasion of host organisms by parasitic fungi (Berghen and Petterson 1974, Jones *et al.* 1974), and perhaps during initial stages of pollen tube growth (Gahan 1973, Knox *et al.* 1970) and remodelling of developing slime molds (Ashworth 1971). The degradation of food reserves during seed germination of barley and other plants is carried out by hydrolases (proteases, amylases, RNAse, glucanases) secreted by the

---

* There are reports from electron microscopic cytochemical studies (*e.g.*, Hislop *et al.* 1974) suggesting that, in at least some cases, hydrolase release is by exocytosis of the contents of lysosome-like, membrane-delimited bodies. The insectivorous plants may be favorable material for future studies (Knox and Heslop-Harrison 1971).

aleurone tissues; secretion is responsive to gibberellins, osmotic factors and other influences (ARMSTRONG and JONES 1973, CHRISPEELS 1974). The secreted hydrolases closely resemble the lysosomal enzymes of the tissues involved (GIBSON and PALEY 1972); of interest for future studies of turnover is the observation that some, at least, are synthesized at the expense of pre-existing proteins (FILNER and VARNER 1967). Since the rates of hydrolase synthesis in aleurone tissue can be altered by osmotic changes, one line of thinking is that there is a kind of self-regulating system wherein the osmotic effects of the products of hydrolysis inhibit synthesis of additional enzymes (CHRISPEELS 1974, JONES and ARMSTRONG 1971).

Other nutritional roles attributed to extracellular hydrolases include, for instance, the provision of phosphates through hydrolysis of diverse substrates by the extracellular acid phosphatase induced in *Euglena* by phosphate-starvation (SOMER and BLUM 1965, we have already pointed out that this enzyme differs, *e.g.*, in its responses to inhibitors from the lysosomal phosphatase of *Euglena*, BLUM 1965). It has also been suggested that the extracellular enzymes secreted in tissues of higher plants may be elements of mechanisms for defense against bacteria, viruses and other potential pathogens.

Finally, a number of investigators have found that lowering the pH of growth media can stimulate growth of plant cells such as those of *Avena* coleoptile. This is interesting in light of the fact that a few plant hormones including auxins seem to participate in control of extracellular pH. One hypothesis now being investigated is that the acidification of the space between cell and wall leads to enhanced activities of acid hydrolases present there—the enzymes could act on the wall to facilitate its expansion (JOHNSON *et al.* 1974).

V.2. Hydrolases in Secretory Cells

V.2.1. Lysosomes and Secretory Processes; the Thyroid Gland

Earlier we outlined one way in which gland cells make "interesting" use of lysosomes—crinophagy permits cells to degrade excess secretion granules. And, if the views expounded in Section II.4.2.2 are correct, a second type of lysosome involvement is in degradation of the membranes responsible for transport and release of secretions. Another important example is provided by the thyroid (BAUER and MEYER 1967, A. B. NOVIKOFF *et al.* 1974, PITT-RIVERS and CAVALIERI 1963, SELJELID 1967, WETZEL *et al.* 1965, WISSIG 1960, WOLLMAN 1969). Thyroid epithelial cells synthesize and store in the extracellular follicle lumen, an iodinated glycoprotein, thyroglobulin. Upon proper stimulation (*e.g.*, by thyroid stimulating hormone) the cells reincorporate thyroglobulin through endocytosis and deliver it to their lysosomes. *In vitro*, extracts of thyroid lysosomes can degrade the globulin and in so doing release tyrosines and thyronines (including triiodothyronine and thyroxine, which constitute the biologically active circulating hormones; DOPHEIDE *et al.* 1969, McQUILLAN and TRIKOJUS 1971, WOLLMAN 1969). It is inferred that lysosomal proteases act similarly *in vivo*. The subsequent steps whereby hormone appears in the blood stream are not well understood. It might be premature to discard entirely the idea that a few of the lysosomes release their contents by exocytosis (see *e.g.*, SMITH 1972) since undegraded thyroglobulin appears in the lymph when thyroid hormone release is stimulated. But of course, there are other possible explanations for this observation (*e.g.*, leakage between cells [WOLLMAN 1969] or the failure of some endocytosis vesicles to fuse with lysosomes and their subsequent fusion with the plasma membrane). Given the likely permeability properties of lysosome surface membranes (Section I.3.1.4) there would seem to be no major barrier to prevent the diffusion of iodinated amino acids out of the organelles. It should also be recalled that when proteins labelled with radioactive iodine are endocytosed by cells such as cultured macrophages, a great deal of the label subsequently appears in the medium still attached to amino acids, probably without the occurrence of much exocytosis of lysosome contents. Perhaps the thyroid hormones, which are somewhat larger than amino acids, can also cross the lysosome membrane. From the cytoplasm the hormones could then diffuse or be transported across the plasma membrane and into the capillaries that abound near the basal regions of the cells.

LIBBIN *et al.* (1974) among others ascribe a central role to lysosomes in the genesis of erythropoietin. Triton-isolated kidney lysosomes can modify serum components to give rise to molecules that behave like erythropoietin in the conventional bioassay, based on iron uptake into red cells (Section IV.3 has summarized evidence that extracellular hydrolases can generate various other active molecules from circulating precursors).

A few gland cells, notably those of the prostate and rodent preputial gland, secrete acid hydrolases such as acid phosphatase, esterase, proteases, and β-glucuronidase (BRANDES 1966, DOTT 1969, HELMINEN and ERICSSON 1970, 1972, MANN 1963, RUENWONGSA and CHULAVATANOL 1974). Exocytosis of Golgi-derived secretory granules appears to be the release mechanism, at least in the prostate (HELMINEN and ERICSSON 1970). Care is needed in analysis of the sources of seminal fluid enzymes since work on vasectomized individuals suggest there may be a contribution from the testes (DINGLE and DOTT 1969, DOTT and DINGLE 1968). The evidence as to whether the secreted enzymes are truly lysosomal hydrolases is inconclusive. Both similarities and differences among comparable enzymes of hepatic lysosomes, prostate cells, seminal fluid and so forth have been noted for such characteristics as saccharide side chains, substrate specificities and responses to inhibitors (ANDRES and PALLAVICICI 1973, BARRETT 1972, BULLOCK and WINCHESTER 1973 a, b, DERECHIN *et al.* 1971, KILSHEIMER and AXELROD 1957, LONDON *et al.* 1955).

In the preputial gland, β-glucuronidase and acid phosphatase are both present in the secretion granules, but, at least from cytochemical observations, other lysosomal hydrolases seem to be absent (MESQUITA-GUIMARAES and COIMBRA 1974). One might speculate that the secretion granules are a sort of evolutionary descendant of more typical lysosomes.

V.2.2. Hydrolases in the Golgi Apparatus and Secretion Granules

There are a number of additional observations on hydrolases in secretory cells that may eventually turn out to be pertinent to lysosome function although their present status is cloudy; there often is no good reason to assume that the enzymes in question are identical to the comparable ones in lysosomes (see *e.g.*, the comments to this effect concerning acid phosphatases by FARQUHAR *et al.* 1967, GREENBERG *et al.* 1967, NYQUIST and MOLLENHAUER 1974).

Cytochemically demonstrable acid phosphatase is found in some secretion granules of various gland cells and in the Golgi sacs (or GERL) that produce these granules [Fig. 55; see *e.g.*, NOVIKOFF (1963), NOVIKOFF and ESSNER (1962), OGAWA *et al.* (1962) for Paneth cells, pancreatic islets and some others; HOLTZMAN and DOMINITZ (1968) and GREENBERG *et al.* (1967) for

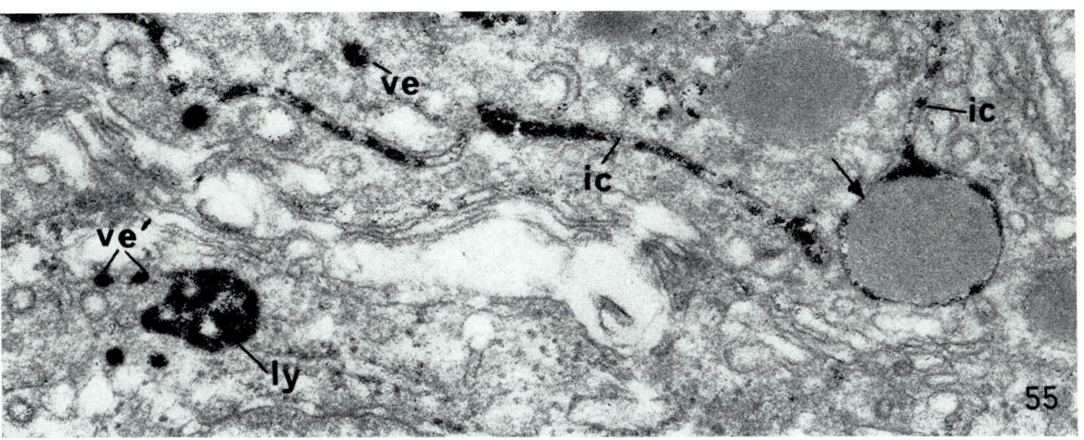

Fig. 55. Golgi region of a somatotrophic cell from a normal rat pituitary incubated to demonstrate acid phosphatase activity. Reaction product is seen in sacs (*ic*) of the Golgi apparatus, and surrounding a secretion granule (arrow) that seems to be forming in a dilated region of one of these sacs. Product also is present in a multivesicular body (*Ly*) and several small vesicles (*ve'*). ×35,000. (From FARQUHAR, M. G., 1969: In: Lysosomes in biology and pathology. [DINGLE, J. T., and H. B. FELL, eds.], Vol. 2, pp. 462—482. Amsterdam: North-Holland Publishing.)

adrenal medulla; SMITH and FARQUHAR (1966), HOPKINS (1969), MASUR and HOLTZMAN (1969) for pituitary cells; FARQUHAR *et al.* (1974) for the very low density lipoproteins of hepatocytes; HORIE *et al.* (1971) for exocrine pancreas]. When appropriate secretion granule-rich fractions are examined biochemically, appreciable levels of hydrolases often are not found [see DEAN and HOPE (1967) for pituitary and SMITH and WINKLER (1969) for adrenal]. Perhaps this relates to the fact that the granules which seem to be forming or newly formed are the most strikingly reactive ones in cytochemical acid phosphatase preparations; older granules often are largely unreactive (HOLTZMAN and DOMINITZ 1968, OSINCHAK 1966, NOVIKOFF 1963, SMITH and FRAQUHAR 1966).

It is not yet known if other acid hydrolases are present in the granules along with the phosphatase. RIECKEN and PEARSE (1966) assert that esterase is detectable in the Paneth cell granules but their work was light microscopic and electron microscopy may be needed to determine whether the activity is truly in the granules as opposed to conventional lysosomes (*cf.*, BEHNKE and NOE 1964). ERLANDSEN *et al.* (1974) have demonstrated lysozyme in the Paneth cell granules; this may be related to anti-microbial functions of the cells (ERLANDSEN and CHASE 1972). Limited cytochemical work by ourselves and others thus far has failed to demonstrate aryl sulfatase or other acid hydrolases in secretion granules of cells such as those of adrenal medulla or pituitary.

One line of thought is that the acid phosphatase in secretion granules gets there by accident (see *e.g.*, DE DUVE and WATTIAUX 1966). Both lysosomes and secretion granules contain proteins made in the rough endoplasmic reticulum and "packaged" near the Golgi apparatus and if the mechanisms that maintain separate transport routes for different proteins are imperfect, there might occur some "unintentional" admixing.

Another possibility is that the phosphatase is playing a modulating or controlling metabolic role. For example, some Golgi apparatus transferases responsible for addition of sugar residues to glycoproteins release nucleotides as a product of the reactions they catalyze; conceivably the hydrolysis of these nucleotides by phosphatases of the Golgi apparatus or GERL can influence the rates of transferase reactions (P. M. NOVIKOFF *et al.* 1971). Some other reported activities of the Golgi apparatus might also be influenced by phosphatases (*e.g.*, the phosphorylation of casein, BINGHAM and FARELL 1974). Or possibly the enzyme reflects the operation of a special mode of crinophagy involving degradation of secretions within GERL (P. M. NOVIKOFF *et al.* 1971).

In general, the conception that some kinds of hydrolases in forming secretion granules can contribute directly to the processing of secretory materials is steadily gaining ground. STEINER and his colleagues have developed a convincing picture of insulin formation, according to which, proinsulin molecules synthesized in the rough endoplasmic reticulum are cleaved to give rise to insulin by a protease (pH optimum, roughly 6.5) that acts during or soon after the packaging of secretion granules in the Golgi region. (See STEINER *et al.* 1974 for review and KEMMLER *et al.* 1973; also note the comments following KEMMLER *et al.* 1972, in which Smith outlines cytochemical evidence for appropriate localizations of proteases.) There is growing evidence for intracellular proteolytic processing of a number of other secretions. (KEMPER *et al.* 1974, LODISH 1973, MILSTEM *et al.* 1972, PASQUINI *et al.* 1974, SCOTT *et al.* 1973; TAGER and STEINER 1974.) For some, such as collagen, the locale of supposed processing has yet to be definitively established—probably procollagen is cleaved to tropocollegen extracellularly (LAZARUS 1972, PONTZ *et al.* 1973). For others, it is highly likely that processing is intracellular. DEWALD and TOUSTER (1973) have speculated that an α-D-mannosidase they find in hepatocyte fractions enriched in Golgi apparatus, may participate in processing of glycoproteins and one might also extend a comparable idea to the glucuronidase activity of the endoplasmic reticulum. Perhaps some of the effects of these or other hydrolases are related to the alterations in the state of aggregation of material that are important for "condensation" of granule contents into their final concentrated form. And, SMITH and WINKLER (1969) conjecture that phospholipases might act on secretion granule mem-

branes to generate lysophospholipids, the latter components could be important in promoting the subsequent exocytic fusions of the granules.

The type II cells of lung contain large inclusion bodies with a characteristic
lamellated ultrastructure. These bodies possess cytochemically demonstrable
acid phosphatase, esterase and aryl sulfatase activities (BALIS and CONEN
1964, CASTLEMAN *et al.* 1973, GOLDFISCHER *et al.* 1968, HATASA and NAKA-
MURA 1965, MEBAN 1973). Biochemical studies with partially purified cell
fractions also indicate the presence of hydrolases such as acid phosphatase,
aryl sulfatase and N-acetyl-β-glucosaminidase but cathepsin D and nuclease
activities are low. The bodies are reported to arise from more conventional
lysosomes such as multivesicular bodies (GOLDENBERG *et al.* 1969, SOROKIN
1973). The lamellated bodies apparently contribute surfactant to the extracellular surfaces of the lung, by exocytosis; as always one must be careful
about drawing conclusions as to direction of movement from electron micrographs but in this case autoradiographic evidence and studies on cells stimulated to release the bodies, have adequately supplemented the initial observations of lamellated bodies fused with the cell surface (ADAMSON and BOW-
DEN 1973, FAULKNER 1969, GOLDENBERG *et al.* 1969, KRASNO *et al.* 1972,
NIDEN *et al.* 1967). Phosphatidic acid phosphatase is thought to participate
in the formation of the surfactant component dipalmitoyl lecithin (ETHERTON
and BOTHAM 1970, MEBAN 1970). Thus one might tentatively surmise that
lysosomal enzymes are here being put to use in the manufacture of the secretory product. The hydrolases released with the other contents of the lamellated bodies might also help combat microorganisms at the alveolar surfaces.

Inadequate supply or improper distribution of surfactant is a source of
problems for premature infants and in some respiratory disorders (KRASNO
et al. 1972, ROBERTSON and ENHORNING 1974). Conceivably some of the
hydrolase deficiencies in storage diseases generate difficulties along these lines
by interfering with surfactant metabolism.

V.2.3. Melanin

Finally, although they are not secretory materials in the usual sense, some
comments on pigment granules are in order. The "tyrosinase" enzyme complex that produces melanin is packaged into melanosomes by sacs near the
Golgi apparatus (EPPIG 1974). In melanoma cells, what are considered to
be the same sacs contain acid phosphatase activity, leading to the suggestion
that they are better referred to as GERL than as Golgi sacs (NOVIKOFF *et al.*
1968). The melanosomes of melanoma cells and of normal melanocytes, frequently contain cytochemically detectable acid phosphatase (NOVIKOFF *et al.*
1968, WOLFF and SHREINER 1971). In light of this and of the discussion in
the previous section it is interesting that tyrosinase may require proteolytic
activation (see *e.g.,* BARISAS and McGUIRE 1974 for amphibian tissue). According to TURNER *et al.* (1975), the melanosomes of goldfish melanophores
acquire tyrosinase through fusion of Golgi vesicles; some of the forming
melanosomes have the appearance of multivesicular bodies, perhaps reflecting
internalization of their surface membrane that balances the addition of Golgi
vesicle membrane.

In melanomas, melanin granules are commonly seen in autophagic vacuoles, suggesting that autophagy may contribute to regulating the numbers of granules in the cells (MISHIMA 1967, NOVIKOFF *et al.* 1968; recall that pigment granules in retinula cells also are subject to lysosomal sequestration [Section II.3.3]). Presumably the enzymes of forming melanin granules would be more susceptible to actual degradation by the hydrolases than are the pigments, which seem to be resistant (see WOLFF *et al.* 1974, WOLFF and SHREINER 1971). Thus inclusion in a lysosome might "turn off" melanin synthesis by a given melanosome but have little immediate effect on the melanin already formed. According to JIMBOW *et al.* (1974) the melanocytes of developing feathers in white chickens show many more melanosomes in autophagic vacuoles than do the corresponding cells in black chickens and the autophagocytosed melanosomes show little if any melanin; the pigmentation differences are genetically based, so it may be especially interesting to determine the nature of the controls governing the lysosomal behavior.

Many melanin-containing cells in skin or feathers are not themselves able to synthesize melanin. Rather they acquire melanosomes from melanocytes, probably by some type of endocytic process (WOLFF and HONIGSMANN 1971). Sometimes this accompanies the death of melanocytes—macrophages can phagocytose the debris, including the melanin granules. But apparently there are less drastic transfer mechanisms that are quantitatively significant but have yet to be elucidated to the satisfaction of all investigators. During the loss of pigment from developing frog retinal cells exocytic release of granules is seen (HOLLYFIELD 1973); the pigment subsequently is phagocytosed. And WOLFF *et al.* (1974) have shown that isolated melanosomes injected into skin can be taken up in patterns quite similar to the normal pigment distribution. But a number of other workers favor the concept that the passage of melanin from one cell to another depends largely on some sort of transfer of melanosomes or small portions of melanocyte cytoplasm from the intact cell, rather than prior release of granules. (In principle, different transfer mechanisms should produce granules with different numbers of delimiting membranes in the recipient cells: exocytosis and subsequent endocytosis would lead to a granule with one membrane [as is often observed]; phagocytosis of bits of cytoplasm might result in granules with two membranes, their own and that of the phagocytic vacuole. But for any of the mechanisms it is easy to devise *ad hoc* explanations of the final morphology, involving digestion of membranes and so forth.)

V.3. Lysosomes in Animal Gametes

We have already mentioned one role of lysosomes in gametogenesis—the phagocytosis by Sertoli cells of the residual cytoplasmic droplet shed by maturing sperm. Before being taken up, the residual droplets show much autophagy, presumably reflecting their unusual metabolic status (DIETERT 1966, STANG-VOSS 1972). Sertoli cells additionally phagocytose degenerating sperm resulting, *e.g.*, from autoimmune phenomena, or testosterone treatment (BARHAM and BERLIN 1974, GARDNER and SHERUEY 1974). Phagocytosis also occurs in the ovary; for example, macrophages invade atretic follicles, and in insects, follicle cells take up degenerating oocytes during oocyte resorption induced by starvation or other conditions (Bell 1971, Davies and King 1972 a, b). We have also alluded to the presence of extracellular hydrolases in seminal fluid (Section V.2.2); comparable enzymes (and some inhibitors) are also present in female genital tracts (*e.g.*, Sukeno 1972). The hydrolases might have various functions including, for example, influencing the viscosity of the fluids encountered by the sperm (*e.g.*, Mann 1967, Ruenwongsa and Chulavatanol 1974). As for lysosomes in the gametes themselves, there still is much to learn.

V.3.1. Sperm

The acrosomes of sperm cells contain hydrolases that facilitate sperm passage through egg coatings—these enzymes vary somewhat from species to species but include neutral proteases and polysaccharidases resembling hyaluronidase (see *e.g.*, Allison and Harree 1970, Dott 1969, Stambaugh and Buckley 1969). The acrosome forms from Golgi apparatus-derived material and is released through events similar to exocytosis. Cytochemical and biochemical studies on various species (mouse, ram, rat, salamander, sea urchin) indicate the presence of a number of acid hydrolases such as acid phosphatase, sulfatase, and esterase (Allison and Hartree 1970, Anderson 1968, Buongiorno and Bertolini 1967, Bryan and Unnithan 1973, Dingle and Dott 1969, Posalski *et al.* 1968). This, plus the finding that acrosomes also share vital-staining properties with lysosomes (Allison and Hartree 1970) sustains the current tendency to regard these organelles as close relatives or modified forms of lysosomes. But some acrosome enzymes seem to differ in specificities, optimal incubation conditions and so forth from usual lysosomal hydrolases (see *e.g.*, the discussions of mammalian sperm proteases by Gilboa *et al.* 1973, Palakoski and Mororie 1973, Zanewald *et al.* 1972, and of testicular hyaluronidase by Barrett 1963, 1972). Furthermore, in some species such as the rabbit, acid phosphatase and several other characteristic lysosomal hydrolases are not detectable in the acrosome (Stambaugh and Buckley 1969). Dingle and Dott (1969) find for bull, that during spermiogenesis most of the RNAse, acid proteinase and acid phosphatase of the spermatid is shed with the cytoplasmic droplet but glucosaminidase and galactosaminidase activities are retained and are detected in the acrosome.

V.3.2. Eggs

For oocytes and ova, an outstanding unresolved issue is the role of lysosomes in yolk deposition and digestion (for review see Pasteels 1973). In many animal species, both vertebrate and invertebrate, a large proportion of yolk constituents are synthesized by cells other than the developing oocytes and reach their final destinations in the yolk bodies through endocytosis (Anderson 1969, Cruiskshank 1971, Paulson and Rosenberg 1972, Roth and Porter 1964, Wolin *et al.* 1973). For example, work by Wallace, Dumont and co-workers has shown that in the amphibian, *Xenopus,* most yolk protein derives from vitellogenin, a phosphoprotein secreted by the liver. Vitellogenin is pinocytosed by growing ooctes from fluids in communication with the circulation (BRUMMETT and DUMONT 1974) and once in the oocytes it is converted to the storage proteins, phosvitin and lipovitellin through limited proteolysis (BERGINK and WALLACE 1974, JARED et al. 1973). Further digestion does not occur until the yolk is mobilized during development. Interestingly, when vitellogenin is injected directly into the oocyte cytoplasm, rather than entering through pinocytosis, the bulk of the protein is degraded to low molecular weight molecules within a number of hours (DEHN and WALLACE 1973). The basis for the differential treatment of molecules entering in different ways is not known.

In lower invertebrates, notably sponges, somewhat different storage processes take place—the forming reproductive cells phagocytose other cell types intact, or in large fragments (DE VOS 1971) and from cytochemical demonstrations of acid phosphatase they are thought to utilize acid hydrolases in processing the endocytosed materials (TESSANOW 1969, TIFFON et al. 1973).

Oocytes and ova do contain apparently conventional lysosomes and for some eggs at some stages acid phosphatase is demonstratable in sacs near the Golgi apparatus (e.g., ANDERSON 1969, KESSEL and DECKER 1972). In addition cytochemical studies and limited biochemical work strongly suggests that acid hydrolases are present in yolk platelets of a number of animals, especially certain invertebrates (molluscs, sea urchins; PASTEELS 1969, 1973, KRISCHER and CHAMBERS 1970, SCHUEL et al. 1972). For some species, the cytochemical reactivity is most evident during initial periods of yolk digestion (Fig. 56; DALAQ 1963, PASTEELS 1969) but, for example GIORGI (1974) reports that in Drosophila, acid phosphatase is demonstrable in the newly formed yolk platelets, although not in ones that are somewhat older.

Fully satisfactory evidence for acid hydrolases in amphibian yolk bodies is very difficult to obtain. For frog yolk platelets there are reports of cytochemically demonstrable acid phosphatase, capable of acting on phosphoproteins (casein was used as the substrate). But in some of the cells of the preparations studied, nuclear staining was noted (DENNIS 1964) suggesting that there might have been diffusion of reaction product. A preliminary report on axolotls by LEMANSKI and ALDOROTY (1974) indicates that acid phosphatase is present in the yolk platelets during development of embryos. BECK (1965) claims that acid phosphatase is evident in chick yolk at a time in the blastoderm stage when yolk is being digested.

The presence of acid hydrolases in yolk constitutes presumptive evidence for a role of lysosomes either in yolk production (e.g., through modifications of blood borne macromolecules; LOCKSCHIN 1969) or in mobilization of nutrients for developing embryos. The mechanisms by which the enzymes are brought into the yolk bodies are not known. Perhaps in some species they enter by fusion of lysosomes during stages of yolk digestion, but for others, the suggestion has been made (PASTEELS 1973) that hydrolases are incorporated in yolk platelets long before the enzymes act and somehow remain inactive until properly triggered. PASTEELS also points out that as yolk digestion procedes, the yolk bodies come to resemble multivesicular bodies (see also DALCQ and PASTEELS 1966); perhaps this reflects the endocytic history of the yolk. In any event, an explanation must be sought for the fact that yolk precursors endocytosed by oocytes are not degraded until much later. Is it the paucity of oocyte lysosomes in comparison to endocytosis structures that "protects" the latter or are these more specific controls? * How is the selective hydrolysis of *Xenopus* vitellogenin brought about—are selective pro-

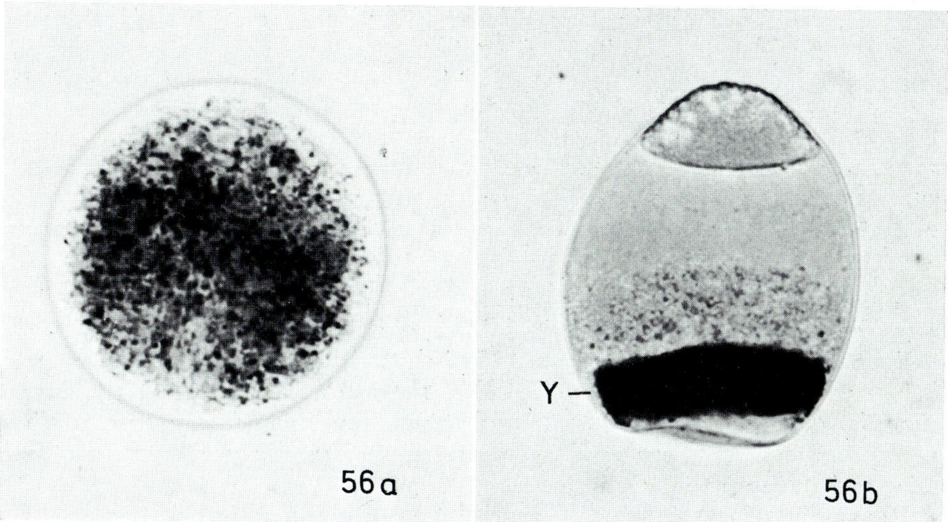

Fig. 56. Eggs of the mollusc, *Barnea* (pronuclei stage) incubated to demonstrate thiolacetic-esterase activity. The egg in *b* was centrifuged, which stratifies the organelles into intracellular layers each of which is enriched in a particular type of structure. *a* Shows an uncentrifuged egg. Esterase reaction product is prominent in yolk platelets; these are normally seen scattered throughout the cell; they occupy the layer designated *Y* in the centrifuged preparation. ×700. (From PASTEELS, J. J., 1973: In: Lysosomes in biology and pathology. [DINGLE, J. T., ed.], Vol. 3, pp. 216—234. Amsterdam: North-Holland Publishers.)

teases responsible or does the protein conformation expose only certain links for hydrolysis by non-specific enzymes (BERGINK and WALLACE 1974)? Perhaps work on insect fat body will provide some useful leads. Proteins endocytosed by the cells of this organ enter typical multivesicular bodies and other lysosomes at some stages of development, and larger "storage granules" at others (COLLINS 1969, THOMASSEN and MITCHELL 1972). According to COLLINS (1972) there is an inverse relationship between hydrolase levels in the cells and storage of proteins. As with oogenetic phenomena in vertebrates, the events in the fat body are under hormonal control (*e.g.,* COLLINS 1969, THOMASSEN and MITCHELL 1972).

The cortical granules that fuse with the egg cell surface during fertilization in many species, are believed to contain proteases, at least in some cases (VAC-

---

* For various species, protease inhibitors reportedly are present in yolk and elsewhere in the embryo (see *e.g.,* BAINTNER and FEHER 1974, for chick embryos and SLAUGHTER and TRIPLETT 1975, for frogs).

14*

QUIER *et al.* 1973). In a few studies, the granules of these enzymes showed cytochemically demonstrable acid phosphatase although this is not invariably true (ANDERSON 1968, KESSEL and DECKER 1972, SCHUEL *et al.* 1972). Relatively pure fractions containing cortical granules do not seem very rich in acid hydrolases (SCHUEL *et al.* 1972) and there is presently no decisive reason to regard these bodies as lysosomes.

The regression of the corpus luteum involves formation of autophagic vacuoles as well as macrophage engulfment (DINGLE *et al.* 1968). There also are some interesting possibilities for roles of lysosomes or their enzymes in release of ova from follicles and in embryonic implantation, but there are only fragments of solid relevant information (BJERSINK and CAJANDER 1974, WOESSNER 1969). Additional information is also needed for definitive evaluation of such interesting proposals as those that envisage a major role for heterophagy, by yolk sac epithelia, in embryonic nutrition (BAINTNER and FEHER 1974, WILLIAMS *et al.* 1973).

V.4. Closing Comments

Even though mysteries abound, none of the major "problem areas" outlined in the prior pages seems intractable. If the tenor of the discussion has at times been overly confusing, this probably is due more to my own lack of perception than to the absence of progress—it is easier to stress the uncertainties than to see a way out.

For studies of certain problems such as the mechanisms of lysosome acidification, plant cells may prove particularly useful. Especially for those cells with large central *vacuoles,* there is much evidence for ionic gradients between the *vacuole* and the extracellular medium and for pertinent active or facilitated transport mechanisms. Further study of *vacuole* enzymes and of the *tonoplast* (*vacuole* membrane) may provide key insights applicable to more "conventional" lysosomes.

Probably the most weighty unresolved questions concern the role(s) of lysosomes in intracellular turnover. The guesses ventured above are being subjected to experimental tests in several laboratories. And the time may be suitable for even more direct attacks. One can directly introduce known and labelled intracellular components into cells and follow their fate. Micro-injection experiments have already been alluded to. Perhaps advantage can also be taken of the recent advances in cell fusion techniques. Through such techniques one might be able to create suitable mixtures of labelled and unlabelled cytoplasmic constituents for detailed study of turnover of organelles; especially since one might be able to utilize cultured cells lacking lysosomal enzyme activities (storage disease material), it should not be difficult to get some new insights into the problem. Unicellular organisms might also be used to advantage.

Perhaps the specific line of reasoning just outlined, is far-fetched, or overlooks some phenomenon well known to experts in cell culture. The point is, that the rapidly expanding range of available techniques, and cells or organisms, has been relatively little exploited for study of lysosomes. There is every reason to expect increasingly diversified and technically sophisticated work on these organelles, and thus, there is sound basis for optimism about the remaining puzzles.

Acknowledgements

Dr. A. B. NOVIKOFF introduced me to lysosomes during my tenure as a post-doctoral fellow in his laboratory. His enthusiasm and insight are much appreciated. He also read and commented valuably upon portions of the manuscript. Dr. S. SILVERSTEIN went over the entire manuscript—his criticisms and his many excellent and original ideas were exceedingly helpful. Dr. C. DE DUVE read through the initial portions of the manuscript and encouraged me to rethink some areas and thereby to avoid several pitfalls. Dr. S. GOLD-FISCHER has repeatedly refreshed my viewpoints with his lively perceptions.

The donors of illustrative materials are acknowledged in the corresponding captions. In addition, for contributions of unpublished material, key bits of information, advice and useful discussions I am grateful to many colleagues including especially: Drs. A. ALLISON, S. BEYCHOK, W. BOWERS, Z. COHN, PH. COULOMB, R. DETER, J. F. DICE, J. T. DINGLE, P. EDELSON, J. L. E. ERICSSON, E. ESSNER, D. M. GILL, A. L. GOLDBERG, J. HIRSCH, S. HOCHSTEIN, P. LEUENBERGER, M. LOCKE, R. LOCKSHIN, PH. MATILE, M. MÜLLER, J. J. PASTEELS, B. POOLE, F. H. SCHNEIDER, A. D. SMITH, J. M. TAGER, D. C. WARHURST, R. WALLACE, G. WEISSMANN, Z. WERB, and D. ZIPSER.

BERDINE ABLER and BRUCE SKINNER put up with my atrocious handwriting and typed the manuscript. Dr. FE REYES, LEONARD KASHNER, IRENE TARR provided able technical and photographic assistance. JANE DORFMAN of the Columbia Biology Library was very helpful in locating materials. Grant support for unpublished research described in the manuscript was from the National Institute of Neurological Diseases and Stroke (NS 09475).

We also thank the publishers of the various journals cited in the Figure legends for permission to reprint material. This includes: Pergamon Press, Gauthier-Villars-Paris, the Federation of American Societies for Experimental Biology, the American Medical Association, Associated Scientific Publishers (North Holland division), Friedr. Vieweg and Sohn, Rockefeller Univ. Press, the Journal of Neuropathology and Experimental Neurology, Williams and Wilkens (Histochemical Society and the U.S. Canadian division of the Int. Acad. Pathology). Little Brown and Co., Cambridge Univ. Press, Academic Press, Macmillan Journals Ltd., Wistar Press, American Soc. Biological Chemists.

Bibliography

Aaronson, S., 1973: Digestion in phytoflagellates. In: Lysosomes in biology and pathology. (Dingle, J. T., ed.), Vol. 3, pp. 18—37. Amsterdam: North-Holland Publishing.

Abelson, H. T., L. F. Johnson, S. Penman, and H. Green, 1974: Changes in RNA in relation to growth of the fibroblast: II. The lifetime of mRNA, rRNA, and tRNA in resting and growing cells. Cell 1, 161—165.

Abraham, R., L. Goldberg, and F. Coulstin, 1972: Uptake and storage of degraded carageenan in lysosomes of reticuloendothelial cells of the rhesus monkey, *Macaca mulatta*. Exp. molec. Path. **17**, 77—93.

— R. Hendy, and P. Grasso, 1968: Formation of myeloid bodies in rat liver lysosomes after chloroquine administration. Exp. molec. Path. **9**, 212—229.

Abrahams, S. J., and E. Holtzman, 1973: Secretion and endocytosis in insulin-stimulated rat adrenal medulla. J. Cell Biol. **56**, 540—558.

Ackerman, G. A., 1968: Ultrastructure and cytochemistry of the developing neutrophil. Lab. Invest. **19**, 290—302.

Ackerman, N. R., and J. R. Beebe, 1974: Release of lysosomal enzymes by alveolar mononuclear cells. Nature **247**, 475—477.

Adachi, M., L. Schenck, and B. W. Volk, 1974: Ultrastructural studies of eight cases of fetal Tay-Sachs' disease. Lab. Invest. **30**, 102—112.

— J. Torii, L. Schneck, and B. Volk, 1971: Fine structures of early Tay Sachs' disease. In: Sphingolipids, sphingolipidoses and allied disorders (Adv. Exp. Med. Biol. **19**). (Volk, B. W., and S. M. Aronson, eds.), pp. 1—13. New York: Plenum Press.

Adams, J. G., W. P. Winter, D. C. Rucknagel, and H. H. Spence, 1972: Biosynthesis of hemoglobin Ann Arbor; evidence for catabolic and feedback regulation. Science **176**, 1427—1429.

Adamson, E. Y. R., and D. M. Bowden, 1973: The intracellular site of surfactant synthesis: autoradiographic studies on murine and avian lung explants. Exp. molec. Path. **18**, 112—124.

Aikawa, M., 1972: High resolution autoradiography of malarial parasites treated with H³-chloroquine. Amer. J. Path. **67**, 277—284.

Albin, R., R. T. Dowell, R. Zak, and M. Rabinowitz, 1973: Synthesis and degradation of mitochondrial components in hypertrophied rat heart. Biochem. J. **136**, 629—635.

Albright, F. R., D. A. White, and W. J. Lennarz, 1973: Studies on enzymes involved in the catabolism of phospholipids in *Escherichia coli*. J. biol. Chem. **248**, 3968—3977.

Ali, S. Y., and L. Evans, 1968: Purification of rabbit kidney cytokinase and comparison of its properties with human urokinase. Biochem. J. **107**, 293—303.

— — 1973: Enzymatic degradation of cartilage in osteoarthritis. Fed. Proc. **32**, 1494—1498.

Allen, J. M., 1969: Lysosomes in bacterial infection. In: Lysosomes in biology and pathology. (Dingle, J. T., and H. B. Fell, eds.), Vol. 2, pp. 41—68. Amsterdam: North-Holland Publishing.

— E. M. Brieger, and R. J. W. Rees, 1965: Electron microscopy of the host cell-parasite relation in murine leprosy. J. Path. Bact. **89**, 301—306.

ALLEN, R. D., 1974: Food vacuole growth with microtubule associated membrane transport in *Paramecium*. J. Cell Biol. **63**, 904—922.
— 1975: Evidence for firm linkages between microtubules and membrane bounded vesicles. J. Cell Biol. **64**, 497—503.
— and R. W. WOLF, 1974: The cytoproct of *Paramecium caudatum*: structure, function, microtubules, and fate of food vacuole membranes. J. Cell Sci. **14**, 611—631.
ALLEN, R. R., J. J. LEE, S. H. HUTNER, and J. STORM, 1966: Prolonged culture of the voracious flagellate *Peranema* in antioxidant containing media. J. Protozool. **13**, 103—108.
ALLEN, T. D., and C. S. POTTEN, 1975: Desmosomal form, fate and function in mammalian epidermis. J. Ultrastruct. Res. **51**, 94—105.
ALLENDE, C. C., J. E. ALLENDE, and R. A. FIRTEL, 1974: The degradation of ribonucleic acids injected into *Xenopus laevis* oocytes. Cell **2**, 189—201.
ALLISON, A. C., 1968: Effects of drugs and toxic agents on lysosomes. In: Interaction of drugs and subcellular components in animal cells. (CAMPBELL, P. N., ed.), pp. 218—236. London: Churchill Ltd.
— 1969: Lysosomes and cancer. In: Lysosomes in biology and pathology. (DINGLE J. T., and H. B. FELL, eds.), Vol. **2**, pp. 178—204. Amsterdam: North-Holland Publishing.
— 1970: On the role of macrophages in some pathological processes. In: Mononuclear phagocytes. (VAN FURTH, R., ed.), pp. 422—440. Philadelphia: F. A. Davis.
— 1972: Types of lysosomal abnormality and the light they throw on normal lysosomal functions. In: Defects in cellular organelles and membranes in relation to mental retardation. (BENSON, P. F., ed.), (CIBA Foundation Study Group #2).
— 1974: The role of microfilaments and microtubules in cell movement, endocytosis and exocytosis. In: Locomotion of tissue cells. (CIBA Foundation symposium 14-New Series), pp. 109—148. Elsevier Press.
— and P. DAVIES, 1972: The control of lysosomal enzyme synthesis and the effects of steroids. In: Effects of drugs on cellular control mechanisms. (RABIN, B. R., and R. B. FREEDMAN, eds.), pp. 49—67. London: Macmillan Press.
— and E. F. HARTREE, 1970: Lysosomal enzymes in the acrosome and their possible role in fertilization. J. Reprod. Fert. **21**, 501—515.
— and L. MALLUCCI, 1964: Uptake of hydrocarbon carcinogens by lysosomes. Nature **203**, 1024—1027.
— and M. R. YOUNG, 1969: Vital staining and fluorescence microscopy of lysosomes. In: Lysosomes in biology and pathology. (DINGLE, J. T., and H. B. FELL, eds.), Vol. **2**, pp. 600—628. Amsterdam: North-Holland Publishing.
— J. S. HARRINGTON, and M. BIRBECK, 1966: An examination of the cytotoxic effects of silica on macrophages. J. exp. Med. **124**, 141—153.
AMSTERDAM, A., I. OHAD, and M. SCHRAMM, 1969: Changes in the ultrastructure of the acinar cells of the rat parotid gland during the secretory cycle. J. Cell Biol. **41**, 753—773.
— M. SCHRAMM, I. OHAD, Y. SALOMON, and Z. SELINGER, 1971: Concomitant synthesis of membrane protein and exportable protein of the secretory granule in rat parotid gland. J. Cell Biol. **50**, 187—200.
ANDERSON, E., 1969: Oogenesis in the cockroach *Periplaneta americana* with special reference to the specialization of the oolemma and the fate of coated vesicles. J. Microscopie **8**, 721—738.
ANDERSON, S. E., and J. S. REMINGTON, 1974: Effect of normal and activated human macrophages on *Toxoplasma gondii*. J. exp. Med. **139**, 1154—1177.
ANDERSON, W. A., 1968: Cytochemistry of sea urchin gametes. III. Acid and alkaline phosphatase activity of spermatozoa and fertilization. J. Ultrastruct. Res. **25**, 1—14.
ANDREWS, A. T., and C. PALLAVICINI, 1973: Bovine milk acid phosphatase. I. Some kinetic studies and other properties using a partially purified preparation. Biochim. biophys. Acta. **321**, 197—209.
ANSORGE, S., P. BOHLEY, M. KIRSCHKE, J. LAGNER, and M. HANSON, 1971: Metabolism of insulin and glucagon: breakdown of radioiodinated insulin and glucagon in rat liver cell fractions. Eur. J. Biochem. **19**, 283—288.

Antoine, J. C., S. Avrameas, N. K. Gonatas, A. Stieber, and J. D. Gonatas, 1964: Plasma membrane and internalized immunoglobulins of lymph node cells studied with conjugates of antibody and its FAB fragments with horseradish peroxidase. J. Cell Biol. **63**, 12—23.

Apte, B. N., P. N. Bhavsar, and O. Siddigi, 1974: The regulation of aryl sulphatase in *Aspergillus nidulans.* J. molec. Biol. **86**, 637—658.

Arborgh, B., T. Berg, and J. L. E. Ericsson, 1973: Quantification of acid phosphatase and aryl sulphatase in rat hepatic parenchymal and Kupffer cells. FEBS Letters **35**, 51—53.

— H. Glaumann, T. Berg, and J. L. E. Ericsson, 1974: Isolation of Kupffer cell lysosomes, with observations on their chemical and enzymic composition. Exp. Cell Res. **88**, 279—288.

Archer, G. T., and J. G. Hirsch, 1963: Isolation of granules from eosinophil leucocytes and study of their enzyme content. J. exp. Med. **118**, 277—285.

Arias, I. M., 1972: Transfer of bilirubin from blood to bile. Semin. Hematol. **9**, 55—70.

— D. Doyle, and R. T. Schimke, 1969: Studies on the synthesis and degradation of proteins of the endoplasmic reticulum of rat liver. J. biol. Chem. **244**, 3303—3315.

Armstrong, D., S. Dimmitt, D. H. Boehme, S. C. Leonberg, and W. Vogel, 1974: Leukocyte peroxidase deficiency in a family with a dominant form of Kuf's disease. Science **186**, 155—156.

Armstrong, J. A., and P. D'Arcy Hart, 1971: Response of cultured macrophages to Mycobacterium tuberculosis with observations on fusion of lysosomes with phagosomes. J. exp. Med. **134**, 713—740.

Armstrong, J. E., and R. L. Jones, 1973: Osmotic regulation of α-amylase synthesis and polyribosome formation in aleurone cells of barley. J. Cell Biol. **59**, 444—455.

Aronson, N. N., and E. A. Davidson, 1968: Digestive activity of lysosomes. II. The digestion of macromolecular carbohydrates by extracts of rat liver lysosomes. J. biol. Chem. **243**, 4564—4573.

Arstila, A., J. Jauregui, J. Chang, and B. Trump, 1971: Studies on cellular autophagocytosis. Lab. Invest. **24**, 162—174.

Arstila, A. U., and B. F. Trump, 1968: Studies on cellular autophagocytosis: the formation of autophagic vacuoles in the liver after glucagon administration. Amer. J. Path. **53**, 687—733.

— M. A. Smith, and B. F. Trump, 1972: Microsomal lipid peroxidation: Morphological characterization. Science **175**, 530—533.

— I. J. M. Nuuja, and B. F. Trump, 1974: Studies on cellular autophagocytosis: vinblastine induced autophagy in the rat liver. Exp. Cell Res. **87**, 249—252.

Aschenbrenner, V., R. Druyan, R. Albin, and M. Rabinowitz, 1970: Haem a, cytochrome c, and total protein turnover in mitochondria from rat heart and liver. Biochem. J. **119**, 157—160.

Ashford, T. P., and K. R. Porter, 1962: Cytoplasmic components in hepatic cell lysosomes. J. Cell Biol. **12**, 198—202.

Ashworth, J. M., 1971: Cell development in the cellular slime mould *Dictyostelium discoideum.* S. E. B. Symp. **25**, 27—49.

— and E. Wiener, 1973: The lysosomes of the cellular slime mould, *Dictyostelium discoideum.* In: Lysosomes in biology and pathology. (Dingle, J. T., ed.), Vol. 3, pp. 38—48. Amsterdam: North-Holland Publishing.

Askonas, B. A., and L. Jaroskova, 1970: Antigen in macrophages and antibody induction. In: Mononuclear phagocytes. (Van Furth, ed.), pp. 595—610. Philadelphia: F. A. Davis.

Aster, R. M., 1965: Studies of the fate of platelets in rats and man. Blood **34**, 117—128.

Atkinson, P. H., 1975: Synthesis and assembly of HeLa cell plasma membrane glycoproteins and proteins. J. Biol. Chem. **250**, 2123—2134.

Austin, J., 1973: Metachromatic Leukodystrophy (Sulfatide Lipidosis). In: Lysosomes and storage diseases. (Hers, H. G., and F. Van Hoof, eds.), pp. 411—437. New York: Academic Press.

Autissier, F., and A. Kepes, 1971: Segregation of membrane markers during cell division in *Escherichia coli*. II. Segregation of lac permease and mel permease studied with a penicillin technique. Biochim. biophys. Acta. **249**, 611—615.

Axline, S., 1968: Isozymes of acid phosphatase in normal and Calmette-guerin bacillus induced rabbit alvelolar macrophages. J. exp. Med. **128**, 1031—1048.

Axline, S. G., and Z. A. Cohn, 1970: *In vitro* induction of lysosomal enzymes by phagocytosis. J. exp. Med. **131**, 1239—1260.

— and E. P. Reavan, 1974: Inhibition of phagocytosis and plasma membrane mobility of the cultivated macrophage by cytochalasin B. J. Cell Biol. **62**, 647—659.

Bach, G., R. Friedman, B. Weissmann, and E. F. Neufeld, 1972: The defect in Hurler and Scheie syndromes: deficiencies of -L-iduronidase. Proc. nat. Acad. Sci. (U.S.A.) **69**, 2048—2051.

Badenoch-Jones, P., and H. Baum, 1973: Progesterone-induced permeability changes in rat liver lysosomes. Nature New Biol. **242**, 123—124.

— 1974: Further evidence for the heterogeneity of liver lysosomes. FEBS Letter **43**, 227—230.

Baggiolini, M., C. de Duve, P. L. Masson, and J. F. Heremans, 1970: Association of lactoferrin with specific granules in rabbit heterophil leukocytes. J. exp. Med. **131**, 559—570.

— J. G. Hirsch, and C. de Duve, 1969: Resolution of granules from rabbit heterophil leukocytes into two distinct populations by zonal sedimentation. J. Cell Biol. **40**, 529—541.

Bailey, E., C. B. Taylor, and W. Bartley, 1967: Turnover of mitochondrial components of normal and essential fatty acid deficient rats. Biochem. J. **104**, 1026—1032.

Baintner, K., and G. Feher, 1974: Fate of egg white trypsin inhibitor and start of proteolysis in developing chick embryo and newly hatched chick. Dev. Biol. **36**, 272—278.

Bainton, D. F., 1972: Origin, content, and fate of PMN granules. In: Phagocytic mechanisms in health and diesease. (Williams, R. C., Jr., and H. M. Judenberg, eds.). New York: Intercontinental Medical Book Co.

— 1973: Sequential degranulation of the two types of polymorphonuclear leukocyte granules during phagocytosis of microorganisms. J. Cell Biol. **58**, 249—264.

— and M. G. Farquhar, 1968: Differences in enzyme content of azurophil and specific granules of polymorphonuclear leukocytes. I. Histochemical staining of bone marrow smears. II. Cytochemistry and electron microscopy of bone marrow cells. J. Cell Biol. **39**, 286—298, 299—317.

— — 1970: Segregation and packaging of granule enzymes in eosinophilic leukocytes. J. Cell Biol. **43**, 54—73.

— J. L. Ullyot, and M. G. Farquhar, 1971: The development of neutrophilic polymorphonuclear leukocytes in human bone marrow. J. exp. Med. **134**, 907—934.

Balduni, C., C. L. Balduni, and E. Ascari, 1974: Membrane glycopeptides from old and young human erythrocytes. Biochem. J. **140**, 557—560.

Balis, J. U., and P. E. Conen, 1964: The role of alveolar inclusion bodies in the developing lung. Lab. Invest. **13**, 1215—1229.

Ballard, F. J., M. F. Hopgood, L. Reshef, and R. W. Hanson, 1974: Degradation of phosphoenolpyruvate carboxykinase (guanosine triphosphate) *in vivo* and *in vitro*. Biochem. J. **140**, 531—538.

Bangham, A. D., and R. M. C. Dawson, 1962: Electrokinetic requirements for the reaction between Cl. perfringens α-toxin phospholipase C and phospholipid substrates. Biochim. biophys. Acta **59**, 103—115.

Barham, S. S., and J. D. Berlin, 1974: Fine structure and cytochemistry of testicular cells in man treated with testosterone propionate. Cell. Tiss. Res. **148**, 159—182.

Barisas, B. G., and J. S. McGuire, 1974: A proteolytically activated tyrosinase from frog epidermis. J. biol. Chem. **249**, 3151—3156.

Barka, T., 1974: Lysosomes and regulation of cell proliferation. In: The liver and its diseases. (Shaffner, Sherlock, and Leevy, eds.), pp. 19—23. Intercontinental Medical Book Corp.

BARKA, T., and P. J. ANDERSON, 1962: Histochemical methods for acid phosphatase using hexazonium pararosanilin as coupler. J. Histochem. Cytochem. **10**, 741—753.

BARNHART, M., 1965: I. Importance of neutrophilic leukocytes in the resolution of fibrin. Fed. Proc. **24**, 846—853.

BARRATT, M. E. J., 1973: The role of soft connective tissue in the response of pig articular cartilage in organ culture to excess of retinol. J. Cell Sci. **13**, 205—219.

BARRETT, A. J., 1969: Properties of lysosomal enzymes. In: Lysosomes in biology and pathology. (DINGLE, J. T., and H. B. FELL, eds.), Vol. 2, pp. 245—312. Amsterdam: North-Holland Publishing.

— 1972: Lysosomal enzymes. In: Lysosomes: A laboratory handbook. (DINGLE, J. T., ed.), pp. 46—135. Amsterdam: North-Holland Publishing.

— and J. T. DINGLE, 1967: A lysosomal component capable of binding cations and a carcinogen. Biochem. J. **105**, 20 P.

— — (ed.), 1971: Tissue Proteinases. Amsterdam: North-Holland Publishing.

— and P. M. STARKEY, 1973: The interaction of α-2-macroglobulin with proteinases. Biochem. J. **133**, 709—724.

BARTHOLOMEW, B. A., and A. L. PERRY, 1973: The properties of synovial fluid β-mannosidase activity. Biochim. biophys. Acta. **315**, 123—127.

BARTON, R., 1966: Fine structure of mesophyll cells in senescing leaves of *Phaseolus*. Planta (Berl.) **71**, 314—325.

BAUDHIN, P., P. EVERARD, and J. BERTHET, 1967: Electron microscope examination of subcellular fractions. I. Preparation of representative samples from suspensions of particles. J. Cell Biol. **32**, 181—191.

BAUER, W. C., and J. S. MEYER, 1965: Origin and fate of colloid droplets of the thyroid gland: A study using iodine-125 electron microscopic autoradiography. Lab. Invest. **14**, 1795—1808.

BEATTIE, D. S., 1969: The turnover of the protein components of the inner and outer membrane fractions of rat liver mitochondria. Biochem. biophys. Res. Commun. **35**, 712—727.

BEAUFAY, H., 1972: Methods for the isolation of lysosomes. In: Lysosomes: A laboratory handbook. (DINGLE, J. T., ed.), pp. 1—45. Amsterdam: North-Holland Publishing.

— A. AMAR-COSTESEC, D. THINES-SEMPOUX, M. WIBO, M. ROBBI, and J. BERTHET, 1974: Analytical studies of microsomes and isolated subcellular membranes from rat liver. III. Subfractionation of the microsome fraction by isopycnic and differential centrifugation in density gradients. J. Cell Biol. **61**, 213—231.

BEAULATON, J., 1967: Localisation d'activite lytique dans la glande prothoracique du ver a soie du chene (Anthera pernyi Guer) au stade prenymphal. II. Les vacuoles autolytiques (cytolysosomes). J. Microscopie **6**, 345—370.

BECK, F., 1965: The distribution of acid phosphatase in the chick blastoderm. Exp. Cell Res. **37**, 504—508.

— J. B. LLOYD, and C. A. SQUIER, 1972: Histochemistry. In: Lysosomes: A laboratory handbook. (DINGLE, J. T., ed.), pp. 200—239. Amsterdam: North-Holland Publishing.

BECKER, F. F., and T. C. CORNWALL, 1971: Phlorizin-induced autophagocytosis during hepatocytic glycogenolysis. Exp. molec. Path. **14**, 103—109.

— and B. P. LANE, 1966: Regeneration of the mammalian liver. IV. Evidence of the role of cytoplasmic alterations in preparation for cell division. Amer. J. Path. **49**, 227—237.

BECKER, N. H., A. HIRANO, and H. M. ZIMMERMAN, 1968: Observations of the distribution of exogenous peroxidase in the rat cerebrum. J. Neuropath. exp. Neurol. **27**, 439—452.

BEGG, K. J., and W. D. DONACHIE: Topography of outer membrane growth in *E. coli*. Nature New Biol. **245**, 38—39.

BEHNKE, O., and H. MOE, 1964: An electron microscope study of mature and differentiating paneth cells in the rat, especially of their endoplasmic reticulum and lysosomes. J. Cell Biol. **22**, 633—652.

BEISEL, W. R., and M. I. RAPOPORT, 1969: Inter-relations between adrenocortical functions and infectious illness. New Eng. J. Med. **48**, 541—546, 596—604.

BELL, W. J., 1971: Starvation induced oocyte resorption and yolk protein salvage in *Periplaneta americana*. J. Insect. Physiol. **17**, 1099—1111.

Benacerraf, B., 1964: Functions of the Kupffer cells. In: The Liver. (Rouiller, Ch., ed.), Vol. 2, pp. 37—62. New York: Academic Press.

Benes, F., J. A. Higgins, and R. J. Barrnett, 1973: Ultrastructural localization of phospholipid synthesis in rat trigeminal nerve during myelination. J. Cell Biol. **57**, 613—629.

Bengt, A. M., H. Glaumann, and J. L. E. Ericsson, 1974: Studies on iron loading of rat liver lysosomes: effects on the liver and distributions and fate of iron. Lab. Invest. **30**, 664—673, 674—680.

Bennett, G., and C. P. Leblond, 1971: Passage of fucose-^{3}H label from the Golgi apparatus into dense and multivesicular bodies in the duodenal columnar cells and hepatocytes of the rat. J. Cell Biol. **51**, 875—881.

— and A. Haddad, 1974: Migration of glycoprotein from the Golgi apparatus to the surface of various cell types as shown by autoradiography after labelled fucose injection into rats. J. Cell. Biol. **60**, 258—284.

Bentfield, M. E., and D. F. Bainton, 1974: Localization of lysosomal enzymes in rat megakaryocytes and platelets. J. Cell Biol. **63**, 22 a.

Berger, J. D., 1973: Selective inhibition of DNA synthesis in macronuclear fragments in *Paramecium aurelia* exconjugants and its reversal during macronuclear regeneration. Chromosoma **44**, 33—48.

— and R. F. Kimball, 1964: Specific incorporation of precursors of DNA feeding labelled bacteria to *Paramecium aurelia*. J. Protozool. **11**, 534—537.

Bergeron, J. J. M., J. H. Ehrenreich, P. Siekevitz, and G. E. Palade, 1973: Golgi fractions prepared from rat liver homogenates. II. Biochemical characterizations. J. Cell Biol. **59**, 73—88.

— W. H. Evans, and I. I. Gerschwind, 1973: Insulin binding to rat liver Golgi fractions. J. Cell Biol. **59**, 771—776.

Berghem, L. E. R., and L. G. Pettersson, 1974: The mechanism of enzymatic cellulose degradation: isolation and some properties of a β-glucosidase from *Trichoderma viride*. Eur. J. Biochem. **46**, 295—305.

Bergink, F. W., and R. A. Wallace, 1974: Precursor product relationship between amphibian vitellogenin and the yolk proteins lipovitellin and phosvitin. J. biol. Chem. **249**, 2897—2903.

Berjack, P., and T. A. Villiers, 1972: Ageing in plant embryos. II. Age induced damage and its repair during early seed germination. New Phytol. **71**, 135—144.

— 1968: A lysosome-like organelle in the root cap of *Zea Mays*. J. Ultrastruct. Res. **23**, 233—243.

— and T. A. Villiers, 1970: Ageing in plant embryos. I. The establishment of the sequence of development and senescence in the root cap during germination. New Phytol. **69**, 929—938.

Berkeley, R. C. W., S. J. Brewer, J. M. Ortiz, and J. B. Gillespie, 1973: An exo-β-N-acetylglucosaminidase from *Bacillus subtilis* B: characterization. Biochim. biophys. Acta. **309**, 157—168.

Berlin, R. D., J. M. Oliver, T. E. Ukena, and H. H. Yin, 1974: Control of cell surface topography. Nature **247**, 45—46.

Bernsohn, J., and H. J. Grossmann (eds.), 1971: Lipid storage diseases: enzymatic defects and clinical implications. New York: Academic Press.

Berther, J., L. Berthet, F. Appelmans, and C. de Duve, 1951: Tissue fractionation studies. 2. The nature of the linkage between acid phosphatase and mitochondria in rat liver tissue. Biochem. J. **50**, 182—189.

Bertini, F., 1969: The effect of low body temperature on the intraparticulate hydrolysis of I^{131}-albumin in toads. Z. Naturforsch. **24 b**, 141—142.

Bessis, M. C., and J. Breton-Gorius, 1962: Iron metabolism in the bone marrow as seen by electron microscopy: a critical review. Blood **19**, 635—663.

Betz, H., H. Hinge, and H. Holzer, 1974: Isolation and purification of two inhibitors of proteinase B from yeast. J. biol. Chem **249**, 4515—4521.

Bhisey, A. N., and J. J. Freed, 1971: Altered movement of endosomes in colchicine-treated cultured macrophages. Exp. Cell Res. **64**, 2130—2138.

Biava, C., and M. West, 1965: Lipofuscin-like granules in vascular smooth muscle and juxtaglomerular cells of human kidneys. Amer. J. Path. **47**, 287—313.

Bibb, C., and R. W. Young, 1974 a: Renewal of fatty acids in the membrane of visual cell outer segments. J. Cell Biol. **61**, 327—343.

— — 1974 b: Renewal of glycerol in the visual cells and pigment epithelium of the frog retina. J. Cell Biol. **62**, 378—389.

Bingham, E. W., and H. M. Farrell, 1974: Casein kinase from the Golgi apparatus of lactating mammary gland. J. biol. Chem. **249**, 3647—3651.

Birks, R. I., and co-workers, 1972: Organelle formation from pinocytotic elements in neurites of cultured sympathetic ganglia. J. Neurocytol. **1**, 311—330.

Bishop, J. M., and G. Koch, 1967: Purification and characterization of poliovirus induced infectious double-stranded ribonucleic acid. J. biol. Chem. **242**, 1736—1743.

Bissell, D. M., L. Hammaker, and R. Schmid, 1972: Liver sinusoidal cells: identification of a subpopulation for erythrocyte catabolism. J. Cell Biol. **54**, 107—119.

Bitensky, L., 1963: The reversible activation of lysosomes in normal cells and the effects of pathological conditions. In: Ciba foundation symposium on lysosomes. (de Reuck, A. V. S., and M. P. Cameron, eds.), pp. 362—383. Boston: Little Brown.

— R. G. Butcher, and J. Chayen, 1973: Quantitative cytochemistry in the study of lysosomal function. In: Lysosomes in biology and pathology. (Dingle, J. T., ed.), Vol. 3, pp. 465—510. Amsterdam: North-Holland Publishing.

Bjersing, L., and S. Cajander, 1974: Ovulation and the mechanism of follicle rupture. III. Transmission electron microscopy of rabbit germinal epithelium prior to induced ovulation. Cell and Tissue Research, (Z. Zellforsch.) **149**, 313—328.

Black, W. J., T. Takano, and T. J. Peters, 1973: Subcellular distribution of cholesteryl-esterase in rabbit aortic smooth muscle cells. J. Cell Biol. **59**, 26 a.

Bloodgood, R. A., 1974: Resorption of organelles containing microtubules. Cytobios. **9**, 143—161.

Blum, J. J., 1965: Observations on the acid phosphatases of *Euglena gracilis*. J. Cell Biol. **24**, 223—234.

Bock, K. W., P. Siekevitz, and G. E. Palade, 1971: Localization and turnover studies of membrane nicotinamide-adenine-dinucleotide glycohydrolase in rat livers. J. biol. Chem. **246**, 188—195.

Bode, F., H. Pockrandt-Hemstedt, K. Baumann, and R. Kinne, 1974: Analysis of the pinocytic process in rat kidney: isolation of pinocytic vesicles from rat kidney cortex. J. Cell Biol. **63**, 998—1008.

Bohley, P., H. Kirschke, J. Langner, S. Ansorge, B. Wiederanders, and H. Hanson, 1971: Intracellular protein breakdown. In: Tissue proteinases. (Barrett, A. J., and J. T. Dingle, eds.), pp. 187—219. Amsterdam: North-Holland Publishing.

Bok, D., and M. O. Hall, 1971: The role of the pigment epithelium in the etiology of inherited retinal dystrophy in the rat. J. Cell Biol. **49**, 664—682.

— and R. W. Young, 1972: The renewal of diffusely distributed protein in the outer segments of rods and cones. Vision Res. **12**, 161—168.

Bolender, R. P., 1974: Sterological analysis of the guinea pig pancreas. I. Analytical model and quantitative description of non-stimulated pancreatic exocrine cells. J. Cell Biol. **61**, 269—287.

— and E. R. Weibel, 1973: A morphometric study of the removal of phenobarbital induced membranes from hepatocytes after cessation of treatment. J. Cell Biol. **56**, 746—761.

Boler, R. K., and R. B. Arhelger, 1966: Microtubules in cytosomes and cytosegresomes of rabbit proximal tubule epithelium. Lab. Invest. **15**, 320—331.

Bond, J. S., 1971: A comparison of the proteolytic susceptibility of several rat liver enzymes. Biochem. biophys. Res. Commun. **43**, 333—339.

Bonsignore, A., A. deFlora, M. A. Margiarotti, I. Lorenzoni, and S. Alema, 1968: A new hepatic protein inactivating glucose-6-phosphate dehydrogenase. Biochem. J. **106**, 147—154.

Boon, T., 1971: Inactivation of ribosomes *in vitro* by colicin E 3. Proc. nat. Acad. Sci. (U.S.A.) **68**, 2421—2425.

Borek, C., N. Grob, and M. M. Berger, 1973: Surface alterations in transformed epithelial and fibroblastic cells in culture: a disturbance of membrane degradation versus biosynthesis. Exp. Cell Res. **77**, 207—215.

Borun, T. W., and G. S. Stein, 1972: The synthesis of acidic chromosomal proteins during the cell cycle of HeLa S-3 cells. II. The kinetics of residual protein synthesis and transport. J. Cell Biol. **52**, 308—315.

Bosmann, H. B., 1974: Release of specific protease during the mitotic cycle of L 5187 murine leukemic cells by sublethal autolysis. Nature **249**, 144—145.

— T. Lockwood, and H. R. Morgan, 1974: Surface biochemical changes accompanying primary infection with Rous Sarcoma virus. II. Proteolytic and glycosidase activity and sublethal autolysis. Exp. Cell Res. **83**, 25—30.

Boutry, J.-M., and A. B. Novikoff, 1975: Cytochemical studies on Golgi apparatus, GERL and lysosomes in neurons of dorsal root ganglia in mice. Proc. Nat. Acad. Sci. (U.S.) **72**, 508—512.

Bowen, I. D., 1973: Glycosidases, β-glucosidases, β-glucuronidase. In: Electron microscopy of enzymes. (Hayat, M. A., ed.), Vol. **1**, pp. 77—103. New York: Van Nostrand Reinhold.

Bowers, B., and T. E. Olszewski, 1972: Pinocytosis in *Acantamoeba*. J. Cell Biol. **53**, 681—694.

Bowers, W. E., 1969: Lysosomes in lymphoid tissues: spleen, thymus, and lymph nodes. In: Lysosomes in biology and pathology. (Dingle, J. T., and H. B. Fell, eds.), Vol. **1**, pp. 167—191. Amsterdam: North-Holland Publishing.

— 1970: A biochemical study of lysosomes in lymphoid tissues. In: Mononuclear phagocytes. (Van Furth, R., ed.), pp. 81—96. Philadelphia: F. A. Davis.

— 1973: Lysosomes in rat thoracic duct lymphocytes fractionated by zonal centrifugation. J. Cell Biol. **59**, 177—184.

Boyer, H. W., L. T. Chow, A. Dugaiczyk, J. Hedgpeth, and H. M. Goodman, 1973: DNA substrate site for the Eco R II restriction endonuclease and modification methylase. Nature New Biol. **244**, 40—44.

Braatz, J. A., and E. C. Heath, 1974: The role of polysaccharide in the secretion of protein by *Micrococcus sodonensis*. J. biol. Chem. **249**, 2536—2547.

Bracker, C. E., and S. N. Grove, 1971: Continuity between cytoplasmic and endomembranes and outer mitochondrial membranes in fungi. Protoplasma **73**, 15—34.

— and C. M. Williams, 1966: Comparative ultrastructure of developing sporangea and asci in fungi. In: Int. Cong. for Electron Microscopy. (Uyeda, R., ed.), Vol. **6** (part 2), pp. 307—308. Tokyo: Maruzen Co.

Bradford, W. D., J. G. Elchlepp, A. N. Arstila, B. F. Trump, and T. D. Kinney, 1969: Iron metabolism and cell membranes. Amer. J. Path. **56**, 201—228.

Bradfute, O. E., Chapman-Andresen, and W. A. Jensen, 1964: Concerning morphological evidence for pinocytosis in higher plants. Exp. Cell Res. **36**, 207—210.

Brady, R. O., 1971: The ups and downs of complex lipid metabolism. In: Lipid storage diseases. (Bersohn, J., and H. J. Grossman, eds.), pp. 275—289. New York: Academic Press.

— P. G. Pentschev, and A. E. Gal, 1975: Investigations in enzyme replacement therapy in lipid storage diseases. Fed. Proc. **34**, 1310—1315.

— 1973: The abnormal biochemistry of inherited disorders of lipid metabolism. Fed. Proc. **32**, 1660—1667.

— and F. M. King, 1973: Gaucher's Disease. In: Lysosomes and storage diseases. (Hers, H. G., and F. Van Hoof, eds.), pp. 381—394. New York: Academic Press.

Braidman, I., and D. Robinson, 1973: Glycosidases in pregnancy and postnatal development in the rabbit. Biochem. Soc. Trans. **1**, 1176—1179.

Brandes, D., 1965: Observations on the apparent mode of formation of "pure" lysosomes. J. Ultrastruct. Res. **12**, 63—80.

— 1966: The fine structure and histochemistry of prostatic glands in relation to sex hormones. Int. Rev. Cytol. **20**, 207—276.

Brandes, D., and E. Anton, 1966: The role of lysosomes in cellular lytic processes. III. Electron histochemical changes in mammary tumors after treatment with cytoxan and vitamin A. Lab. Invest. **15**, 987—1006.

— — 1969: Lysosomes in uterine involution: intracytoplasmic degradation of myofilaments and collagen. J. Gerontol. **24**, 55—69.

— D. E. Buetow, R. Bertini, and D. B. Malkoff, 1964: Roles of lysosomes in cellular lytic processes. I. Effect of carbon starvation in *Euglena gracilis*. Exp. molec. Path. **3**, 583—609.

— R. Bertini, and E. W. Smith, 1965: Role of lysosomes in cellular lytic processes. II. Cell death during holocrine secretion in sebaceous glands. Exp. molec. Path. **4**, 245—265.

Brandt, E. J., R. Zeigel, and R. Swank, 1974: Abnormal control of a lysosomal enzyme in mice with Chediak-Higashi syndrome. J. Cell Biol. **63**, 35 a.

Brandt, P. W., and A. R. Freeman, 1967: The role of surface chemistry in the biology of pinocytosis. J. Colloid and Interface Science **25**, 47—56.

Brandt, P., and K. B. Hendil, 1972: Plasma membrane permeability and pinocytosis in *Chaos chaos*. C. R. Lab. Carlsberg **38**, 423—443.

Braten, T., 1973: Autoradiographic evidence for the rapid disintegration of one chloroplast in the zygote of the green alga, *Ulva mutabilis*. J. Cell Sci. **12**, 385—389.

Braunsteiner, M., and F. Schmalzl, 1970: Cytochemistry of monocytes and macrophages. In: Mononuclear phagocytes. (Van Furth, R., ed.), pp. 62—81. Philadelphia: F. A. Davis.

Breton-Gorius, J., and J. Guichard, 1972: Ultrastructural localization of peroxidase activity in human platelets and megakaryocytes. Amer. J. Path. **66**, 277—286.

Bretz, M., and M. Baggiolini, 1973: Association of the alkaline phosphatase of rabbit polymorphonuclear leukocytes with the membrane of the specific granules. J. Cell Biol. **59**, 696—707.

Bretz, U., and M. Baggiolini, 1974: Biochemical and morphological characterization of azurophil and specific granules of human polymorphonuclear leukocytes. J. Cell Biol. **63**, 251—269.

Briggs, R. T., M. L. Karnovsky, and M. J. Karnovsky, 1975: Cytochemical demonstration of hydrogen peroxide in polymorphonuclear leukocyte phagosomes. J. Cell Biol. **64**, 254—259.

Brizzee, K. R., P. A. Cancilla, N. Sherwood, and P. S. Timiras, 1969: The amount and distribution of pigments in neurons and glia of the cerebral cortex. J. Gerontol. **24**, 127—135.

Brock, M. A., 1970: Ultrastructural studies on the life cycle of a short-lived metazoan, *Campanularia flexura*. J. Ultrastruct. Res. **32**, 118—141.

— and R. J. Hay, 1971: Comparative ultrastructure of chick fibroblasts *in vitro* at early and late stages during their growth span. J. Ultrastruct. Res. **36**, 391—411.

Brockman, H. C., J. M. Law, and F. J. Kezdy, 1973: Catalysis by adsorbed enzyme: the hydrolysis of tripropionin by pancreatic lipase adsorbed to siliconized glass beads. J. biol. Chem. **24**, 4965—4970.

Brostoff, S. W., W. Reuter, M. Michens, and E. M. Eylar, 1974: Specific cleavage of the A protein from myelin with cathepsin D. J. biol. Chem. **249**, 559—567.

Brostrom, C. P., and H. Jeffay, 1970: Protein catabolism in rat liver homogenates; a revaluation of the energy requirements for protein catabolism. J. biol. Chem. **245**, 4001—4008.

Brot, F. E., J. M. Glaser, K. J. Roozen, W. S. Sly, and P. D. Stahl, 1974: *In vitro* correction of deficient human fibroblasts by β-glucuronidase from different human sources. Biochem. biophys. Res. Commun. **57**, 1—8.

Brownstone, Y. S., and C. Chapman-Andresen, 1971: Degradation of endocytosed albumin by *Chaos chaos*. C. R. Lab. Carlsberg **38**, 297—313.

Brummett, A. R., and J. N. Dumont, 1974: Localization of negative charges on the surface of *Xenopus oocytes*. Anat. Rec. **178**, 317—326.

Brun, A., and U. Brunk, 1973: Heavy metal localization and age-related accumulations in the rat nervous system. Histochemie **34**, 333—342.

BRUNK, U., 1973: Distribution and shifts of ingested marker particles in residual bodies and other lysosomes. Exp. Cell Res. **79**, 15—27.

— A. BRUN, and G. SKOLD, 1968: Histochemical demonstration of heavy metals with the sulfide-silver method. Acta. Histochem. **31**, 345—357.

— — 1972: Histochemical evidence for lysosomal uptake of lead in tissue cultured fibroblasts. Histochemie **29**, 140—146.

— and J. L. E. ERICSSON, 1972 a: The demonstration of acid phosphatase in *in vitro* cultured tissue cells: studies on the significance of fixation, tonicity, and permeability. Histochem. J. **4**, 349—363.

— — 1972 b: Cytochemical evidence for the leakage of acid phosphatase through ultrastructurally intact lysosomal membranes. Histochem. J. **4**, 479—491.

— — J. PONTEN, and R. WESTERMARK, 1973: Residual bodies and aging in cultured human glia cells. Effects of entrance into phase III and prolonged periods of confluence. Exp. Cell Res. **79**, 1—14.

BRUNS, G. A. P., and P. S. GERALD, 1974: Human acid phosphatase in somatic cell hybrids. Science **184**, 480—481.

BRUSH, J. S., 1971: Purification and characterization of a protease with specificity for insulin from rat muscle. Diabetes **20**, 151—155.

BRYAN, J. M. D., and R. R. UNNITHAN, 1973: Cytochemical localization of non specific esterase and acid phosphatase in spermatozoa of the mouse, *Mus musculus*. Histochemie **33**, 169—180.

BUCKLEY, I. K., 1973: The lysosomes of cultured chick embryo cells: a correlated light and electron microscopic study. Lab. Invest. **29**, 411—421.

BUJAK, J. S., and R. K. ROOT, 1974: The role of peroxidase in the bactericidal activity of human blood eosinophils. Blood **43**, 727—736.

BUKHARI, A. I., and D. ZIPSER, 1973: Mutants of *Echerichia coli* with a defect in the degradation of nonsense fragments. Nature New Biol. **243**, 238—241.

BULLOCK, S., and B. WINCHESTER, 1973: The N-acetylhexosaminidase components of the ram testis and epididymis. Biochem. J. **133**, 593—599.

BUNGE, M. B., 1973: Fine structure of nerve fibers and growth cones of isolated sympathetic neurons in culture. J. Cell Biol. **56**, 713—735.

BUNN, H. F., 1972: Erythrocyte destruction and hemoglobin catabolism. Semin. Hematol. **9**, 3—17.

BUNT, A. H., 1969: Formation of coated and synaptic vesicles within neurosecretory axon terminals of the crustacean sinus gland. J. Ultrastruct. Res. **28**, 411—421.

BUONGIORNO, M., and B. BERTOLINI, 1967: Subcellular localization of some acid hydrolases in *Triturus cristatus* sperm. Histochemie **8**, 34—44.

BURLEIGH, M. C., A. J. BARRETT, and G. S. LAZARUS, 1974: Cathepsin B-1, a lysosomal enzyme that degrades native collagen. Biochem. J. **137**, 387—398.

BUTLER, R. D., and E. W. SIMON, 1971: Ultrastructural aspects of senescence in plants. Adv. Gerontol. Res. **3**, 73—146.

BUTTERWORTH, F. M., and B. L. LA TENDRESSE, 1973: Quantitative studies of cytochemical and cytological changes during cell death in the larval fat body of *Drosophila melanogaster*. J. Insect Physiol. **19**, 1487—1499.

CABIB, E., R. ULANE, and B. BOWERS, 1973: Yeast chitin synthase: separation of the zymogen from its activating factor and recovery of the latter in the vacuole fraction. J. biol. Chem. **248**, 1451—1458.

CAMERON, I. C., 1971: Cell proliferation and renewal in the mammalian body. In: Cellular and molecular renewal in the mammalian body. (CAMERON, I. C., and J. D. THRASHER, eds.), pp. 45—85. New York: Academic Press.

CAMPBELL, F. R., 1968: Nuclear elimination from the normoblast of fetal guinea pig liver as studied with electron microscopy and serial sectioning techniques. Anat. Rec. **160**, 539—554.

— 1972: Electron microscope studies on the fate of erythroblast ribosomes. Anat. Rec. **174**, 513—520.

Cantz, M., and M. Kresse, 1974: Sandhoff disease: Defective glycosaminoglycan catabolism in cultured fibroblasts and its correction by β-N-acetylhexosaminidase. Eur. J. Biochem. **47**, 581—590.

Capecchi, M. R., N. E. Capecchi, S. H. Hughes, and G. M. Wahl, 1974: Selective degradation of abnormal proteins in mammalian tissue culture cells. Proc. Nat. Acad. Sci. (U.S.) **71**, 4732—4736.

Carasso, N., P. Favard, et S. Goldfischer, 1964: Localisation a l'échelle des ultrastructures, d'activites de phosphatases en rapport avec les processus digestifs chez un cilié (*Campanella umbellaria*). J. Microscopie **3**, 297—322.

Cardella, C. J., P. Davies, A. C. Allison, 1974: Immune complexes induce selective release of lysosomal hydrolases from macrophages. Nature **247**, 46—48.

Carlson, P., 1973: The use of protoplasts for genetic research. Proc. nat. Acad. Sci. (U.S.A.) **70**, 598—602.

Carothers, Z. B., 1972: Membrane continuity between plasmalemma and nuclear envelope in spermatogenic cells of *Blasia*. Science **175**, 652—654.

Carr, I., 1973: The Macrophage: A review of ultrastructure and function. New York: Academic Press.

Carroll, M., and D. Robinson, 1974: A low molecular weight protein cross reacting with human liver N-acetyl β-D-glucosaminidase. Biochem. J. **137**, 217—221.

Casley-Smith, J. R., 1964: The Brownian movement of pinocytic vesicles. J. roy. micr. Soc. B **82**, 257—261.

— 1969: Endocytosis: the different energy requirements for the uptake of particles by small and large vesicles into peritoneal macrophages. J. Microscopy **90**, 15—30.

Castle, J. D., J. D. Jamieson, and G. E. Palade, 1974: Secretion granules of the rabbit parotid gland: contamination analysis of the membrane subfraction. J. Cell Biol. **63**, 52 a.

Castleman, W. L., D. L. Dungworth, and W. S. Tyler, 1973: Cytochemically demonstrable alterations of lung acid phosphatase following ozone exposure. Lab. Invest. **29**, 310—319.

Catesson, A. M., and Y. Czaninski, 1968: Localisation ultrastructurale de la phosphatase acide et cycle saissoner dans les tissus conducteurs de quelques arbres. Bull. Soc. Franc. Physiol. Veget. **14**, 65—73.

Catherwood, B., and F. R. Singer, 1974: Generation of a carboxyl terminal fragment of bovine parathyroid hormone by canine renal plasma membrane. Biochem. biophys. Res. Commun. **57**, 469—475.

Ceccarelli, B., W. P. Hurlburt, and A. Mauro, 1973: Turnover of transmitter and synaptic vesicles at the frog neuromuscular junction. J. Cell Biol. **57**, 499—524.

Chapman-Andresen, C., and H. Holter, 1955: Studies on the ingestion of C 14 glucose by pinocytosis in the amoeba *Chaos chaos*. Exp. Cell Res. Suppl. **3**, 52—63.

— 1964: Measurements of material uptake by cells: pinocytosis. In: Methods in cell physiology. (Prescott, D., ed.), Vol. I, pp. 277—304. Academic Press.

— and D. Lagunoff, 1966: The distribution of acid phosphatase in the amoeba *Chaos chaos*. C. R. Lab. Carlsberg **35**, 419.

Chen, L.-T., and L. Weiss, 1972: Electron microscopy of the red pulp of human spleen. Amer. J. Anat. **134**, 425—458.

— — 1973: The role of the sinus wall in the passage of erythrocytes through the spleen. Blood **41**, 529—537.

Chesney, C. McI., E. Harper, and R. W. Coleman, 1974: Human platelet collagenase. J. Clin. Invest. **53**, 1647—1654.

Chevalier, G., and A. J. Collet, 1972: *In vivo* incorporation of choline-³H, leucine-³H and galactose-³H in alveolar type II pneumocytes in relation to surfactant synthesis; a quantitative radioautographic study in mouse by electron microscopy. Anat. Rec. **174**, 289—310.

Chien, S., G. W. Cooper, K.-M. Jan, L. H. Miller, C. Howe, S. Usami, and P. Lalezart, 1974: N-acetylneuraminic acid deficiency in erythrocyte membranes: biophysical and biochemical correlates. Blood **43**, 445—460.

Chio, K. S., U. Reiss, B. Fletcher, and A. L. Tappel, 1969: Peroxidation of subcellular organelles: formation of lipofuscin-like fluorescent pigments. Science **166**, 1535—1536.

CHLAPOWSKI, F. J., and R. N. BAND, 1971: Assembly of lipids into membranes in *Acanthamoeba palestinensis.* J. Cell Biol. **50**, 625—633; 634—651.

CHRISPEELS, M. J., 1974: Mechanism of osmotic regulation of hydrolase synthesis in aleurone cells of barley: inhibition of protein synthesis. Biochem. biophys. Res. Commun. **53**, 99—104.

CHRISTIANSEN, R. G., and J. M. MARSHALL, 1965: A study of phagocytosis in the amoeba *Chaos chaos.* J. Cell Biol. **25**, 443—457.

CHRISTMAN, J. K., G. ACS, S. SILAGI, E. W. NEWCOMB, and S. SILVERSTEIN, 1974: Correlation of decreased release of plasminogen activator with loss of tumorigenicity in bromodeoxyuridine-grown melanoma cells. J. Cell Biol. **63**, 61 a.

CHUA, N. H., G. BLOBEL, P. SIEKEVITZ, and G. E. PALADE, 1973: Attachment of chloroplast polysomes to thylakoid membranes in *Chlamydomonas reinhardti.* Proc. nat. Acad. Sci. (U.S.A.) **70**, 1554—1558.

CIARANELLO, R. D., and J. AXELROD, 1973: Genetically controlled alterations in the rate of degradation of phenylethanolamine N-methyl transferase. J. biol. Chem. **248**, 5616—5623.

COCKING, E. C., 1966: Electron microscope studies in isolated plant protoplasts. Z. Naturforsch. **21 B**, 581—584.

— 1970: Virus uptake, cell wall regeneration and virus multiplication in isolated plant protoplasts. Int. Rev. Cytol. **28**, 89—124.

COFFEY, J. W., and C. DE DUVE, 1968: Digestive activity of lysosomes. I. The digestion of proteins by extracts of rat liver lysosomes. J. biol. Chem. **243**, 3255—3263.

— and Q. PLETSCH, 1971: Association of an alkaline 5-nucleotidase with the lysosomes of rat liver. Fed. Proc. **30**, 1228 abs.

COHEN, M. M., and R. HIRSCHHORN, 1971: Lysosomal and non-lysosomal factors in chemically-induced chromosome breakage. Exp. Cell Res. **64**, 209—217.

COHN, Z. A., 1970 a: Lysosomes in mononuclear phagocytes. In: Mononuclear Phagocytes. (VAN FURTH, R., ed.), pp. 50—58. Philadelphia: F. A. Davis.

— 1970 b: Endocytosis and intracellular digestion. In: Mononuclear phagocytes. (VAN FURTH, R., ed.), pp. 121—128. Philadelphia: F. A. Davis.

— and B. BENSON, 1965: The *in vitro* differentiation of mononuclear phagocytes. J. exp. Med. **122**, 455—466.

— and M. E. FEDORKO, 1969: The formation and fate of lysosomes. In: Lysosomes in biology and pathology. (DINGLE, J. T., and H. B. FELL, eds.), Vol. 1, pp. 43—63. Amsterdam: North-Holland Publishing.

— and E. PARKS, 1967: The regulation of pinocytosis in mouse macrophages. IV. The immunological induction of pinocytosis vesicles, secondary lysosomes and hydrolytic enzymes. J. exp. Med. **125**, 1091—1104.

COHNEN, G., S. D. DOUGLAS, E. KUNIG, and G. BRITTINGER, 1973: Acid phosphatase cytochemistry of mitogen-transformed normal and chronic lymphocytic leukemia lymphocytes. Exp. Cell Res. **80**, 297—304.

COLBEAU, A., J. NACHBAUR, and P. M. BIGNAIS, 1971: Enzyme characterization and lipid composition of rat liver subcellular membranes. Biochim. biophys. Acta. **249**, 462—492.

COLLARD, J. G., and L. A. SMETS, 1974: Effects of proteolytic inhibitors on growth and surface architecture of normal and transformed cells. Exp. Cell Res. **86**, 75—80.

COLLIER, R. J., 1975: Diphtheria toxin: mode of action and structure. Bact. Rev. **39**, 54—85.

COLLINS, J. V., 1969: The hormonal control of fat body development in *Calpodes ethlius* (*Lepidoptera, hesperiidae*). J. Insect Physiol. **15**, 341—352.

— 1973: Soluble acid phosphatases and the control of protein storage in insect fat body. J. Cell Biol. **59**, 61 a.

CONCHIE, J., A. J. MAY, and G. A. LEVVY, 1961: Mammalian glycosidases. 3. The intracellular localization of β-glucuronidase in different mammalian tissues. Biochem. J. **79**, 324—330.

CONE, R. E., J. J. MARCHALONIS, and R. T. ROLLEY, 1971: Lymphocyte membrane dynamics: metabolic release of cell surface proteins. J. exp. Med. **134**, 1373—1384.

CONTRACTOR, S. F., and B. SHANE, 1972: Purification and characterization of lysosomal β-glucuronidase from human placenta. Biochem. J. **128**, 11—18.

Conway, T. P., W. T. Morgan, H. H. Liem, and U. Muller-Eberhard, 1975: Catabolism of photo-oxidized and desialylated hemopexin in the rabbit. J. Biol. Chem. **250**, 3067—3073.

Cook, G. M. W., 1973: The Golgi apparatus: Form and function. In: Lysosomes in biology and pathology. (Dingle, J. T., ed.), Vol. **3**, pp. 237—277. Amsterdam: North-Holland Publishing.

Cook, J. D., W. E. Barry, C. Gershko, G. Fillet, and C. A. Finch, 1973: Iron kinetics with emphasis on iron overload. Amer. J. Path. **72**, 337—343.

Coombs, R. R. A., and H. B. Fell, 1969: Lysosomes in tissue damage mediated by allergic reaction. In: Lysosomes in biology and pathology. (Dingle, J. T., and H. B. Fell, eds.), Vol. **2**, pp. 3—18. Amsterdam: North-Holland Publishing.

Cooper, R. A., and J. H. Jandl, 1969: The role of membrane lipids in the survival of red cells in hereditary spherocytosis. J. Clin. Invest. **48**, 736—744.

Cope, G. H., and M. A. Williams, 1973: Quantitative analysis of the constituent membranes of parotid acinar cells and of the changes evident after exocytosis. Z. Zellforsch. **145**, 311—330.

Corpe, W. A., and H. Winters, 1972: Hydrolytic enzymes of some periphytic marine bacteria. Canad. J. Microbiol. **18**, 1483—1490.

Cotran, R. S., and L. H. M. Cotran, 1970: Ultrastructural localization of horseradish peroxidase and endogenous peroxidase activity in guinea pig peritoneal macrophages. J. Immunol. **105**, 1536—1546.

— and M. J. Karnovsky, 1967: Vascular leakage induced by horseradish peroxidase in the rat. Proc. Soc. Exp. Biol. Med. **126**, 557—568.

— and M. Litt, 1969: The entry of granule associated peroxidase into phagocytic vacuoles of eosinophils. J. exp. Med. **129**, 1291—1299.

Coulomb, C., 1973: Diversite des corps multivesiculaire et notion d'heterophagie dans le meristeme radiulaire de *Scorsonere*. J. Microscopie **16**, 345—360.

— et P. Coulomb, 1973: Participation des structures Golgiennes a la formation des vacuoles autolytiques et a leur approvisionnement enzymatique. C. R. Acad. Sci. (Paris) D **277**, 2685—2688.

Coulomb, P., 1971: Phytolysomes dans les fronds d'*Asplenium fontanum* (Filicinees, polypodiacees). Isolement sur gradient dosages de quelques hydrolases et controle des culots obtenu par la microscope electronique. J. Microscopie **11**, 299—318.

— et J. Coulon, 1971: Fonctions de l'appareil de Golgi dans les meristemes radiculaire de la courge (*Curcurbita pepo* L. *curcurbitacee*). J. Microscopie **10**, 203—214.

— C. Coulomb et J. Coulon, 1972: Origine et fonctions des phytolysosomes dans le meristeme radiculaire de la Courge (*Cucurbita pepo* L. *cucurbitacee*). Relations reticulum endoplasmique-dictosomes-phytolysosomes. J. Microscopie **13**, 263—279.

— and N. Vicente (editors): Concept du systeme lysosomal chez les vegetaux et chez les animaux. (Suppl. to Bulletin de l'Observatoire de la mer.) Le Brusc-Var, France: Fondation Scientifique Ricard 1974.

Cowan, N. J., and C. Milstein, 1974: Stability of cytoplasmic ribonucleic acid in a mouse myeloma: estimation of the half-life of the messenger RNA coding for an immunoglobulin light chain. J. molec. Biol. **82**, 469—482.

Cox, J. P., and M. L. Karnovsky, 1973: The depression of phagocytosis by exogenous cyclic nucleotides, prostaglandins, and theophylline. J. Cell Biol. **59**, 480—490.

Creemers, J., and P. J. Jacques, 1971: Endocytotic uptake and vesicular transport of ingested horseradish peroxidase in the vacuolar apparatus of rat liver cells. Exp. Cell Res. **67**, 188—203.

Crosby, W. M., 1957: Siderocytes and the spleen. Blood **12**, 165—170.

Crossley, A. C., 1968: The fine structure and mechanism of breakdown of larval intersegmental muscles in the blowfly, *Calliphora erythrocephala*. J. Insect Physiol. **14**, 1389—1407.

Crotty, W. J., and M. C. Ledbetter, 1973: Membrane continuities involving chloroplasts and other organelles in plant cells. Science **182**, 839—841.

Cruickshank, W. J., 1971: Follicle cell protein synthesis in moth oocytes. J. Insect. Physiol. **17**, 217—232.

CURNUTTE, J. T., and B. M. BABIOR, 1974: Biological defense mechanisms: the effect of bacteria and serum on superoxide production by granulocytes. J. Clin. Invest. **53**, 1662—1672.

CUZNER, M. L., A. N. DAVISON, and N. A. GREGSON, 1966: Turnover of brain mitochondrial membrane lipids. Biochem. J. **107**, 618—626.

CZECH, M. P., D. G. LYNN, and W. S. LYNN, 1973: Cytochalasin B-sensitive, 2-deoxy-D-glucose transport in adipose cell and ghosts. J. Biol. Chem. **248**, 3636—3641.

DACREMENT, G., J. A. KINT, and G. COCQUYT, 1974: Brain sphingolipids in I cell disease mucolipidosis II). J. Neurochem. **22**, 599—601.

DAEMS, W. TH., and P. BREDEROO, 1973: Electron microscopic studies on the structure, phagocytic properties and peroxidative activity of resident and exudate peritoneal macrophages in the guinea pig. Z. Zellforsch. **144**, 247—297.

— E. WISSE, and P. BREDEROO, 1969: Electron microscopy of the vacuolar apparatus. In: Lysosomes in biology and pathology. (DINGLE, J. T., and H. B. FELL, eds.), Vol. **1**, pp. 64—112. Amsterdam: North-Holland Publishing.

— — — 1972: Electron microscopy of the vacuolar apparatus. In: Lysosomes: A laboratory handbook. (DINGLE, J. T., ed.), pp. 150—199. Amsterdam: North-Holland Publishing.

DALCQ, A. M., 1963: The relation to lysosomes of the *in vivo* metachromatic granules. In: Ciba foundation symposium on lysosomes. (DE REUCK, A. V. S., and M. P. CAMERON, eds.), pp. 226—263. Boston: Little Brown.

— and J. J. PASTEELS, 1966: Remarks on the formation of the multivesicular bodies from yolk platelets: a rectification of authorship. Z. Zellforsch. **72**, 125.

DALES, S., 1969: Role of lysosomes in cell-virus interactions. In: Lysosomes in biology and pathology. (DINGLE, J. T., and H. B. FELL, eds.), Vol. **2**, pp. 69—86. Amsterdam: North-Holland Publishing.

— 1973: Early events in cell-animal virus interactions. Bact. Rev. **37**, 103—135.

— and E. H. MOSSBACH, 1968: Vaccinia as a model for membrane biogenesis. Virology **35**, 564—583.

D'ARCY-HART, P., J. A. ARMSTRONG, C. A. BROWN, and P. DRAPER, 1972: Ultrastructural study of the behavior of macrophages toward parasitic mycobacteria. Infect. Immun. **5**, 803—807.

DARZYNKIEXICZ, Z., and B. A. W. ARNASON, 1974: Suppression of RNA synthesis in lymphocytes by inhibitors of proteolytic enzymes. Exp. Cell Res. **85**, 95—104.

DAUWALDER, M., W. G. WHALEY, J. E. KEPHART, 1969: Phosphatases and differentiation of the Golgi apparatus. J. Cell Sci. **4**, 455—497.

— — — 1972: Functional aspects of the Golgi apparatus. Subcell. Biochem. **1**, 225—275.

DAVEY, M. R., and E. C. COCKING, 1972: Uptake of bacteria by isolated higher plant protoplasts. Nature **239**, 455—456.

DAVID, J. R., 1970: The role of macrophages in delayed hypersensitivity reactions. In: Mononuclear phagocytes. (VAN FURTH, R., ed.), pp. 486—493. Philadelphia: F. A. Davis.

DAVIDSON, S. J., 1973: Protein absorption by renal cells. II. Very rapid lysosomal digestion of exogenous ribonuclease *in vitro*. J. Cell Biol. **59**, 213—222.

— W. K. HUGHES, and A. BARNWELL, 1971: Renal protein absorption into subcellular particles. I. Studies with intact kidneys and fractionated homogenates. Exp. Cell Res. **67**, 171—187.

— and S. W. SONG, 1975: A thermally-induced alteration in lysosome membrane salt permeability at 0° and 37 °C. Biochim. Biophys. Acta **375**, 274—285.

DAVIES, M., 1973: The effect of Triton WR-1339 on the subcellular distribution of Trypan blue and I 125 labelled albumin in rat liver. Biochem. J. **136**, 57—65.

DAVIES, P., and P. E. KING, 1972 a: The effects of host deprivation on the calyx cells of *Nasonia ultrapennis* (Walke) (*Hymenoptera: Pteromalidae*). Z. Zellforsch. **134**, 529—538.

— — 1972 b: The ultrastructure of oosorption in *Locusta migratoria migratorioides*. Z. Zellforsch. **135**, 275—286.

Davies, P., G. A. Rita, K. Krakauer, and G. Weissmann, 1971: Characterization of a neutral protease from lysosomes of rabbit polymorphonuclear leucocytes distinct from collagenase. Biochem. J. **123**, 559—569.
— A. C. Allison, and A. D. Haswell, 1973: Selective release of lysosomal hydrolases from phagocytic cells by cytochalasin B. Biochem. J. **134**, 33—41.
— R. I. Fox, M. Polyzonis, A. C. Allison, and A. D. Haswell, 1973: The inhibition of phagocytosis and facilitation of exocytosis in rabbit polymorphonuclear leukocytes by Cytochalasin B. Lab. Invest. **28**, 16—22.
— A. C. Allison, J. Ackerman, A. Butterfield, and S. Williams, 1974 a: Asbestos induces selective release of lysosomal enzymes from mononuclear phagocytes. Nature **251**, 423—425.
— R. C. Page, and A. C. Allison, 1974 b: Changes in cellular enzymes levels and extracellular release of lysosomal acid hydrolases in macrophages exposed to group A streptococcal cell wall substance. J. exp. Med. **139**, 1262—1282.
Davis, A. T., R. Estensen, and P. G. Quie, 1971: Cytochalasin B. III. Inhibition of human polymorphonuclear leukocyte phagocytosis. Proc. Soc. exp. Biol. Med. **137**, 161—167.
Davis, W. C., S. S. Spicer, W. B. Greene, and Padgett, 1971: Ultrastructure of cells in bone marrow and peripheral blood of normal mink and mink with the homologue of the Chediak-Higashi trait of humans. II. Cytoplasmic granules in eosinophils, basophils, mononuclear cells, and platelets. Amer. J. Path. **63**, 411—432.
Davis, W. L., and S. D. Douglas, 1972: Defective granule formation and function in the Chediak-Higashi syndrome in man and animals Semin. Hematol **9**, 431—450.
Dawson, G., and A. O. Stein, 1970: Lactosyl ceramidosis; catabolic enzyme defect of glycosphingolipid metabolism. Science **170**, 556—558.
— and C. C. Sweeley, 1970: *In vivo* studies on glycosphingolipid metabolism in porcine blood. J. biol. Chem. **245**, 410—416.
— A. C. Stoolmiller, and N. S. Raden, 1974: Inhibition of β-glucosidase by N-n-Hexyl-O-glucosylspingosine in cell strains of neurological origin. J. biol. Chem. **249**, 4683—4646.
Dawson, R. M. C., 1973: The exchange of phospholipids between cell membranes. Sub. Cell Biochem. **2**, 69—89.
Dean, C. R., and D. B. Hope, 1967: The isolation of purified neurosecretory granules from bovine pituitary posterior lobes: comparison of granule protein constituents with those of neurophysin. Biochem. J. **104**, 1082—1088.
De Both, N. J., J. M. van Dongen, B. van Hofwegen, J. Keulmans, W. J. Visser, and H. Galjaard, 1974: The influences of various cell kinetic conditions on functional differentiation in the small intestine of the rat. Develop. Biol. **38**, 119—137.
De Camilli, P., D. Peluchetti, and J. Meldolesi, 1974: Structural difference between luminal and lateral plasmalemma in pancreatic acinar cells. Nature **248**, 245—247.
De Chatelet, L. R., C. E. McCall, L. C. McPhail, and R. B. Johnston, 1974: Superoxide dismutase activity in leukocytes. J. clin. Invest. **53**, 1197—1201.
Decker, R. S., 1974: Lysosomal packaging in differentiating and degenerating anuran lateral motor column neurons. J. Cell Biol. **61**, 599—612.
— 1974 b: Lysosomal behavior during lateral motor column neurogenesis. Dev. Biol. **41**, 146—161.
De Duve, C., 1963: The lysosome concept. In: Ciba foundation Symposium on Lysosomes. (de Reuck, A. V. S., and M. P. Cameron, eds.), pp. 1—28. Boston: Little Brown.
— 1965: The separation and characterization of subcellular particles. Harvey Lectures **59**, 49—87.
— 1968: Lysosomes as targets for drugs. In: Interaction of drugs and subcellular components in animal cells. (Campbell, P. N., ed.), pp. 155—169. London: Churchill Ltd.
— 1969: The lysosome in retrospect. In: Lysosomes in biology and pathology. (Dingle, J. T., and H. B. Fell, eds.), Vol. **1**, pp. 3—40. Amsterdam: North-Holland Publishing.
— 1971: Tissue fractionation, Past and Present. J. Cell Biol. **50**, 20 D—55 D.
— 1972: Lysosomes in pathology and therapeutics. Abstracts, International Symposium on Lysosomes, Hakone, Japan (Aug.-Sept. 1972), p. 4.

De Duve, C., 1973: Biochemical studies on the occurence, biogenesis, and life history of mammalian peroxisomes. J. Histochem. Cytochem. **21**, 941—948.

— T. de Barsy, B. Poole, A. Trouet, P. Tulkens, and F. van Hoof, 1974: Lysosomotropic agents. Biochem. Pharmacol. **23**, 2495—2531.

— and R. Wattiaux, 1966: Functions of lysosomes. Ann. Rev. Physiol. **28**, 435—492.

De Heer, D. H., M. S. Olson, and R. N. Pinckard, 1974: Characterization of rat liver subcellular membranes: demonstration of membrane specific autoantigens. J. Cell Biol. **60**, 460—472.

Dehlinger, P. J., and R. T. Schimke, 1971: Size distribution of membrane proteins of rat liver and their relative rates of degradation. J. biol. Chem. **246**, 2574—2583.

Dehn, P. F., and R. A. Wallace, 1973: Sequestered and injected vitellogenin: alternate routes of protein processing in *Xenopus oocytes*. J. Cell Biol. **58**, 721—724.

De Jong, W. W., F. S. M. van Kleef, and H. Bloemendal, 1974: Intracellular carboxy-terminal degradation of the α A Chain of α-crystallin. Eur. J. Biochem. **48**, 271—276.

De La Iglesia, F. A., G. Fewer, G. Takada, and Y. Matsuda, 1974: Morphological studies on secondary phospholipidosis in human liver. Lab. Invest. **30**, 539—549.

Dellman, H. D., 1973: Degeneration and regeneration of neurosecretory systems. Int. Rev. Cytol. **36**, 215—216.

De Lores-Arnais, G. R., M. A. de Canal, and E. de Robertis, 1971: Turnover of proteins in subcellular fractions of rat cerebral cortex. Brain Res. **31**, 179—184.

Denis, M., 1964: Phosphoproteine phosphatase et resorption du vitellus chez les amphibiens: une etude cytochemique electrophorétique et immunologique. J. Embryol. exp. Morph. **12**, 197—217.

De Petris, S., and M. C. Raff, 1973: Normal distribution, patching, and capping of lymphocyte surface immunoglobulins studied by electron microscopy. Nature New Biol. **241**, 257—259.

De Petrocellis, B., P. Siekevitz, and G. E. Palade, 1970: Changes in chemical composition of thylakoid membranes during greening of the Y-1 mutants of *Chlamydomonas*. J. Cell Biol. **44**, 618—635.

De Pierre, J. W., and M. L. Karnovsky, 1973: Plasma membranes of mammalian cells: a review of methods for their characterization and isolation. J. Cell Biol. **56**, 275—303.

Derechin, M., W. Ostrowski, M. Galka, and E. A. Barnard, 1971: Acid phosphomonosterase of human prostate: molecular weight, dissociation and chemical composition. Biochim. biophys. Acta. **25**, 143—154.

De Reuck, A. V. S., and M. P. Cameron (eds.), 1963: Ciba foundation symposium on lysosomes. Boston: Little Brown.

Desbuquois, B., and P. Cuatrecorsas, 1972: Independence of glucagon receptors and glucagon inactivation in liver cell membranes. Nature New Biol. **237**, 202—204.

Deshmukh, K., and M. E. Nimni, 1973: Effects of lysosomal enzymes on the type of collagen synthesized by articular cartilage. Biochem. biophys. Res. Commun. **53**, 424—431.

Desnick, R. J., W. Krivit, and M. L. Sharp, 1973: In utero diagnosis of Sandhoff's disease. Biochem. biophys. Res. Commun. **51**, 20—26.

Desser, R. K., R. Himmelboch, W. H. Evans, M. Januska, M. Mage, and E. Shelton, 1972: Guinea pig heterophil and eosinophil peroxidase. Arch. biochem. biophys. **148**, 452—465.

Deter, R. L., 1971: Quantitative characterization of dense bodies, autophagic vacuoles and acid phosphatase-bearing particle populations during the early phases of glucagon induced autophagy in rat liver. J. Cell Biol. **48**, 473—489.

— P. Baudhuin, and C. de Duve, 1967: Participation of lysosomes in cellular autophagy induced in rat liver by glucagon. J. Cell Biol. **35**, C11—C16.

— and C. de Duve, 1967: Influences of glucagon, an inducer of cellular autophagy, on some properties of rat liver lysosomes. J. Cell Biol. **33**, 437—449.

De Vos, L., 1971: Etude ultrastructurale de la gemmulogenese chez Ephydatia fluviatilis. J. Microscopie **10**, 283—304.

Devreotes, P. N., and D. M. Fambrough, 1975: Acetylcholine receptor turnover in membranes of developing muscle fibers. J. Cell Biol. **65**, 353—358.

230 Bibliography

De Wald, B., R. Rindler-Ludwig, U. Bretz, and M. Baggiolini, 1975: Subcellular localization of neutral proteases in neutrophilic polymorphonuclear leukocytes. J. Exp. Med. **141**, 709—723.

Dewald, B., and O. Touster, 1973: A new α-D-mannosidase occurring in Golgi membranes. J. biol. Chem. **248**, 7223—7233.

Diamond, R. L., M. Brenner, and W. F. Loomis, 1973: Mutations affecting N-acetyl-glucosaminidase in *Dictyostelium discoideum*. Proc. nat. Acad. Sci. (U.S.) **70**, 3356—3360.

Di Augustine, R. P., 1974: Lung concentric laminar organelle: Hydrolase activity and compositional analysis. J. biol. Chem. **249**, 584—593.

Dice, J. F., P. J. Dehlinger, and R. T. Schimke, 1973: Studies on the correlation between size and relative degradation rate of soluble proteins. J. biol. Chem. **248**, 4220—4228.

— and R. T. Schimke, 1972: Turnover and exchange of ribosomal proteins from rat liver. J. biol. Chem. **272**, 98—111.

Dickerson, R. E., and I. Geis, 1969: The structure and action of Proteins. New York: Harper and Row.

Dietert, S. E., 1966: Fine structure of the formation and fate of the residual bodies of the mouse spermatozoa with evidence for the participation of lysosomes. J. Morph. **120**, 317—345.

Dietert, S. C., and T. J. Scallan, 1969: An ultrastructural and biochemical study of the effects of three inhibitors of cholesterol biosynthesis upon murine adrenal gland and testis: histochemical evidence for a lysosome response. J. Cell Biol. **40**, 44—60.

Dingle, J. T., 1963: Action of vitamin A on the stability of lysosomes *in vivo* and *in vitro*. In: Ciba Foundation Symposium on Lysosomes. (de Reuck, A. V. S., and M. P. Cameron, eds.), pp. 384—427. Boston: Little Brown.

— 1968: Vacuoles, vesicles, and lysosomes. Brit. Med. Bull. **24**, 141—145.

— 1969: The extracellular secretion of lysosomal enzymes. In: Lysosomes in biology and pathology. (Dingle, J. T., and H. B. Fell, eds.), Vol. **2**, pp. 421—436. Amsterdam: North-Holland Publishing.

— (ed.), 1972: Lysosomes: a laboratory handbook. Amsterdam: North-Holland Publishing.

— (ed.), 1973 a: Lysosomes in biology and pathology. Vol. **3**. Amsterdam: North-Holland Publishing.

— 1973 b: Lysosomal enzymes in skeletal tissues. In: Hard tissue growth, repair, and remineralization. (Ciba Found. Symp. 11), pp. 295—313.

— and A. J. Barrett, 1969: Uptake of biologically active substances by lysosomes. Proc. roy. Soc. B. **173**, 85—93.

— and H. M. Dott, 1969: Lysosomal enzymes in bull and ram sperm. Biochem. J. **111**, 35 p.

— and H. B. Fell (eds.), 1969: Lysosomes in biology and pathology. Vols. **1** and **2**. Amsterdam: North-Holland Publishing.

— M. F. Hay, and R. M. Moor, 1968: Lysosomal function in the corpus luteum of the sheep. J. Endocr. **40**, 325—336.

— H. B. Fell, and A. M. Glauert, 1969: Endocytosis of sugars in embryonic skeletal tissues in organ culture. IV. Lysosomal and other biochemical effects. General discussion. J. Cell Sci. **4**, 139—154.

— A. J. Barrett, A. R. Poole, and P. Stovin, 1972: Inhibition by pepstatin of human cartilage degradation. Biochem. J. **127**, 443—444.

— A. R. Poole, G. S. Lazarus, and A. J. Barrett, 1973: Immunoinhibition of intracellular protein digestion in macrophages. J. exp. Med. **137**, 1124—1141.

Distler, J. J., and G. W. Jourdain, 1973: The purification and properties of β-galactosidase from bovine testis. J. biol. Chem. **248**, 6772—6780.

D'Monte, B., N. Marks, K. Dalta, and A. Lajtha, 1970: Protein turnover in membranous fractions. In: Protein metabolism of the nervous system. (Lajtha, A., ed.), pp. 185—217. New York: Plenum.

Dogiel, V. A., 1965: General Protozoology, 2nd. ed. London: Oxford Univ. Press.

Dopheide, Ta. A., C. A. Mfnzies, M. T. McQuillan, and V. M. Trikojus, 1969: Studies with purified pig thyroglobulin and thyroid enzymes. Biochim. biophys. Acta **181**, 105—115.

DORFMAN, A., R. MATALON, J. A. CIFONELLI, J. THOMPSON, and G. DAWSON, 1971: The degradation of acid mucopolysaccharide and the mucopolysaccharidoses. In: Sphingolipids, sphingolipidoses and allied disorders (Adv. Exp. med. Biol. 19). (VOLK, B. W., and S. M. ARONSON, eds.), pp. 195—210. New York: Plenum Press.

DOTT, H. M., 1969: Lysosomes and lysosomal enzymes in the reproductive tract. In: Lysosomes in biology and pathology. (DINGLE, J. T., and H. B. FELL, eds.), Vol. 1, pp. 330 to 360. Amsterdam: North-Holland Publishing.

— and J. T. DINGLE, 1968: Distribution of lysosomal enzymes in the spermatozoa and cytoplasmic droplets of bull and ram. Exp. Cell Res. 52, 523—540.

DOUCE, R., 1974: Site of biosynthesis of galactolipids in spinach chloroplasts. Science 183, 852—853.

DOUGLAS, W. W., J. NAGASAWA, and R. SCHULZ, 1971: Electron microscopic studies on the mechanism of secretion of posterior pituitary hormones and significance of microvesicles ("synaptic vesicles"): evidence of secretion by exocytosis and formation of microvesicles as a by-product of this process. Mem. Soc. Endocrinol. 19, 353—378.

DRAPER, P., and R. J. W. REES, 1970: Electron transparent zone of mycobacteria may be a defense mechanism. Nature 228, 860—861.

DRESSER, D. W., and N. A. MITCHISON, 1968: The mechanism of immunological paralysis. Adv. Immunol. 8, 129—181.

DROLLER, M. J., 1974: An electron microscopic study of the time course of platelet nucleotide, calcium and acid phosphatase secretion. Lab. Invest. 31, 197—205.

DRUTZ, D. J., M. J. CLINE, and L. LEVY, 1974: Leukocyte antimicrobial function in patients with leprosy. J. clin. Invest. 53, 380—386.

DRUYAN, R., B. DE BERNARD, and M. RABINOWITZ, 1969: Turnover of cytochromes labelled with δ-aminolevulinic acid. J. biol. Chem. 244, 5874—5880.

DRYSDALE, J. W., and H. N. MUNRO, 1968: Regulations of synthesis and turnover of ferritin in rat liver. J. biol. Chem. 241, 3630—3637.

DUBOIS, P., 1972: Origine et developpment de l'appareil de Golgi an cours de la differenciation cellulaire dans une glande endocrine chez l'homme: l'antehypophyse foetale. J. Microscopie 13, 193—206.

DUBOWSKY, N., 1974: Selectivity of ingestion and digestion in the chrysomonad flagellate Ochromonas malhamensis. J. Protozool. 21, 295—298.

DUCKWORTH, W. C., M. A. HEINEMANN, and A. E. KITAOCHI, 1972: Purification of insulin specific protease by affinity chromatography. Proc. nat. Acad. Sci. (U.S.) 69, 3698—3702.

DUNHAM, P. B., I. M. GOLDSTEIN, and G. WEISSMANN, 1974: Potassium and amino acid transport in human leukocytes exposed to phagocytic stimuli. J. Cell Biol. 63, 215—226.

DUNN, W. B., J. M. HARDIN, and S. S. SPICER, 1968: Ultrastructural localization of myeloperoxidase in hyman neutrophil and rabbit heterophil and eosinophil leukocytes. Blood 32, 835—844.

DUTTA, G. P., 1973: Recent advances in the cytochemistry and ultrastructure of cytoplasmic inclusions in *Mastigophora* and *Opalinata* (protozoa). Int. Rev. Cytol. 36, 93—135.

DUTTON, G. R., and S. BARONDES, 1969: Microtubular protein synthesis, and metabolism in developing brain. Science 166, 1637—1638.

DURPHY, M., P. N. MANLEY, and E. C. FRIEDBERG, 1974: A demonstration of several deoxyribonuclease activities in mammalian cell mitochondria. J. Cell Biol. 62, 695—706.

EAGLE, H., K. A. PIEZ, R. FLEISCHMAN, and V. I. OYAMA, 1959: Protein turnover in mammalian cell cultures. J. biol. Chem. 234, 592—597.

EBNER, E., T. L. MASON, and G. SCHATZ, 1973: Mitochondrial assembly in respiration deficient mutants of Saccharomyces cerevisiae. II. Effects of nuclear and extrachromosomal mutations on the formation of cytochrome c oxidase. J. biol. Chem. 248, 5369—5378.

EDELMAN, G. M., and G. MOLLER, 1973: The immune system as a model for cellular maturation and differentiation. (Neurosciences Research Program Bulletin, Vol. 11, No. 2.) Neurosci. Res. Prog. Brookline, Mass.

EDELSON, P. J., and Z. A. COHN, 1973: Peroxidase medicated mammalian cell cytotoxicity. J. exp. Med. 138, 318—323.

EDELSON, P. J., 1974: Effects of concanavalin A on mouse peritoneal macrophages. I. Stimulation of endocytic activity and inhibition of phago-lysosome formation. II. Metabolism of endocytised proteins and reversibility of the effects by mannose. J. exp. Med. **140**, 1364—1386, 1387—1398.

EECKHAUT, Y., 1973: Digestion and lysosomes in zooflagellates. In: Lysosomes in biology and pathology. (DINGLE, J. T., ed.), Vol. 3, pp. 3—17. Amsterdam: North-Holland Publishing.

EGUCHI, E., and T. H. WATERMAN, 1968: Cellular basis for polarized light perception in the spider crab Libinium. Z. Zellforsch. **84**, 87—101.

EHRENREICH, B. A., and Z. A. COHN, 1967: The uptake and digestion of iodinated human serum albumin by macrophages *in vitro*. J. exp. Med. **129**, 941—958.

— — 1969: The fate of peptides pinocytosed by macrophages *in vitro*. J. exp. Med. **129**, 227—245.

EISEN, A. Z., E. A. BAUER, and J. J. JEFFREY, 1971: Human skin collagenase. The role of serum alpha globulins in the control of activity *in vivo* and *in vitro*. Proc. nat. Acad. Sci. (U.S.) **68**, 248—251.

ELLIOT, A. M., and I. J. BAK, 1964: The fate of mitochondria during aging in *Tetrahymena*. J. Cell Biol. **20**, 113—136.

— and G. L. CLEMMONS, 1966: An ultrastructural study of ingestion and digestion in *Tetrahymena pyriformis*. J. Protozool. **13**, 311—323.

ELSBACH, P., D. ZUKER-FRANKLIN, and C. SANSARICQ, 1969: Increased lecithin synthesis during phagocytosis by normal leukocytes and by leukocytes of a patient with chronic granulomatous disease. New Engl. J. Med. **280**, 1319—1322.

— P. PATRIARCA, P. PETTIS, T. P. STOSSEL, R. J. MASON, and M. VAUGHAN, 1972: The appearance of lecithin 32 P synthesized from lysolecithin 32 P in phagosomes of polymorphonuclear leukocytes. J. clin. Invest. **51**, 1910—1914.

— P. PETTIS, S. BECKERDITE, and R. FRANSON, 1973: Effect of phagocytosis by rabbit granulocytes on macromolecular synthesis and degradation in different species of bacteria. J. Bact. **115**, 490—497.

EPPIG, J. J., 1974: Tyrosinase. In: Electron microscopy of enzymos. (HAYAT, M. A., ed.), Vol. 2, pp. 79—89. New York: Van Nostrand Reinhold.

ERICSSON, J. L. E., 1965: Transport and digestion of hemoglobin in the proximal tubule. Lab. Invest. **14**, 1—15.

— 1969 a: Mechanism of cellular autophagy. In: Lysosomes in biology and pathology. (DINGLE, J. T., and H. B. FELL, eds.), Vol. 2, pp. 345—394. Amsterdam: North-Holland Publishing.

— 1969 b: Studies on induced cellular autophagy. I. Electron microscopy of cells with *in vivo* labelled lysosomes. II. Characterization of the membrane bordering autophagosomes in parenchymal liver cells. Exp. Cell Res. **55**, 95—106; Exp. Cell Res. **56**, 393—405.

— U. T. BRUNK, and B. ARBORGH, 1974: Demonstration of acid phosphatase in *in vitro* cultured cells: normal and subjected to lysosomal damage. Histochem. J. **6**, 63—64.

ERIKSSON, L. C., and G. DALLNER, 1972: Membrane biogenesis and rough microsomal fractions. J. Cell Biol. **55**, 70 A.

ERLANDSON, S. L., and D. G. CHASE, 1972: Paneth cell function: phagocytosis and intracellular digestion of intestinal microorganisms. I. Hexamita muris. II. Spiral microorganism. J. Ultrastruct. Res. **41**, 291—318, 319—333.

— J. A. PARSONS, and T. B. TAYLOR, 1974: Ultrastructural immunocytochemical localization of lysozyme in the Paneth cells of man. J. Histochem. Cytochem. **22**, 401—413.

ESSNER, E., 1960: Electron microscopic study of erythrophagocytosis. J. Biophys. Biochem. Cytol. **7**, 329—334.

— 1973: Phosphatases. In: Electron microscopy of enzymes. (HAYAT, M. A., ed.), Vol. **1**, pp. 44—76. New York: Van Nostrand Reinhold.

— and H. HAIMES, 1975: Structure and functions of GERL in macrophages. J. Histochem. Cytochem. **23**, 314.

— and A. B. NOVIKOFF, 1960: Human hepatocellular pigments and lysosomes. J. Ultrastruct. Res. **3**, 374—391.

ESSNER, E., and C. OLIVER, 1973: A hereditary alteration in kidneys of mice with Chediak-Higashi syndrome. Amer. J. Path. **73**, 217—232; 1974: Fate of exogenous peroxidase in renal lysosomes of mice with Chediak-Higashi syndrome. Amer. J. Path. **77**, 407—422.

— — 1974: Lysosome formation in hepatocytes of mice with Chediak-Higashi syndrome. Lab. Invest. **30**, 596—607.

— A. B. NOVIKOFF, and N. QUINTANA, 1965: Nucleoside phosphatase activities in rat cardiac muscle. J. Cell Biol. **25**, 201—215.

ESTENSEN, R. D., J. G. WHITE, and B. HOLMES, 1974: Specific degranulation of human PMN leukocytes. Nature **248**, 347—348.

ESTEVE, J.-C., 1970: Distribution of acid phosphatase in *Paramecium caudatum:* its relations with the process of digestion. J. Protozool. **17**, 24—35.

ETHERINGTON, D. J., 1974: The purification of bovine cathepsin B 1 and its mode of action on bovine collagen. Biochem. J. **137**, 547—557.

ETO, Y., U. WIESMANN, and N. N. HERSCHOKOWITZ, 1974: Sulfogalactosyl and sphingosine sulfatase: characteristics of the enzyme and its deficiency in metachromatic leuko-dystrophy in human cultured skin fibroblasts. J. biol. Chem. **249**, 4955—4960.

EVANS, G. W., 1973: Copper homeostasis in the mammalian system. Physiol. Rev. **53**, 535—570.

— R. S. DUBOIS, and K. M. HAMBRIDGE, 1973: Wilson's disease: identification of an abnormal copper binding protein. Science **181**, 1175—1176.

EVANS, W. H., and J. W. GURD, 1971: Biosynthesis of liver membranes: incorporation of H^3-leucine into proteins and C^{14}-glucosamine into proteins and lipids of liver microsomal and plasma membrane fractions. Biochem. J. **125**, 615—624.

EVANSON, J. M., 1971: Mammalian collagenses and their role in connective tissue breakdown. In: Tissue proteinases. (BARRETT, A. J., and J. T. DINGLE, eds.), pp. 327—345. Amsterdam: North-Holland Publishing.

EYTAN, G., and I. OHAD, 1972: Biogenesis of chloroplast membranes. VII. The preservation of membrane homogeneity during development of the photosynthetic lamellar system in an algal mutant (*Chlamydomonas reinhardii* y-I). VIII. Modulation of chloroplast lamellae composition and function induced by discontinuous illumination and inhibition of ribonucleic acid and protein synthesis during greening of *Chlamydomonas reinhardii* y-I mutant cells. J. biol. Chem. **247**, 112—121, 122—262.

— R. C. JENNINGS, G. FORTI, and I. OHAD, 1974: Biogenesis of chloroplast membranes. J. biol. Chem. **249**, 738—744.

FAHIMI, H. D., 1970: The fine structural localization of endogenous and exogenous per-oxidase activity in Kupffer cells of rat liver. J. Cell Biol. **47**, 247—262.

FAHRENBACH, W. H., 1969: The morphology of the eyes of Limulus. II. Ommatidia of the compound eye. Z. Zellforsch. **93**, 451—483.

FAKHER, A., and D. SCHLESSINGER, 1966: Synthesis and breakdown of ribonucleic acid in *Escherichia coli* starving for nitrogen. Biochim. biophys. Acta **119**, 183—191.

FARKAS, V., P. BIELY, and S. BAUER, 1973: Extracellular β-glucanase of the yeast *Saccharonyces cerevisiae*. Biochim. biophys. Acta **321**, 246—255.

FARQUHAR, M. G., 1969: Lysosome function in regulating secretion: disposal of secretory granules in cells of the anterior pituitary gland. In: Lysosomes in biology and pathology. (DINGLE, J. T., and H. B. FELL, eds.), Vol. 2, pp. 462—482. Amsterdam: North-Holland Publishing.

— 1971: Processing of secretory products by cells of the anterior pituitary gland. Mem. Soc. Endocrinol. **19**, 79—124.

— D. F. BAINTON, M. BAGGIOLINI, and C. DE DUVE, 1972: Cytochemical localization of acid phosphatase activity in granule fractions from rabbit polymorphonuclear leuko-cytes. J. Cell Biol. **54**, 141—156.

— J. J. M. BERGERON, and G. E. PALADE, 1974: Cytochemistry of Golgi fractions prepared from rat liver. J. Cell Biol. **60**, 8—25.

FAULKNER, C. S., 1969: The role of the granular pneumocyte in surfactant metabolism. Arch. Path. **87**, 521—525.

FAUVE, R. M., 1970: The effects of corticorteroids on some functions of macrophages. In: Mononuclear phagocytes. (VAN FURTH, R., ed.), pp. 265—280. Philadelphia: F. A. Davis.

FAVARD, P., et N. CARASSO, 1963: Mise en evidence d'un process de micropinocytose interne au niveau des vacuoles digestives d'*Epistylis anastatica* (*Cilie peritriche*). J. Microscopie 2, 495—498.

FEDORKO, J., and S. U. MORSE, 1965: Isolation, characterization, and distribution of acid mucopolysaccharide in rabbit leucocytes. J. exp. Med. **121**, 139—148.

FEDORKO, M. E., 1974: Loss of iron from mouse peritoneal macrophages *in vitro* after uptake of Fe-55-ferritin and Fe-55-ferritin-rabbit antiferritin complexes. J. Cell Biol. **62**, 802—814.

— and J. G. HIRSCH, 1966: Cytoplasmic granule formation in myelocytes: an electron microscope radioautographic study on the mechanism of formation of cytoplasmic granules in rabbit heterophilic myelocytes. J. Cell Biol. **29**, 307—316.

— — and Z. A. COHN, 1968: Autophagic vacuoles produced *in vitro*. I. Studies on cultured macrophages exposed to chloroquine. II. Studies on the mechanism of formation of autophagic vacuoles produced by chloroquine. J. Cell Biol. **38**, 377—391, 392—402.

— N. L. CROSS, and J. G. HIRSCH, 1973: Appearance and distribution of ferritin in mouse peritoneal macrophage *in vitro* after uptake of heterologous erythrocytes. J. Cell Biol. **57**, 289—305.

FELDMANN, M., and G. J. V. NOSSAL, 1972: Cellular basis of antibody production. Quart. Rev. Biol. **47**, 269—302.

FELIG, PL., and J. WAHREN, 1974: Protein turnover and amino acid metabolism in the regulation of gluconeogenesis. Fed. Proc. **33**, 1092—1104.

FELTON, J., M. MEISLER, and K. PAIGEN, 1974: A locus determining β-galactosidase in the mouse. J. biol. Chem. **249**, 3267—3272.

FERRANS, V. J., L. M. BUJA, W. C. ROBERTS, and D. S. FREDRICKSON, 1973: Chlyomicrons and the formation of foam cells in Type I hyperlipoproteinemia. Amer. J. Path. **70**, 253—272.

FESSAS, P., D. LOUKOPOULOS, and A. KALBOYA, 1966: Peptide analysis of the inclusions of erythroid cells in β-thalassemia. Biochim. biophys. Acta **300**, 430—432.

FIEDLER, F., and L. GLASER, 1973: Assembly of bacterial cell walls. Biochim. biophys. Acta **300**, 467—485.

FIELDING, J., and B. E. SPEYER, 1974: Iron transport intermediates in human reticulocytes and the membrane binding site of irontransferrin. Biochim. biophys. Acta **263**, 387—396.

FILNER, P., and J. E. VARNER, 1967: A test for *de novo* synthesis of enzymes: density labelling with H_2O^{18} of barley α-amylase induced by gibberellic acid. Proc. nat. Acad. Sci. (U.S.) **58**, 1520—1526.

— J. L. WRAY, and J. E. VARNER 1969: Enzyme induction in higher plants. Science **165**, 358—367.

FINERAN, B. A., 1972: Ultrastructure of vacuolar inclusions in root tips. Protoplasma **72**, 1—18.

FISCHER, C. A., and P. MORELL, 1974: Turnover of proteins in myelin and myelin-like material of mouse brain. Brain Res. **74**, 51—66.

FISER-SZAFARZ, and D. SZAFARZ, 1973: Lysosomal hyaluronidase activity in normal rat liver and in chemically induced hepatomas. Cancer Res. **33**, 1104—1108.

FISHMAN, M., and F. L. ADLER, 1970: Heterogeneity of macrophage functions in relation to the immune response. In: Mononuclear phagocytes. (VAN FURTH, R., ed.), pp. 581—592. Philadelphia: F. A. Davis.

FLEISCHER, B., and S. FLEISCHER, 1970: Preparation and characterization of Golgi membranes from rat liver. Biochim. biophys. Acta **219**, 301—319.

— — 1971: Comparison of cellular membranes of liver with emphasis on the Golgi complex as a discrete organelle. In: Biomembranes. (MANSON, L. A., ed.), Vol. II, pp. 75—94. New York: Plenum Press.

FLETCHER, M. J., and D. R. SANADI, 1961: Turnover of rat-liver mitochondria. Biochim. biophys. Acta **51**, 356—360.

FLICKINGER, C. J. 1969: The pattern of growth of the Golgi complex during the fetal and post-natal development of the rat epididymis. J. Ultrastruct. Res. **27**, 344—360.

Fouquet, J.-P., 1974: La spermiation et la formation des corps residuels chez le Hamster: rôle des cellules de Sertoli. J. Microscopie **19**, 161—168.

Fowler, S., and C. de Duve, 1969: Digestive activity of lysosomes. III. The digestion of lipids by extracts of rat liver lysosomes. J. biol. Chem. **244**, 471—481.

Fox, H., 1972: Tissue degeneration: an electron microscopic study of the tail skin of *Rana temporaria* during metamorphosis. Arch. Biol. (Liège) **83**, 373—394.

— 1973: Degeneration of the tail notochord of *Rana temporaria* at metamorphic climax. Z. Zellforsch. **139**, 371—386.

Frank, A. L., and A. K. Christensen, 1968: Localization of acid phosphatase in lipofuscin granules and possible autophagic vacuoles in interstitial cells of the guinea pig testis. J. Cell Biol. **36**, 1—14.

Franke, E. E., and J. Kartenbeck, 1971: Outer mitochondrial membrane continuous with endoplasmic reticulum. Protoplasma **73**, 35—41.

Franke, W. W., D. J. Morre, B. Deumling, R. D. Cheetham, J. Kartenbeck, E.-D. Jarasch, and H.-W. Zentgraf, 1971: Synthesis and turnover of membrane proteins in rat liver: an examination of the membrane flow hypothesis. Z. Naturforsch. **26 B**, 1031—1039.

Franson, R. C., and M. Waite, 1973: Lysosomal phospholipase A_1 and A_2 of normal and Bacillus Calmette-Guerin induced alveolar macrophages. J. Cell Biol. **56**, 621—627.

Freed, J. J., and M. M. Leibowitz, 1970: The association of a class of saltatory movements with microtubules in cultured cells. J. Cell Biol. **45**, 334—354.

Freychet, P., R. Kahn, J. Roth, and D. M. Neville, 1972: Insulin interactions with liver plasma membranes: independence of binding of the hormone and its degradation. J. biol. Chem. **247**, 3953—3961.

Friend, D. S., 1969: Cytochemical staining of multivesicular bodies and Golgi vesicles. J. Cell Biol. **41**, 267—279.

— and M. G. Farquhar, 1967: Functions of coated vesicles during protein absorption in the rat vas deferens. J. Cell Biol. **35**, 357—376.

Friis, R. R., 1972: Interaction of L-cells and *Chlamydia psittaci*. Entry of the parasite and host responses to its development. J. Bact. **110**, 706—721.

Fristrom, D., 1969: Cellular degeneration in the production of some mutant phenotypes in *Drosophila melanogaster*. Molec. gen. Genetics **103**, 363—379.

Fritz, P. J., E. S. Vessell, L. White, and K. M. Pruitt, 1969: The roles of synthesis and degradation in determining tissue concentrations of lactate dehydrogenase. Proc. nat. Acad. Sci. (U.S.) **62**, 558—565.

Fromme, M. G., 1968: Studies on intracellular degeneration of phagocytosed microorganisms by electron microscopic autoradiography. Lab. Invest. **18**, 211—214.

Fujiwara, K., T. Sahai, T. Oda, S. Igarashi, 1973: The presence of collagenase in Kupffer cells of the rat liver. Biochem. biophys. Res. Commun. **54**, 531—537.

Fulks, R. M., J. B. Li, and A. L. Goldberg, 1975: Effects of insulin, glucose and amino acids on protein turnover in rat diaphragm. J. Biol. Chem. **250**, 290—298.

Gahan, P. B., 1965: Reversible activation of lysosomes in rat liver. J. Histochem. Cytochem. **13**, 334—338.

— 1967: Histochemistry of lysosomes. Int. Rev. Cytol. **21**, 2—54.

— 1973: Plant lysosomes. In: Lysosomes in biology and pathology. (Dingle, J. T., ed.), Vol. **3**, pp. 69—85. Amsterdam: North-Holland Publishing.

— and A. J. Maple, 1966: The behaviour of lysosome-like particles during cell differentiation. J. exp. Bot. **17**, 151—155.

Galjaard, H., A. Hoogeveen, H. A. de Wit-Verbeck, A. J. J. Reuser, W. Keigzer, A. Westerveld, and D. Bootsma, 1974: Tay-Sachs and Sandhoff's disease; intergenic complementation after somatic cell hybridization. Exp. Cell Res. **87**, 444—448.

Galle, P., 1974: Rôle des lysosomes et des mitochondries dans les phénomènes de concentration et d'elimination d'elements mineraux (uranium et or) par le rein. J. Microscopie **19**, 17—24.

Gallin, J. I., J. S. Bujak, E. Patten, and S. M. Wolff, 1974: Granulocyte formation in the Chediak-Higashi syndrome of mice. Blood **43**, 201—206.

GAN, J. C., and H. JEFFAY, 1967: Origins and metabolism of the intracellular amino acid pools in rat liver and muscle. Biochim. biophys. Acta **148**, 448—459.

GANS, H., V. SUBRAMANIAN, and B. M. TAN, 1968: Selective phagocytosis: a new concept in protein catabolism. Science **159**, 107—110.

GANSCHOW, R. E., and R. T. SCHIMKE, 1969: Independent genetic control of the catalytic activity and the rate of degradation of catalase in mice. J. biol. Chem. **244**, 4649—4658.

GARDNER, P. J., and P. D. SHERVEY, 1974: An ultrastructural and histochemical study of autoimmune aspermatogenesis in testis. Z. Zellforsch. **147**, 191—198.

GASKO, O., and D. DANON, 1972: Deterioration and disappearance of mitochondria during reticulocyte maturation. Exp. Cell Res. **75**, 159—169.

GATT, S., Y. BARENHOLZ, I. BORKOVSKI-KUBILER, Z. BEN-GERSHON, 1971: Interaction of enzymes with lipid substrates. In: Sphingolipids, sphingolipidoses and allied disorders (Adv. Exp. Med. Biol. **19**). (VOLK, B. W., and S. M. ARONSON, eds.), pp. 237—256. New York: Plenum Press.

GEAR, A. R. L., 1970: Inner and outer membrane enzymes of mitochondria during liver regeneration. Biochem. J. **120**, 577—587.

— A. D. ALBERT, and J. M. BEDNAREK, 1974: The effect of the hypocholestolemic drug Clofibrate on liver mitochondrial biogenesis: a role for neutral mitochondrial proteases. J. biol. Chem. **249**, 6495—6504.

GERSTEN, D. M., T. W. KIMMERER, and H. B. BARMANN, 1974: The lysosome periphery: biochemical and electrokinetic properties of the tritosome surface. J. Cell Biol. **60**, 764—774.

GERVIN, S. W., and E. HOLTZMAN, 1972: The fate of exogenous peroxidase in the thymus of newborn and young adult mice. J. Histochem. Cytochem. **20**, 445—462.

GESSNER, T. P., S. R. HIMMELHOCH, and E. SHELTON, 1973: Partial characterization of the protein component of eosinophil granules isolated from guinea pig exudates. Arch. biochem. biophys. **156**, 383—389.

GESTRELIUS, S., B. MATTIASSON, and K. MOSBACH, 1973: On the regulation of the activity of immobilized enzymes: microenvironmental effects of enzyme generated pH changes. Eur. J. Biochem. **36**, 89—96.

GEUZE, J. J., and C. POORT, 1973: Cell membrane resorption in the rat exocrine pancreas cell after *in vivo* stimulation of the secretion as studied by *in vitro* incubation with extracellular space markers. J. Cell Biol. **57**, 159—174. See also Cell Tissue Res. **156**, 1—20.

GIBSON, R. A., L. G. PALEG, 1972: Lysosomal nature of hormonally induced enzymes in wheat aleurone cells. Biochem. J. **128**, 367—375.

GIELEN, J. E., F. M. GOUJON, and D. W. NEBERT, 1972: Genetic regulation of aryl hydro-carbon hydroxylase induction. II. Simple mendelian expression in mouse tissues *in vivo*. J. biol. Chem. **247**, 1125—1137.

GILBERT, B. E., and T. C. JOHNSON, 1972: Protein turnover during maturation of mouse brain tissue. J. Cell Biol. **53**, 143—147.

GILBERT, F., R. KUCHERLAPATI, R. P. CREGAN, M. J. MURNAME, G. J. DARLINGTON, and F. H. RUDDLE, 1975: Tay-Sach's and Sandhoff's diseases: the assignment of genes for hexosaminidases A and B to individual human chromosomes. Proc. nat. Acad. Sci. (U.S.) **72**, 263—267.

GILBOA, E., Y. ELKANA, and M. RIGHI, 1973: Purification and properties of human acrosin. Eur. J. Biochem. **39**, 85—92.

GILL, D. M., A. M. PAPPENHEIMER, and T. UCHEDA, 1973: *Diphtheria* toxin, protein synthesis and the cell. Fed. Proc. **32**, 1508—1515.

GIORGI, F., 1974: Acid phosphatase localization in ovarian cells of *Drosophila melanogaster*: role of the Golgi apparatus. J. Histochem. **6**, 71—78.

GITHENS, S., and M. L. KARNOVSKY, 1973: Biochemical changes during growth and encyst-ment of the cellular slime mold, *Polysphondylium pallidum*. J. Cell Biol. **58**, 522—535.

GLASS, R. D., and D. DOYLE, 1972: On the measurement of protein turnover in animal cells. J. biol. Chem. **247**, 5234—5242.

GLAUERT, A. M., H. B. FELL, and J. T. DINGLE, 1969: Endocytosis of sugars in embryonic skeletal tissues in organ culture. II. Effect of sucrose on cellular fine structure. J. Cell Sci. **4**, 104—131.

GLEICH, G. J., D. A. LOEGERING, F. KUEPPERS, S. P. BAJOJ, and K. G. MANN, 1974: Physiochemical and biological properties of the basic protein from guinea pig eosinophil granules. J. exp. Med. **140**, 313—332.

GLENN, A. R., G. W. BOTH, J. L. McINNER, B. K. MAY, and W. H. ELLIOTT, 1973: Dynamic state of the messenger RNA pool specific for extracellular protease in *Bacillus amyloliquefaciens:* its relevance to the mechanism of enzyme secretion. J. molec. Biol. **73**, 221—230.

GLINSMANN, W., and J. L. E. ERICSSON, 1966: Observations on the subcellular organization of hepatic parenchymal cells. II. Evolution of reversible alterations induced by hypoxia. Lab. Invest. **15**, 762—777.

GOLDBERG, A. L., 1971 a: A role of aminoacyl-tRNA in the regulation of protein breakdown in *Escherichia coli.* Proc. nat. Acad. Sci. (U.S.) **68**, 362—366.

— 1971 b: Effects of protease inhibitors on protein breakdown and enzyme induction in starving *Escherichia coli.* Nature New Biol. **234**, 51—52.

— 1972 a: Degradation of abnormal proteins in *E. coli.* Proc. nat. Acad. Sci. (U.S.) **69**, 422—426.

— 1972 b: Correlation between rates of degradation of bacterial proteins *in vivo* and their sensitivity to proteases. Proc. nat. Acad. Sci. (U.S.) **69**, 2640—2644.

— 1973: Regulation and importance of intracellular protein degradation. In: The Neurosciences. (SCHMITT, F. O., ed.), Vol. **3**. M. I. T. Press (in press).

— and J. F. DICE, 1974: Intracellular protein degradation in mammalian and bacterial cells. Ann. Rev. Biochem. **43**, 835—838.

— E. M. HOWELL, S. B. MARTEL, W. F. PROUTY, and J. B. LI, 1974: The physiological significance of protein degradation in animal and bacterial cells. Fed. Proc. **33**, 1112—1120.

— C. JABLECKI, and J. B. LI, 1974: Effects of use and disuse on amino acid transport and protein turnover in muscle. Ann. (N. Y.) Acad. Sci. **228**, 190—201.

GOLDBERG, I., and I. OHAD, 1970: Biogenesis of chloroplast membranes. J. Cell Biol. **44**, 563—571, 572—592.

GOLDENBERG, V. E., S. BUCKINGHAM, and S. C. SOMMERS, 1969: Pilocarpine stimulation of granular pneumocyte secretion. Lab. Invest. **20**, 147—158.

GOLDFISCHER, S., 1965: The cytochemical demonstration of lysosomal arylsulfatase activity by light and electron microscopy. J. Histochem. Cytochem. **13**, 520—523.

— 1967: Demonstration of copper and acid phosphatase activity in hepatocyte lysosomes in experimental copper toxicity. Nature **215**, 74—75.

— and J. MOSKAL, 1966: Electron probe microanalysis of liver in Wilson's disease. Amer. J. Path. **48**, 304—315.

— and I. STERNLIEB, 1968: Changes in the distribution of hepatic copper in relation to the progression of Wilson's disease (Hepatolenticular degeneration). Amer. J. Path. **53**, 883—901.

— and J. Bernstein, 1969: Lipofuscin (aging) pigment granules of the newborn human liver. J. Cell Biol. **42**, 253—261.

— N. CARASSO, and P. FAVARD, 1963: The demonstration of acid phosphatase activity by electron microscopy in the ergastoplasm of the ciliate *Campanella umbellaria* L. J. Microscopie 2, 611—628.

— E. ESSNER, and A. B. NOVIKOFF, 1973: The localization of phosphatase activities at the level of ultrastructure. J. Histochem. Cytochem. **12**, 72—95.

— H. VILLAVERDE, and R. FORSCHIRM, 1966: The demonstration of acid hydrolase, thermostable reduced diphosphophyridine nucleotide-tetrazolium reductase and peroxidase activities in human lipofuscin pigment granules. J. Histochem. Cytochem. **14**, 641—652.

— Y. KIKKAWA, and L. HOFFMANN, 1968: The demonstration of acid hydrolase activity in the inclusion bodies of type II alveolar cells and other lysosomes in the rabbit lung. J. Histochem. Cytochem. **16**, 102—109.

GOLDFISCHER, S., A. B. NOVIKOFF, A. ALBALA, and C. BIEMPICA, 1970: Hemoglobin uptake by rat hepatocytes and its breakdown in lysosomes. J. Cell Biol. **44**, 513—530.

— B. SCHILLER, I. STEINLIEB, 1970: Copper in hepatocyte lysosomes of the toad *Bufo marinus* L. Nature **228**, 172—173.

GOLDMAN, R., 1973: Dipeptide hydrolysis within intact lysosomes *in vitro*. FEBS Letters **33**, 208—212.

— 1974 a: Induction of vacuolation in the mouse peritoneal macrophage by concanavalin A. FEBS Letters **46**, 203—208.

— 1974 b: Effect of concanavalin A on phagocytosis by macrophages. FEBS Letters **46**, 209—213.

— and H. ROTTENBERG, 1973: Ion distribution in lysosome suspensions. FEBS Letters **33**, 233—238.

GOLDSCHMIDT, R., 1970: *In vivo* degradation of nonsense fragments in *E. coli*. Nature **228**, 1151—1154.

GOLDSPINK, D. F., and A. L. GOLDBERG, 1973: The apparent stimulation of proteolysis by adenosine triphosphate in tissue homogenates. Biochem. J. **134**, 829—832.

GOLDSTEIN, I. M., J. K. HORN, H. B. KAPLAN, and G. WEISSMANN, 1974: Calcium induced lysozyme secretion from human polymorphonuclear leukocytes. Biochem. biophys. Res. Commun. **60**, 807—812.

GOLDSTONE, A., and H. KOENIG, 1972: Biosynthesis of lysosomal glycoproteins in rat kidney. Life Sci. **11**, 511—523.

— — 1973: Physicochemical modifications of lysosomal hydrolases during intracellular transport. Biochem. J. **132**, 267—282.

— — 1974: Autolysis of glycoproteins in rat kidney lysosomes *in vitro*. Effects on the isoelectric focusing behaviour of glycoproteins, aryl sulphatase, and β-glucuronidase. Biochem. J. **41**, 527—535.

— E. SZABO, and H. KOENIG, 1970: Isolation and characterization of acidic lipo-protein in renal and hepatic lysosomes. Life Sciences **9**, 607—616.

— P. KONECNY, and H. KOENIG, 1971: Lysosomal hydrolases; conversion of acidic to basic forms by neuraminidase. FEBS Letters **13**, 68—77.

— H. KOENIG, R. NAYYAR, C. HUGHES, and C. Y. LU, 1973: Isolation and characterization of a rough microsomal fraction from rat kidney that is enriched in lysosomal enzymes. Biochem. J. **132**, 259—266.

GOMORI, G., 1952: Microscopic histochemistry, principles and practice. Chicago: University of Chicago Press.

GORDON, A. H., 1973: The role of lysosomes in protein catabolism. In: Lysosomes in biology and pathology. (DINGLE, J. T., ed.), Vol. **3**, pp. 89—137. Amsterdam: North-Holland Publishing.

GORDON, G. B., L. R. MILLER, and K. G. BENSCH, 1965: Studies on the intracellular digestive process in mammalian tissue culture cells. J. Cell Biol. **25**, 41—55.

GORDON, M. K., K. G. BENSCH, G. G. DEANIN, M. W. GORDON, 1968: Histochemical and biochemical study of synaptic lysosomes. Nature **217**, 523—525.

GORDON, S., and Z. A. COHN, 1973: The macrophage. Int. Rev. Cytol. **36**, 171—214.

— J. TODD, and Z. A. COHN, 1974 a: *In vitro* synthesis and secretion of lysozyme by mononuclear phagocytes. J. exp. Med. **139**, 1228.

— J. C. UNKELESS, and Z. A. COHN, 1974 b: Induction of macrophage plasminogen activator by endotoxin stimulation and phagocytosis: evidence for a two stage process. J. exp. Med. **140**, 995—1010.

GOSS, R. J., 1970: Turnover in cells and tissues. Adv. Cell Biol. **1**, 233—296.

GOTTLICH-RIEMANN, W., J. O. YOUNG, and A. L. TAPPEL, 1971: Cathepsins D, A, and B and the effect of pH in the pathway of protein hydrolysis. Biochim. biophys. Acta **243**, 137—146.

GOULD, A. R., B. U. MAY, and W. H. ELLIOTT, 1973: Accumulation of messenger RNA for extracellular enzymes as a general phenomenon in *Bacillus amyloliquefaciens*. J. molec. Biol. **73**, 213—219.

GRAHAM, R. C., and M. J. KARNOVSKY, 1966: The early stages of absorption of injected horseradish peroxidase in the proximal tubules of the mouse kidney; ultrastructural cytochemistry by a new technique. J. Histochem. Cytochem. **14**, 291—302.

— S. LIMPERT, R. W. KELLERMEYER, 1969: The uptake and transport of exogenous proteins in mouse liver: ultrastructural cytochemical studies with peroxidase tracers. Lab. Invest. **20**, 298—304.

GRANT, L., M. H. ROSS, J. MOSES, P. PROSE, B. W. ZWEIFACH, and R. H. EBERT, 1967: The extravascular nature of Arthus reactions elicited by ferritin: A combined light and electron microscope analysis of immune states in rabbit ear chambers and mesenteries. Z. Zellforsch. **77**, 554—588.

GRASSO, J. A., 1973: Erythropoiesis in the newt *Triturus cristatus* (Laur). J. Cell Sci. **12**, 463—489, 491—523.

GRAY, E. G., and R. A. WILLIS, 1970: On synaptic vesicles, complex vesicles and dense projections. Brain Res. **24**, 147—168.

GRAY, R. H., R. K. BRABEC, S. G. COX, P. W. FOSTER, and I. A. BERNSTEIN, 1974: The ultrastructural localization of glucose-6-phosphatase and possible origin of concentric membranes of hepatic autophagic vacuoles. J. Cell Biol. **63**, 120 a.

GREBNER, E. E., and J. TUCKER, 1973: Human urinary N-acetyl-β-hexosaminidases. Biochim. biophys. Acta **321**, 228—233.

GREEN, H. S., and J. R. CASLEY-SMITH, 1972: Calculations on the passage of small vesicles across endothelial cells by Brownian movement. J. theor. Biol. **35**, 103—111.

GREENBAUM, L. M., 1972: Leukocyte kininogenases and leukokinins from normal and malignant cells. Amer. J. Path. **68**, 613—623.

GREENBERG, R., G. S. FALK, and C. A. KOLEN, 1967: Origin of acid phosphatase associated with adrenal medullary granules. Fed. Proc. **26**, 493.

GREGORIADIS, G., 1973: Drug entrapment in liposomes. FEBS Letters **36**, 292—296.

— and R. A. BUCKLAND, 1973: Enzyme containing liposomes alleviate a model for storage disease. Nature **244**, 170—172.

— A. G. MORELL, I. STEINLIEB, and I. H. SCHEINBERG, 1970: Catabolism of disialylated ceruloplasmin in the liver. J. biol. Chem. **245**, 5833—5837.

— and D. E. NEERUNJUN, 1974: Control of the rate of hepatic uptake and catabolism of liposome entrapped proteins injected into rats; possible therapeutic applications. Eur. J. Biochem. **47**, 179—185.

— D. PUTNAM, L. LOUIS, and D. NEERUNJUN, 1974: Comparative effects of non-entrapped and liposome entrapped neuraminidase injected into rats. Biochem. J. **140**, 323—330.

GRIFFIN, F. M., and S. C. SILVERSTEIN, 1973: Segmental response of the macrophage plasma membrane to a phagocytic stimulus. J. Cell Biol. **59**, 123 a.

GRIMES, G. W., H. R. MAHLER, and P. S. PERLMAN, 1974: Mitochondrial morphology. Science **185**, 630—631.

GRISOLIA, S., 1964: The catalytic environment and its biological implications. Physiol. Rev. **44**, 657—712.

GROSS, G., and C. M. LAPIERE, 1962: Collagenolytic activity in amphibian tissues: a tissue culture assay. Proc. nat. Acad. Sci. (U.S.) **48**, 1014—1022.

GROSS, N. J., G. S. GETZ, and M. RABINOWITZ, 1969: Apparent turnover of mitochondrial deoxyribonucleic acid and mitochondrial phospholipids in the tissues of the rat. J. biol. Chem. **244**, 1552—1559.

GURD, J. W., and W. H. EVANS, 1973: Relative rates of degradation of mouse liver surface membrane proteins. Eur. J. Biochem. **36**, 273—279.

GURDON, J. B., J. B. LINGREL, and A. MARBAIX, 1973: Message stability in injected oocytes. J. Molec. Biol. **80**, 539—551.

HAIDER, M., and H. L. SEGAL, 1972: Some characteristics of the alanine-amino-transferase and arginase-inactivating system of lysosomes. Arch. biochem. biophys. **148**, 228—237.

HALL, M. O., D. BOK, and A. D. E. BACHARACH, 1969: Biosynthesis and assembly of the rod outer segment membrane system. Formation and fate of visual pigment in the frog retina. J. molec. Biol. **45**, 397—406.

240   Bibliography

HAMBERG, M., and B. SAMUELSSON, 1974: Prostaglandin endoperoxides: Novel transforma-
tions of arachidonic acid by human platelets. Proc. nat. Acad. Sci. (U.S.) **71**,
3400—3404.

HAMERMAN, D., R. JANIS, and C. SMITH, 1967: Cartilage matrix depletion by rheumatoid
synovial cells in tissue culture. J. exp. Med. **126**, 1005—1012.

HAMES, B. D., and J. M. ASHWORTH, 1974: The metabolism of macromolecules during
the differentiation of myxamebae of the cellular slime mould *Dictyostelium discoideum*
containing different amounts of glycogen. Biochem. J. **142**, 301—305.

HANNA, M. G., and A. K. SZAKAL, 1968: Localization of I^{125}-labelled antigen in germinal
centers of mouse spleen: histological and ultrastructural autoradiographic studies of the
secondary immune reaction. J. Immunol. **101**, 949—962.

HANOUNE, J., and P. FEIGELSON, 1970: Turnover of protein and RNA of liver ribosomal
components in normal and cortisol-treated rats. Biochim. biophys. Acta **199**, 214—223.

HARDIN, J. M., and S. S. SPICER, 1971: Ultrastructural localization of dialyzed iron-
reactive mucosubstance in rabbit heterophils, basophils, and eosinophils. J. Cell Biol. **48**,
268—286.

HARPER, E., K. J. BLOCH, and J. GROSS, 1971: The zymogen of tadpole collagens. Biochemistry **10**, 3035—3041.

HARTMAN, R., S. B. BOCK-HENNIG, and U. SCHWARZ, 1974: Murein hydrolases in the
envelope of *Escherichia coli*. Eur. J. Biochem. **41**, 203—208.

HASAN, M., P. GLEES, and P. E. SPOERRI, 1974: Dissolution and removal of neuronal
lipofuscin following dimethylaminoethyl p-chlorophenoxyacetate administration to Guinea
pigs. Cell Tiss. Res. **150**, 369—375.

HASILIK, A., and H. HOLZER, 1974: Participation of the tryptophan synthase inactivating
system from yeast in the activation of chitin synthase. Biochem. biophys. Res.
Commun. **53**, 552—559.

— H. MULLER, and H. HOLZER, 1974: Compartmentation of the tryptophan synthase
proteolyzing system in *Saccharomyces cerevisiae*. Eur. J. Biochem. **48**, 111—117.

HATASA, K., and T. NAKAMURA, 1965: Electron microscopic observations of lung alveolar
epithelial cells of normal young mice with special reference to formation and secretion
of osmiophilic lamellar bodies. Z. Zellforsch. **68**, 266—277.

HAUSMANN, E., und W. STOCKEM, 1972: Pinocytose und Bewegung von Amöben. VIII.
Endocytose und intrazelluläre Verdauung bei *Hyalodiscus simplex*. Cytobiologie **5**,
281—300.

— — 1973: Pinocytose und Bewegung von Amöben. XI. Das intrazelluläre Verdauungs-
system verschiedener Amöbenarten. Cytobiologie **7**, 55—75.

— — und K. E. WOHLFARTH-BOTTERMANN, 1972: Pinocytose und Bewegung von Amöben.
VII. Quantitative Untersuchungen zum Membran turnover bei *Hyalodiscus simplex*.
Z. Zellforsch. **127**, 270—286.

HAWIGER, J., and S. TIMMONS, 1973: Dynamic changes in the membranes of leukocytic
lysosomes detected with fluorescent probes and accompanied by release of lysosomal
enzymes. Biochem. biophys. Res. Commun. **55**, 1278—1284.

HAWKINS, D., 1972: Neutrophilic leukocytes in immunologic reactions: evidence for the
selective release of lysosomal constituents. J. Immunol. **108**, 310—317.

— and S. PEETERS, 1971: The response of polymorphonuclear leukocytes to immune complexes *in vitro*. Lab. Invest. **24**, 483—491.

HAWKINS, H. K., J. L. E. ERICSSON, P. BIBERFELD, and B. F. TRUMP, 1972: Lysosome and
phagosome stability in lethal cell injury. Amer. J. Path. **68**, 255—288.

HAYASHI, M., 1965: Histochemical demonstration of N-acetyl-β-glucosaminidase employing
naphthol AS BI N-acetyl-β-glucosamine as substrate. J. Histochem. Cytochem. **13**,
355—360.

— T. SHIRAHAMA, A. S. COHEN, 1968: Combined cytochemical and electron microscopic
demonstration of β-glucuronidase in mouse liver with the use of a simultaneous coupling
azo dye technique. J. Cell Biol. **36**, 289—297.

— Y. HIRAI, and Y. NATORI, 1973: Effects of ATP on protein degradation in rat liver
lysosomes. Nature New Biol. **242**, 163—166.

HAYES, L. W., and A. R. LARRABEE, 1971: Relative turnover rates of proteins and peptides of rat liver fatty acid synthetase. Biochem. biophys. Res. Commun. **45**, 966—963.

HEATH, D. F., and R. N. BARTON, 1973: The design of experiments using isotopes for the determination of the rates of disposal of blood borne substance *in vivo* with special reference to glucose ketone bodies, free fatty acids, and proteins. Biochem. J. **136**, 503—518.

HEFTMAN, E., 1971: Lysosomes in tomatos. Cytobios. **3**, 129—136.

HEINIGER, U., and P. MATILE, 1974: Protease secretion in *Neurospora crassa*. Biochem. biophys. Res. Commun. **60**, 1425—1432.

HELMINEN, H. J., and J. L. E. ERICSSON, 1968: Studies on mammary gland involution (4 articles). J. Ultrastruct. Res. **25**, 193—252.

— — 1970: On the mechanism of lysosomal enzyme secretion. Electron microscopic and histochemical studies on the epithelial cells of the rat's ventral prostate lobe. J. Ultrastruct. Res. **33**, 528—549.

— — 1971: Ultrastructural studies on prostatic involution in the rat. Mechanism of autophagy in epithelial cells with special reference to the rough surfaced endoplasmic reticulum. J. Ultrastruct. Res. **36**, 708—724.

— — 1972: Ultrastructural studies on prostatic involution in the rat. Changes in the secretory pathways. J. Ultrastruct. Res. **40**, 152—166.

HEMMAPLARDH, D., and E. H. MORGAN, 1974: Transferrin and iron uptake by human cells in culture. Exp. Cell Res. **87**, 207—212.

HEMMINKI, K., 1973: Relative turnover of tubulin subunits in rat brain. Biochim. biophys. Acta **310**, 285—288.

HENDLEY, D. D., and B. L. STREHLER, 1965: Enzymatic activities of lipofuscin age pigments: comparative histochemical and biochemical studies. Biochim. biophys. Acta **99**, 406—417.

HENNING, R., and W. STOFFEL, 1972: Ubiquinone in the lysosomal membrane fraction of rat liver. Hoppe-Seyler Z. Physiol. Chem. **353**, 75—78.

— — 1973: Glycosphingolipids in lysosomal membranes. Hoppe-Seyler Z. Physiol. Chem. **354**, 760—770.

— and H. G. HEIDRICH, 1974: Membrane lipids of rat liver lysosomes prepared by free flow electrophoresis. Biochim. biophys. Acta **345**, 326—335.

— H. D. KAULEN, and W. STOFFEL, 1970: Isolation and characterization of the lysosomal and the plasma membrane of the rat liver cell. Hoppe-Seyler Z. Physiol. Chem. **351**, 1191—1199.

— H. PLATTNER, and W. STOFFEL, 1973: Nature and localization of acidic groups on lysosomal membranes. Biochim. biophys. Acta **330**, 61—75.

HENRIKSON, P. A., and U. CLEVER, 1972: Protease activity and cell death during metamorphosis in the salivary gland of *Chironomus tentans*. J. Insect. Physiol. **18**, 1981—2004.

HENSON, P. M., 1970: Mechanisms of release of constituents from rabbit platelets by antigen antibody complexes and complement: lytic and nonlytic reactions. J. Immunol. **105**, 475—489.

— 1971: Interaction of cells with immune complexes: adherence, release of constituents, and tissue injury. J. exp. Med. **134**, 114s—135s.

— 1972: Pathologic mechanisms in neutrophil mediated injury. Amer. J. Path. **68**, 593—605.

HEREWARD, F. V., 1974: Rough membranes in *Schizosaccharomyces pombe* protoplasts. Exp. Cell Res. **87**, 213—218.

HERRON, W. L., B. W. RIEGEL, and M. C. RUBIN, 1971: Outer segment production and removal in the degenerating retina of the dystrophic rat. Invest. Ophthalmol. **10**, 54—63.

HERRON, W. Z., B. W. RIEGEL, O. F. MEYERS, and M. L. RUBIN, 1969: Retinal dystrophy in the rat. A pigment epithelial disease. Invest. Ophthalmol. **8**, 595—604.

HERS, H. G., 1973: The concept of inborn lysosomal disease. In: Lysosomes and storage diseases. (HERS, H. G., and F. VAN HOOF, eds.), pp. 147—171. New York: Academic Press.

HERS, H. G., and T. DE BARSY, 1973: Type II Glycogenosis (acid maltase deficiency). In: Lysosomes and storage diseases. (HERS, H. G., and F. VAN HOOF, eds.), pp. 197—216. New York: Academic Press.

— and F. VAN HOOF (eds.), 1973: Lysosomes and storage diseases. New York: Academic Press.

HERSHKO, A., and G. M. TOMPKINS, 1971: Studies on the degradation of tyrosine amino transferase in hepatoma cells in culture. J. biol. Chem. **246**, 710—714.

HEUSER, J. E., and T. E. REESE, 1973: Evidence for recycling of synaptic vesicle membrane during transmitter release at the frog neuromuscular junction. J. Cell Biol. **57**, 315—344.

HIBBS, J. B., 1974: Heterocytolysis by macrophages activated by Bacillus Calmette-Guerin: lysosome exocytosis into tumor cells. Science **184**, 468—471.

HICKMAN, S., and E. F. NEUFELD, 1972: A hypothesis for I-cell disease: defective hydrolases that do not enter lysosomes. Biochem. biophys. Res. Commun. **49**, 992—997.

HICKS, R. M., 1974: Phil. Trans. Roy. Soc. Lond. **B 268**, 23—68.

HIGGINS, J. A., 1974: J. Cell Biol. **62**, 635—646.

HILL, J. K., and P. A. WARD, 1969: C_3 leukotactic factors produced by a tissue protease. J. exp. Med. **130**, 505—518.

HILLE, M. B., A. J. BARRETT, J. T. DINGLE, and H. B. FELL, 1970: Microassay for cathepsin D shows an unexpected effect of Cycloheximide on limb bone rudiments in organ culture. Exp. Cell Res. **61**, 470—472.

HIPKISS, A. R., and M. KOGUT, 1973: Stimulation of protein breakdown in *Escherichia coli* by dehydrostreptomycin. Biochem. Soc. Trans. **1**, 594—596.

HIRSCH, C. A., and H. H. HIATT, 1966: Turnover of rat liver ribosomes in fed and in fasted rats. J. biol. Chem. **241**, 5936—5940.

HIRSCH, H. E., 1968: Acid phosphatase localization in individual neurons by a quantitative histochemical method. J. Neurochem. **15**, 123—130.

HIRSCH, J. G., 1972: The phagocytic defense system. Symp. Soc. Gen. Microbiol. **22**, 59—74.

— M. E. FEDORKO, 1970: Morphology of mouse mononuclear phagocytes. In: Mononuclear phagocytes. (VAN FURTH, R., ed.), pp. 7—28. Philadelphia: F. A. Davis.

— — and Z. A. COHN, 1968: Vesicle fusion and formation at the surface of pinocytic vesicles in macrophages. J. Cell Biol. **38**, 629—632.

HIRSCHHORN, R., J. M. KAPLAN, A. F. GOLDBERG, K. HIRSCHHORN, and G. WEISSMANN, 1965: Acid phosphatase-rich granules in human lymphocytes induced by phytohemagglutinin. Science **147**, 55—57.

— G. BRITTENGER, K. HIRSCHHORN, and G. WEISSMANN, 1968: Studies on lysosomes. XIII. Redistribution of acid hydrolases in human lymphocytes stimulated by phytohemagglutinin. J. Cell Biol. **37**, 412—423.

— J. GROSSMAN, W. TROLL, and G. WEISSMANN, 1971: The effect of epsilon-amino-caproic acid and other inhibitors of proteolysis upon the response of human peripheral blood lymphocytes to phytohemagglutinin. J. Clin. Invest. **50**, 1206—1217.

HISLOP, E. C., V. M. BARNABY, C. SHELLIS, and F. LABORDA, 1974: Localization of α-L-arabinofuranosidase and acid phosphatase in mycelium of *Sclerotinia fructigena*. J. gen. Microbiol. **81**, 79—99.

HO, M. W., 1973: Hydrolysis of ceramide trihexoside by a specific galactosidase from human liver. Biochem. J. **133**, 1—10.

— 1975: Specificity of low molecular weight glycoprotein effector of lipid glycosidase. FEBS letters **53**, 243—247.

— and N. D. LIGHT, 1973: Glucocerebrosidase: reconstitution from macromolecular components depends on acidic phosphilipids. Biochem. J. **136**, 821—823.

— P. CHEETHAM, and D. ROBINSON, 1973: Hydrolysis of GM_1-ganglioside by human liver β-galactosidase isoenzymes. Biochem. J. **136**, 351—359.

HOCHSCHILD, R., 1971: Lysosomes, membranes, and aging. Exp. Gerontol. **6**, 153—166.

HOFFENBERG, R., A. H. GORDON, E. G. BLACK, and L. N. LOUIS, 1970: Plasma protein catabolism by perfused rat liver: the effect of alteration of albumin concentration and dietary protein depletion. Biochem. J. **118**, 401—404.

HOFFMAN, E. K., L. RASMUSSEN, and E. ZENTHEN, 1974: Cytochalasin B: aspects of phagocytosis in nutrient uptake in *Tetrahymena*. J. Cell Sci. **15**, 403—406.

HOFFMAN, H. P., and C. AVERS, 1973: Mitochondrion of yeast: ultrastructural evidence for one giant branched organelle per cell. Science **181**, 749—751.

— — (Response to GRIMES *et al.*), Science **185**, 631.

HOFFSTEIN, S., R. B. ZURIER, I. M. GOLDSTEIN, and G. WEISSMANN, 1973: Mechanisms of lysosomal enzyme secretion; effects of autonomic antagonists and a component of complement. J. Cell Biol. **59**, 145 A.

— F. STREULI, J. HIRSCH, A. C. FOX, D. E. GENNARO, and G. WEISSMANN, 1974: Cytochemical localization of lysosomal enzyme activity in normal and ischemically injured dog myocardium. J. Cell Biol. **63**, 142 a. (Full account, 1975; Amer. J. Path. **79**, 193—206.)

HOGAN, M., and I. WOOD, 1974: Phagocytosis by pigment epithelium of human retinal cones. Nature **252**, 305—306.

HOHN, D. C., and R. I. LEHRER, 1975: NADPH-oxidase deficiency in X-linked chronic granulomatous disease. J. Clin. Invest. **55**, 707—713.

HOKIN, L. E., 1968: Dynamic aspects of phospholipids during protein secretion. Int. Rev. Cytol. **23**, 187—208.

HOLDSWORTH, G., and R. COLEMAN, 1975: Enzyme profiles of mammalian bile. Biochim. Biophys. Acta **389**, 47—50.

HOLLYFIELD, J. G., 1973: Elimination of egg pigment from developing ocular tissues in the frog *Rana pipiens*. Develop. Biol. **30**, 125—128.

— and A. WARD, 1974: Phagocytic activity in the retinal pigment epithelium of the frog *Rana pipiens*. I. Uptake of polystyrene spheres. II. Exclusion of *Saraner subflava*. J. Ultrastruct. Res. **46**, 327—338, 339—350.

HOLMSEN, H., H. J. DAY, and M. STORMORKEN, 1965: The blood platelet release reaction. Scand. J. Hemotol. **6**, 1—26 (Suppl. #8).

HOLROYDE, C. P., and F. H. GARDNER, 1970: Acquisition of autophagic vacuoles by human erythrocytes: physiological role of the spleen. Blood **36**, 566—575.

HOLT, S. J., 1969: Factors governing the validity of staining methods for enzymes and their bearing upon the Gomori acid phosphatase technique. Exp. Cell Res. Suppl. **7**, 1—27.

— 1963: Some observations on the occurrence and nature of esterases in lysosomes. In: Ciba foundation symposium on lysosomes. (DE REUCK, A. V. S., and M. P. CAMERON, eds.), pp. 114—119. Boston: Little Brown.

— and P. C. BARROW, 1972: Esterases of lysosomes and endoplasmic reticulum of liver. Abstracts, International Symposium on Lysosomes, Hakone, Japan (Aug.—Sept. 1972), pp. 48—49.

— and R. M. HICKS, 1966: The importance of osmiophilia in the production of stable azoindoxyl complexes of high contrast for combined enzyme cytochemistry and electron microscopy. J. Cell Biol. **29**, 361—366.

HOLTZMAN, E., 1969: Lysosomes in the physiology and pathology of neurons. In: Lysosomes in biology and pathology (DINGLE, J. T., and H. B. FELL, eds.), Vol. **1**, pp. 192—216. Amsterdam: North-Holland Publishing.

— 1971: Cytochemical studies of protein transport in the nervous system. Phil. Trans. Roy. Soc. (Lond.) B **261**, 407—421.

— 1974: The biogenesis of organelles. Hospital Practice **9** (3), 75—88.

— and A. B. NOVIKOFF, 1965: Lysosomes in the rat sciatic nerve following crush. J. Cell Biol. **27**, 651—669.

— and R. DOMINITZ, 1968: Cytochemical studies of lysosomes, Golgi apparatus and endoplasmic reticulum in secretion and protein uptake by adrenal medulla cells of the rat. J. Histochem. Cytochem. **16**, 320—336.

— A. B. NOVIKOFF, and H. VILLAVERDE, 1967: Lysosomes and GERL in normal and chromatolytic neurons of the rat ganglion nodosum. J. Cell Biol. **33**, 419—436.

— and E. R. PETERSON, 1969: Protein uptake by mammalian neurons. J. Cell Biol. **40**, 863—869.

— A. R. FREEMAN, and L. A. KASHNER, 1973: Stimulation dependent alterations in peroxidase uptake by lobster neuromuscular junctions. Science **173**, 733—736.

16*

Holtzman, E., S. Teichberg, S. J. Abrahams, E. Citkowitz, S. M. Crain, N. Kawai, and E. R. Peterson, 1973: Notes on synaptic vesicles and related structures, endoplasmic reticulum, lysosomes, and peroxisomes in nervous tissue and the adrenal medulla. J. Histochem. Cytochem. **21**, 349—385.

Holtzman, J. L., T. E. Gram, and J. R. Gillett, 1970: The kinetics of P^{32} incorporation into the phospholipids of hepatic rough and smooth microsomal membranes of male and female rats. Arch. biochem. biophys. **138**, 199—207.

Homewood, L. A., D. C. Warhurst, W. Peters, and V. C. Baggaley, 1972: Lysosomes, pH and the anti-malarial action of chloroquine. Nature **235**, 50—52.

Hook, G. E. R., K. S. Dodgson, F. A. Rose, and M. Worwood, 1973: Relative distribution of arylsulfatase A and B in rat liver parenchymal and other cells. Biochem. J. **134**, 191—195.

Hopkins, C. R., 1969: The fine structural localization of acid phosphatase in the prolactin cell of the teleost pituitary following the stimulation and inhibition of secretory activity. Tissue and Cell **1**, 653—671.

Hopsu-Havu, V. K., A. U. Arstila, M. J. Helminen, H. O. Kalimo, and G. G. Glenner, 1967: Improvements in the method for the electron microscopic localization of aryl sulfatase activity. Histochemie **8**, 54—64.

Horie, A., T. Takino, L. Herman, and P. J. Fitzgerald, 1971: Pancreas acinar cells regeneration. VII. Relationship of acid phosphatase and B-glucuronidase to intracellular organelles and ethionine lesions. Amer. J. Path. **63**, 299—309.

Houdry, J., 1971: Étude histochimique de quelques hydrolases lysosomiques de l'épithelium intestinel au cours du developpement de la larve de *Discoglossus picthus* Otth., Amphibien, Anoure. II. Observation au microscope électronique. Histochemie **26**, 142—159.

Hsu, L., and A. L. Tappel, 1964: Lysosomal enzymes of rat intestinal mucosa. J. Cell Biol. **23**, 233—240.

Hubbard, A. L., and Z. A. Cohn, 1975: Fate of externally disposed plasma membrane proteins of mouse fibroblast following phagocytosis. J. Cell Biol. **64**, 461—479.

Hudgin, R. L., and G. Ashwell, 1974: Studies on the role of glycosyltransferases in the hepatic binding of asialoglycoproteins. J. biol. Chem. **249**, 7369—7372.

— W. E. Pricer, G. Ashwell, R. J. Stockert, and A. G. Morell, 1974: The isolation and properties of a rabbit liver binding protein specific for asialoglycoproteins. J. biol. Chem. **249**, 5536—5543.

Huffer, W. E., 1973: Degradation of chondromucoprotein by macrophage enzymes. Lab. Invest. **29**, 304—309.

Hughes, R. C., B. Sanford, and R. W. Jeanloz, 1972: Regeneration of the surface glycoproteins of a transplantable mouse tumour cell after treatment with neuraminidase. Proc. nat. Acad. Sci. (U.S.) **69**, 942—945.

— and J. Clark, 1974: Growth of baby hamster kidney cells in media containing neuraminidase. Exp. Cell Res. **85**, 362—366.

Hugon, J., and M. Borgers, 1966: Fine structural changes and localization of phosphatases in the epithelium of the duodenal crypt of X-irradiated mice. Histochemie **6**, 209—225.

Huisman, W., L. Lanting, J. M. W. Bouma, and M. Gruber, 1974 a: Proteolysis at neutral pH in a lysosomal cytosol system cannot be attributed to the uptake of proteins into lysosomes. FEBS Letters **45**, 129—131.

— J. M. W. Bouma, and M. Gruber, 1974 b: Facilitation of lysosome disruption by ATP at low pH. Nature **250**, 428—429.

Humphrey, J. A., 1970: Summing up—A layman's view. In: Mononuclear phagocytes. (Van Furth, R., ed.), pp. 625—634. Philadelphia: F. A. Davis.

Humphrey, J., and J. Alward, 1975: A new method for measurement of protein turnover. Biochem. J. **148**, 119—127.

Ide, M., and W. M. Fishman, 1969: Dual localization of B-glucuronidase and phosphatase in lysosomes and in microsomes. Histochemie **20**, 300—321.

Ignarro, L. J., and C. Columbo, 1973: Enzyme release from polymorphonuclear leukocyte lysosomes: regulation by autonomic drugs and cyclic nucleotides. Science **180**, 1181—1182.

IGNARRO, L. J., and W. J. GEORGE, 1974: Hormonal control of lysosomal enzyme release from human neutrophils: elevation of cyclic nucleotide levels by autonomic neurohormones. Proc. nat. Acad. Sci. (U.S.) 71, 2027—2031.
— T. F. LINT, and W. J. GEORGE, 1974: Hormonal control of lysosomal enzyme release from human neutrophils: effects of autonomic agents on enzyme release, phagocytosis and cyclic nucleotide levels. J. exp. Med. 139, 1395—1414.
— R. J. PADDOCK, W. J. GEORGE, 1974: Hormonal control of neutrophil lysosomal enzyme release: effect of epinephrine on adenosine 3,5-monophosphate. Science 183, 855—857.
IHLER, G. M., R. H. GLEW, and F. W. SCHNURE, 1973: Enzyme loading of erythrocytes. Proc. nat. Acad. Sci. (U.S.) 70, 2663—2666.
IKONNE, J. U., and R. B. ELLIS, 1973: N-acetyl-β-D-hexosaminidase component A. Biochem. J. 135, 457—462.
ILES, G. H., P. SEEMAN, D. NAYLOR, and B. CINADER, 1973: Membrane lesions in immunolysis: surface rings, globule aggregates and transient openings. J. Cell Biol. 56, 528—539.
ILLINGWORTH, D. R., O. W. PORTMAN, A. L. ROBERTSON, and W. A. MAGYAR, 1973: The exchange of phospholipids between plasma lipoproteins and rapidly dividing human cells grown in tissue culture. Biochim. biophys. Acta 306, 422—436.
INOUYE, M., N. ARNHEIM, and R. STERNGLANZ, 1973: Bacteriophage T 7 lysozyme is an N-acetyl-muramyl alanine amidase. J. biol. Chem. 248, 7247—7252.
ISHIKAWA, J., and E. YAMADA, 1970: The degradation of the photoreceptor outer segment within the pigment epithelial cell of the rat retina. J. Electron Microscopy 19, 85—91.
ITEN, W., and P. MATILE, 1970: Role of chitinase and other lysosomal enzymes of *Coprinus lagopus* in the autolysis of fruiting bodies. J. gen. Microbiol. 61, 301—309.
IWANIJ, V., N.-H. CHUA, and P. S. SIEKEVITZ, 1975: Synthesis and turnover of ribulose phosphate carboxylase and of its subunits during the cell cycle of *Chlamydomonas reinhardtii.* J. Cell Biol. 64, 572—585.

JACOBSEN, J. V., and J. E. VARNER, 1967: Gibberellic acid induced synthesis of protease by isolated aleurone layers of barley. Plant Physiol. 42, 1596—1600.
JACQUES, P. J., 1969: Endocytosis. In: Lysosomes in biology and pathology. (DINGLE, J. T., and H. B. FELL, eds.), Vol. 2, pp. 395—420. Amsterdam: North-Holland Publishing.
JAMES, W. P. T., D. H. ALPERS, J. E. GERBER, and ISSELBACHER, 1971: The turnover of disaccharidases and brush border proteins in rat intestine. Biochim. biophys. Acta. 230, 194—203.
JAMIESON, J. D., and G. E. PALADE, 1967: Intracellular transport of secretory proteins in the pancreatic exocrine cell. II. Transport to condensing vacuoles and zymogen granules. J. Cell Biol. 34, 597—615.
— — 1968: Intracellular transport of secretory proteins in the pancreatic exocrine cell. IV. Metabolic requirements. J. Cell Biol. 39, 589—603.
— — 1971 a: Condensing vacuole conversion and zymogen granule discharge in pancreatic exocrine cells: metabolic studies. J. Cell Biol. 48, 503—522.
— — 1971 b: Synthesis, intracellular transport and discharge of secretory proteins in stimulated pancreatic exocrine cells. J. Cell Biol. 50, 135—157.
JANCIK, J., and R. SCHAUER, 1974: Sialic acid, a determinant of the life time of rabbit erythrocytes. Hoppe-Seyler Z. Physiol. Chem. 355, 395—400.
JANOFF, A., 1972: Human granulocyte elastase. Amer. J. Path. 68, 579—591.
— and J. D. ZELIGS, 1968: Vascular injury and lysis of basement membrane *in vitro* by neutral protease of human leukocytes. Science 161, 703—704.
— and J. BLONDIN, 1973: The effects of human granulocyte elastase on bacterial suspensions. Lab. Invest. 29, 454—457.
JARED, D. W., J. N. DUMONT, and R. A. WALLACE, 1973: Distribution of incorporated and synthesized proteins among cell fractions of *Xenopus oocytes.* Develop. Biol. 35, 19—28.
JATZKEWITZ, H., and E. MEHL, 1969: Cerebroside sulphatase and aryl sulphatase. A deficiency in metachromatic leukodystrophy. J. Neurochem. 16, 19—28.

JEFFAY, H., and R. J. WINZLER, 1958: The metabolism of serum proteins. I. The turnover rate of rat serum proteins. J. biol. Chem. **231**, 101—109.

JEFFERSON, L. S., D. E. RANNELS, B. L. MUNGER, and H. E. MORGAN, 1974: Insulin in the regulation of protein turnover in heart and skeletal muscle. Fed. Proc. **33**, 1098—1104.

JENNINGS, J. B., 1957: Studies on feeding, digestion and food storage in free-living flatworms (*Platyhelminthes, Turbellaria*). Biol. Bull. **112**, 63—80.

JENSEN, M. S., and D. F. BAINTON, 1973: Temporal changes in pH within phagocytic vacuoles of the polymorphonuclear neutrophilic leukocyte. J. Cell Biol. **56**, 379—388.

JICK, H., and L. SHUSTER, 1966: The turnover of microsomal reduced nicotinamide adenine dinucleotide phosphate-cytochrome-C-reductase in the livers of mice treated with phenobarbital. J. biol. Chem. **241**, 5363—5369.

JIMBOW, K., G. SZABO, and T. B. FITZPATRICK, 1974: Ultrastructural investigation of autophagocytosis of melanosomes and programmed death of melanocytes in white leghorn feathers: a study of morphogenetic events leading to hypomelanosis. Develop. Biol. **36**, 8—23.

JOHNSON, K. D., D. DANIELS, M. J. DOWLER, and D. L. RAYLE, 1974: Activation of *Avena coleoptile* cell wall glycosidases by hydrogen ions and auxin. Plant Physiol. **53**, 224—228.

JOHNSON, R. W., and F. T. KENNEY, 1973: Regulation of amino transferase in rat liver. XI. Studies on the relationship of enzyme stability to enzyme turnover in cultured hepatoma cells. J. biol. Chem. **248**, 4528—4531.

JOHNSON, S. M., 1975: The inability of macrophages to digest liposomes containing a high proportion of cholesterol. Biochem. Soc. Trans. **3**, 160—161.

JONES, D., A. H. GORDON, and J. S. D. BACON, 1974: Cooperative action by endo and exo-β-(1-73)glucanases from parasitic fungi in the degradation of cell wall glucans by *Sclerotininia selerotiorum*. Biochem. J. **140**, 47—55.

JONES, D. G., and R. F. BREARLEY, 1973: An analysis of some aspects of synaptosomal ultrastructure by the use of serial sections. Z. Zellforsch. **140**, 481—496.

JONES, E. G., and A. J. ROCKEL, 1973: Observations on complex vesicles, neurofilamentous hyperplasia, and increased electron density during terminal degeneration in the inferior colliculus. J. comp. Neurol. **147**, 93—118.

JONES, R. E., and J. E. ARMSTRONG, 1971: Evidence for osmotic regulation of hydrolytic enzyme production in germinating barley seed. Plant Physiol. **48**, 137—142.

JONES, R. E., and W. A. MEMMINGS, 1971: Proteolytic enzymes of the yolk-sac splanchnopleur. Biochim. Biophys. Acta **242**, 278—287.

JONES, R. F., S. KATES, and S. J. KELLER, 1968: Protein turnover and macromolecular synthesis during growth and gamete differentiation in *Chlamydomonas reinhardtii*. Biochim. Biophys. Acta **157**, 589—598.

JONES, T. C., and J. G. HIRSCH, 1972: The interaction between *Toxoplasma gondii* and mammalian cells. II. The absence of lysosomal fusion with phagocytic vacuoles containing living parastes. J. exp. Med. **136**, 1173—1194.

— S. YEH, and J. G. HIRSCH, 1972: The interaction between *Toxoplasma gondii* and mammalian cells. I. Mechanism of entry and intracellular fate of the particles. J. exp. Med. **136**, 1157—1172.

JOURNEY, L. J., 1964: Cytoplasmic microtubules in mouse peritoneal macrophages during rejection of MCIM ascites tumor cells. Cancer Res. **24**, 1391—1403.

JUNGALWALA, F. B., 1974: The turnover of myelin phosphatidylcholine and sphingomyelin in the adult rat brain. Brain Res. **78**, 99—108.

KABACK, M. M., and R. R. HOWELL, 1973: Heterozygote detaction and pre-natal diagnosis of lysosomal diseases. In: Lysosomes and storage diseases. (HERS, H. G., and F. VAN HOOF, eds.), pp. 599—616. New York: Academic Press.

KADLUBOWSKI, M., and J. R. HARRIS, 1974: The appearance of a protein in the human erythrocyte membrane during ageing. FEBS Letters **47**, 252—254.

KAHANE, I., and S. RAZIN, 1969: Synthesis and turnover of membrane protein and lipid in *Mycoplasma laidlarvii*. Biochim. biophys. Acta **183**, 79—89.

KALINA, M., Y. KLETTER, and M. ARONSON, 1974: The interaction of phagocytes and the large sized parasite *Cryptococcus neoformans:* a cytochemical and ultrastructural study. Cell Tiss. Res. **152**, 165—174.

KANAI, Y., T. SUGIMURA, T. MATSUSHIMA, and A. KAWAMURA, 1974: Studies on *in vivo* degradation of rat hepatic catalase with or without modification by 3-Amino-1,2,4-triazole. J. biol. Chem. **249**, 6505—6511.

KANESAKI, T., and K. KADOTA, 1968: The vesicle in a basket; a morphological study of the coated vesicle isolated from the nerve endings of the guinea pig brain with special reference to the mechanism of membrane movement. J. Cell Biol. **42**, 202—220.

KAPELLER, M., R. GALOZ, N. B. GROVER, and F. DOLJANSKI, 1973: Natural shedding of carbohydrate containing macromolecules from cell surfaces. Exp. Cell Res. **79**, 152—158.

KAPLAN, R., and D. APIRION, 1974: The involvement of ribonuclease. I. Ribonuclease. II. And polynucleotide phosphorylase in the degradation of stable ribonucleic and during carbon starvation in *Escherichia coli.* J. biol. Chem. **249**, 149—151.

— — 1975: Decay of ribosomal ribonucleic acid in *Escherichia coli* cells starved for various nutrients. J. Biol. Chem. **250**, 3174—3178.

KARAKASHIAN, M. W., and S. J. KARAKASHIAN, 1973: Intracellular digestion and symbiosis in *Paramecium bursaria.* Exp. Cell Res. **81**, 111—119.

KARNOVSKY, M. C., 1973: Chronic granulomatous disease—pieces of a cellular and molecular puzzle. Fed. Proc. **32**, 1527—1533.

KARNOVSKY, M. J., H. D. ENGERS, M. LEVENTHAL, W. D. PERKINS, and E. R. UNANUE, 1972: Topography, migration, and endocytosis of lymphocyte surface macromolecules. Acta Histochem. Cytochem. **5**, 275—277.

KARNOVSKY, M. L., S. SIMMONS, E. A. GLASS, A. W. SHAFER, and P. D. A. HART, 1970: Metabolism of macrophages. In: Mononuclear phagocytes. (VAN FURTH, R., ed.), pp. 103—120. Philadelphia: F. A. Davis.

KATAYAMA, I., C. Y. KI, and L. T. YAM, 1972: Ultrastructural cytochemical demonstration of tartrate resistant acid phosphatase isoenzyme activity in "Hairy cells" of leukemic reticuloendothelosis. Amer. J. Path. **69**, 471—482.

KATCHALSKY, A., 1964: Polyelectrolytes and their biological interaction. Biophys. J. **4** (suppl.), 9—41.

KATO, K., I. HIROHATA, W. M. FISHMAN, and M. ISUKAMOTO, 1972: Intracellular transport of mouse kidney β-glucuronidase induced by gonadotrophin. Biochem. J. **127**, 425—435.

KATSUNUMA, T., E. SCHÖTT, S. ELSÄSSER, and H. HOLZER, 1972: Purification and properties of tryptophan synthase inactivating enzyme from yeast. Eur. J. Biochem. **27**, 520—527.

KATTLOVE, M. E., J. G. WILLIAMS, E. GAYNOR, M. SPIVACK, R. M. BRADLEY, and R. O. BRADY, 1969: Gaucher cells in chronic myeloctic leukemia: an acquired abnormality. Blood **33**, 379—390.

KATUNUMA, N., 1973: Enzyme degradation and its regulation by group specific proteases in various organs of the rat. Curr. Top. Cell. Reg. **7**, 175—203.

— E. KOMINAMI, K. KOBAYASHI, Y. BANNO, K. SUZUKI, K. CHICHIBU, Y. HAMAGUCHI, and T. KATSUNUMA, 1975: Studies on new intracellular proteases in various organs of rat, 1. purification and comparison of their properties. Eur. J. Biochem. **52**, 37—50.

KAULEN, H. D., R. HENNING, and W. STOFFEL, 1970: Comparison of some enzymes of the lysosomal and the plasma membrane of rat liver cell. Hoppe-Seyler Z. Physiol. Chem. **351**, 1555—1563.

KAWASKI, T., and I. YAMASHINA, 1971: Metabolic studies of rat liver plasma membrane using D-1-C^{14}-glucosamine. Biochim. biophys. Acta **225**, 234—248.

KEENE, W. R., and J. H. JANDL, 1959: Studies of the reticuloendothelial mass and sequestering function of rat bone marrow. Blood **26**, 157—175.

KEMMLER, W., J. D. PETERSON, A. M. RUBENSTEIN, and D. R. STEINER, 1972: On the biosynthesis, intracellular transport and mechanisms of conversion of proinsulin to insulin and C-peptide. Diabetes **21**, 572—581.

— D. F. STEINER, J. O. BORG, 1973: Studies on the conversion of proinsulin to insulin. III. Studies *in vitro* with a crude secretion granule fraction isolated from rat islets of Langerhans. J. biol. Chem. **248**, 4544—4551.

KEMPER, B., J. E. HABENER, R. E. MULLIGAN, J. T. POTTS, and A. RICH, 1974: Pre-parathyroid hormone: a direct translation product of parathyroid messenger RNA. Proc. nat. Acad. Sci. (U.S.) **71**, 3731—3735.

KEMSHEAD, J. T., and A. R. HIPKISS, 1974: Degradation of abnormal proteins in *Escherichia coli:* relative susceptibility of canavanyl proteins and puromycin peptides to proteolysis *in vitro*. Eur. J. Biochem. **45**, 535—540.

KENNEDY, E. P., 1970: The lactose permease system of *Escherichia coli*. In: The lactose operon. (BECKWITH, J. B., and D. ZIPSER, eds.), pp. 49—52. New York: Cold Spring Harbor Laboratory.

KENNEY, F. T., 1967: Turnover of rat liver tyrosine amino transferase; stabilization after inhibition of protein synthesis. Science **186**, 525—527.

KENT, G., O. T. MINICK, F. I. VOLINI, and E. ORFEI, 1966: Autophagic vacuoles in human red cells. Amer. J. Path. **48**, 831—857.

KERR, J. F. R., 1970: Liver cell defecation; an electron microscope study of the discharge of lysosomal residual bodies into the intercellular space. Amer. J. Path. **100**, 99—103.

— 1973: Some lysosome functions in liver cells reacting to sublethal injury. In: Lysosomes in biology and pathology. (DINGLE, J. T., ed.), Vol. 3, pp. 365—394. Amsterdam: North-Holland Publishing.

KESSEL, R. G., and R. S. DECKER, 1972: Cytodifferentiation in *Rana pipiens* oocytes. V. Ultrastructural localization of acid phosphatase activity in diplotene oocytes and their associated follicle envelope. J. Microscopie **14**, 169—182.

KIEHN, E. D., and J. J. HOLLAND, 1970: Membrane and non-membrane proteins of mammalian cells. Synthesis, turnover, and size distribution. Biochemistry **9**, 1716—1728.

KILSHEIMER, G. S., and B. AXELROD, 1957: Inhibition of prostatic acid phosphatase by α-hydroxycarboxylic acid. J. biol. Chem. **227**, 879—890.

KINT, J. A., 1974: *In vitro* restoration of deficient β-galactosidase activity in liver of patients with Hurler and Hunter diseases. Nature **250**, 424—425.

— and D. CARTON, 1973: Fabry's Disease. In: Lysosomes and storage diseases. (HERS, H. G., and F. VAN HOOF, eds.), pp. 357—380. New York: Academic Press.

— G. DACREMONT, D. CARTON, E. OYRE, and C. HOOFT, 1973: Mucopolysaccharidosis; secondarily induced abnormal distribution of lysosomal isoenzymes. Science **181**, 352—354.

KITCHING, J. A., 1956: Food vacuoles. Protoplasmatologia 3 D 3 b (54 p.). Wien: Springer.

KLAUS, G. C., 1973: Cytochalasin B dissociation of pinocytosis and phagocytosis by peritoneal macrophages. Exp. Cell Res. **79**, 73—78.

KLEBANOFF, S. J., 1971: Intraleukocytic microbicidal defects. Ann. Rev. Med. **22**, 39—62.

— 1974: Role of the superoxide anion in the myeloperoxidase mediated antimicrobial system. J. biol. Chem. **249**, 3724—3728.

— and J. GREEN, 1973: Degradation of thyroid hormone by phagocytosing human leukocytes. J. Clin. Invest. **52**, 60—72.

KLEVECZ, R. R., 1971: Rapid protein catabolism in mammalian cells is obscured by reutilization of amino acid. Biochem. biophys. Res. Commun. **43**, 76—81.

KLOCKERS, M., and E. F. OSSERMAN, 1974: Localization of lysozyme in normal rat tissues by an immunoperoxidase method. J. Histochem. Cytochem. **2**, 139—146.

KLOETZEL, J. A., 1974: Feeding in ciliated protozoa. I. Pharyngeal disks in *Euplotes*. I. A source of membrane for food vacuole formation? J. Cell Sci. **15**, 379—402.

KNOWLES, S. E., J. M. GUNN, R. W. HANSON, and F. J. BALLARD, 1975: Increased degradation rates of protein synthesized in hepatoma cells in the presence of amino acid analogues. Biochem. J. **146**, 595—600.

KNOX, R. B., and J. HESLOP-HARRISON, 1970: Pollen wall proteins: Localization and enzymic activity. J. Cell Sci. **6**, 1—27.

— — 1971: A cytochemical study of the leaf gland enzymes of insectivorous plants of the genus *Pinquicula*. Planta **96**, 183—211.

KOENIG, H., 1969: Lysosomes in the nervous system. In: Lysosomes in biology and pathology. (DINGLE, J. T., and H. B. FELL, eds.), Vol. 2, pp. 11—162. Amsterdam: North-Holland Publishing.

Kolb, W. P., and H. J. Muller-Eberhard, 1973: The membrane attack mechanism of complement: verification of a stable C 5—9 complex in free solution. J. exp. Med. **138**, 438—451.

Kolodny, E. H., 1973: Clinical and biochemical genetics of the lipidoses. Semin. Hematol. **9**, 251—271.

Kolodny, G. M. K., 1975: Turnover of ribosomal RNA in mouse fibroblasts (3 T 3) in culture. Exp. Cell Res. **91**, 101—106.

Kolody, G. M., 1973: Transfer of cytoplasmic and nuclear proteins between cells in culture. J. molec. Biol. **78**, 197—210.

Kolsch, E., 1970: Patterns of handling and presenting antigen by macrophages. In: Mononuclear phagocytes. (van Furth, R., ed.), pp. 548—561. Philadelphia: F. A. Davis.

— and N. A. Mitchison, 1968: The subcellular distribution of antigen in macrophages. J. exp. Med. **128**, 1059—1079.

Kominami, E., K. Kobayashi, S. Kominami, and W. Katunama, 1972: Properties of a specific protease for pyroxidal enzymes and its biological role. J. biol. Chem. **247**, 6848—6855.

— Y. Banno, K. Chichibu, T. Shiotani, Y. Hamaguchi, and N. Katunuma, 1975: Studies on new intracellular proteases in various organs of rat, 2. mode of limited proteolysis. Eur. J. Biochem. **52**, 51—57.

Konig, E., G. Brittenger, G. Cohnen, 1973: Relation of lysosomal fragility in CLL lymphocytes to PHS reactivity. Nature New Biol. **244**, 247—248.

Kotoulas, O. B., and M. J. Phillips, 1971: Fine structural aspects of the mobilization of hepatic glycogen. I. Acceleration of glycogen breakdown. Amer. J. Path. **63**, 1—22.

— J. Ho, F. Adachi, B. I. Weigensberg, and M. J. Phillips, 1971: Fine structural aspects of the mobilization of glycogen. II. Inhibition of glycogen breakdown. Amer. J. Path. **63**, 23—40.

Kraehenbuhl, J. P., and M. A. Campiche, 1969: Early stages of intestinal absorption of specific antibodies in the newborn: An ultrastructural, cytochemical, and immunological study in the pig, rat, and rabbit. J. Cell Biol. **42**, 345—358.

Krasno, J. R., J. M. Knelson, and F. G. Dalldorf, 1972: Changes in the alveolar lining with onset of breathing. Amer. J. Path. **66**, 471—482.

Kresse, H., 1973: Mucopolysaccharidosis III A (San Fillippo A Disease): deficiency of a heparin sulfamidase in skin fibroblasts and leucocytes. Biochem. biophys. Res. Commun. **54**, 1111—1118.

Krischer, K. N., and E. L. Chambers, 1970: Proteolytic enzymes in sea urchin eggs: characterization, localization and activity before and after fertilization. J. Cell Comp. Physiol. **76**, 23—26.

Kuettner, R., R. Eisenstein, L. W. Soble, and C. Arsenis, 1971: Lysozyme in epiphyseal cartilage. IV. Embryonic chick cartilage lysozyme-its localization and partial characterization. J. Cell Biol. **49**, 450—458.

Kuriyama, Y., and D. J. L. Luck, Membrane associated ribosomes in mitochondria of *Neurospora crassa*. J. Cell Biol. **59**, 776—784.

— T. Omura, P. Siekevitz, and G. E. Palade, 1969: Effects of phenobarbital on the synthesis and degradation of the protein component of rat liver microsomal membranes. J. biol. Chem. **244**, 2017—2026.

Lack, Charles H., 1959: Lysosomes in relation to arthritis. In: Lysosomes in biology and pathology. (Dingle, J. T., and H. B. Fell, eds.), Vol. 1, pp. 493—508. Amsterdam: North-Holland Publishing.

Lagunoff, D., B. A. Nicol, P. Pritzl, 1973: Uptake of β-glucuronidase by deficient human fibroblasts. Lab. Invest. **29**, 449—453.

Lalley, P. A., M. C. Rattaxxi, and T. B. Shows, 1974: Human β-D-N acetylhexosaminidases A and B; expression and linkage relationships in human somatic hybrids. Proc. nat. Acad. Sci. (U.S.) **71**, 1569—1573.

— and T. B. Shows, 1974: Lysosomal and microsomal glucuronidase: genetic variant alters electrophoretic mobility of both hydrolases. Science **185**, 442—444.

Lampen, J. O., G. E. Beltinger, and L. J. Sharkey, 1971: The membrane bound forms of penicillinase in *Bacillus lichenformis* and their significance for the secretion process. Biomembranes, Vol. II. (Manson, C. A., ed.), pp. 211—220. New York: Plenum Press.

Landing, B. H., H. B. Neustein, and S. Kamoshita, 1971: Biopsy diagnosis of lipidoses: background consideration, general concepts and practical aspects. In: Sphingolipids, sphingolipidoses and allied disorders. (Adv. Exp. Med. Biol. **19**) (Volk, B. W., and S. M. Aronson, eds.), pp. 15—35. New York: Plenum Press.

Landry, J., and N. Marceau, 1975: The relative rates of degradation of the plasma membrane glycoproteins from normal rat liver. Biochim. Biophys. Acta **389**, 154—161.

Lane, N. J., and A. B. Novikoff, 1965: Effects of arginine deprivation, ultraviolet radiation and X-radiation on cultured KB cells. A cytochemical and ultrastructural study. J. Cell Biol. **27**, 603—630.

Lanzillo, J. J., and B. L. Fanburg, 1974: Membrane bound angiotension-converting enzyme from rat lung. J. biol. Chem. **249**, 2313—2318.

Lauduron, P., 1974: In: Dynamics of degeneration and growth in neurons. (Fuxe, K. *et al.,* eds.), pp. 245—256. New York: Pergamon Press.

Laurent, G., J. Hildebrand, and O. Thys, 1975: Alterations of rat liver lysosomes and smooth endoplasmic reticulum induced by the diazafluoranthen derivative AC-3579. Lab. Invest. **32**, 580—584.

La Vail, M. M., 1973: Kinetics of rod outer segment renewal in the developing mouse retina. J. Cell Biol. **58**, 650—661.

— R. L. Sidman, and D. O'Neil, 1972: Photoreceptor-pigment epithelial cell relationships in rats with inherited retinal degeneration: radioautographic and electron microscopic evidence for a dual source of extra lamellar material. J. Cell Biol. **53**, 185—209.

— and J. La Vail, 1974: Organelles involved in retrograde axonal transport in chick retinal ganglion cells. Abstracts 4th Annual Meet. Soc. Neuroscience **300** (Fuller account in J. Comp. Neurol. **157**, 303).

Lazarides, E., and K. Weber, 1974: Actin antibody: the specific visualization of actin filaments in non-muscle. Proc. nat. Acad. Sci. (U.S.) **71**, 2268—2272.

Lazarow, P. B., and C. de Duve, 1973: The synthesis and turnover of rat liver peroxisomes. IV. Biochemical pathway of catalase synthesis. V. Intracellular pathway of catalase synthesis. J. Cell Biol. **59**, 491—506, 507—524.

— 1974: The subcellular location of catalase messenger RNA. J. Cell Biol. **63**, 187 a.

Lazarus, G. S., 1972: Collagens, collogenases, and clinicians. Br. J. Dermatol. **86**, 193—199.

— 1973: Studies on the degradation of collagen by collagenases. In: Lysosomes in biology and pathology. (Dingle, J. T., ed.), Vol. **3**, pp. 338—364. Amsterdam: North-Holland Publishing.

— J. R. Parsels, J. Lian, and M. C. Burleigh, 1972: Role of granulocyte collagenase in collagen degradation. Amer. J. Path. **68**, 565.

Lazarus, S. S., B. W. Volk, and H. Barden, 1966: Localization of acid phosphatase activity and secretion mechanism in rabbit pancreatic B-cells. J. Histochem. Cytochem. **14**, 233—246.

Leake, F. S., and Q. N. Myrvik, 1964: Differential release of lysozyme and acid phosphatase from subcellular granules of normal rabbit alveolar macrophages. Brit. J. Exp. Path. **45**, 384—392.

Ledbetter, M. C., and K. R. Porter, 1970: Introduction to the fine structure of plant cells. New York: Springer-Verlag.

Ledeen, R. W., and R. K. Yu, 1973: Structure and enzymic degradation of sphingolipids. In: Lysosome and storage diseases. (Hers, H. G., and F. van Hoof, eds.), pp. 105—145. New York: Academic Press.

Lee, D., 1970: The relative permeability of lysosomes from *Tetrahymena pyriformis* to carbohydrates, lactate, and the cryoprotective nonelectrolytes, glycerol and dimethylsulfoxide. Biochim. biophys. Acta **271**, 550—554.

— 1971: The relative permeability of lysosomes from *Tetrahymena pyriformis* to some amino acids and dipeptides. Biochim. biophys. Acta **225**, 108—112.

— 1972: The effect of glycerol, ethanol, and dimethylsulfoxide on rat liver lysosomes. Biochem. biophys. Acta **266**, 50—55.

LEGRAND, Y., J. CAEN, F. M. BOOYSE, M. E. RAFELSON, B. ROBERT, and L. ROBERT, 1973: Studies on a human blood platelet protease and elastolytic activity. Biochem. biophys. Acta **309** (E), 406—413.

LEHRER, R. I., and M. J. CLINE, 1969: Leukocyte myeloperoxidase deficiency and disseminated candidiasis: the role of myeloperoxidase in resistance to Candida infection. J. Clin. Invest. **48**, 1478—1488.

LEMANSKI, K. F., and R. ALDOROTY, 1974: Role of acid phosphatase in amphibian yolk platelet degradation during early embryogenesis. J. Cell Biol. **63**, 190 a.

LENNY, J. F., P. MATILE, A. WIEMKEN, M. SCHELLENBERG, and J. MEYER, 1974: Activities and cellular localization of yeast proteases and their inhibitors. Biochem. biophys. Res. Commun. **60**, 1378—1383.

LESCA, P., and C. PAOLETTI, 1969: A protein inhibitor of acid deoxyribonuclease. Proc. nat. Acad. Sci. (U.S.) **64**, 913—918.

LESKES, A., P. SIEKEVITZ, and G. E. PALADE, 1971: Differentiation of endoplasmic reticulum in hepatocytes. I. Glucose-6-phosphatase distribution *in situ*. II. Glucose-6-phosphatase in rough microsomes. J. Cell Biol. **49**, 264—287; **49**, 288—302.

LEVINE, A. M., J. A. HIGGINS, and R. J. BARRNETT, 1972: Biogenesis of plasma membranes in salt glands of salt-stressed domestic ducklings: localization of acyltransferase activity. J. Cell Sci. **11**, 855—873.

LEVY, M. R., and S. C. CHOU, 1973: Activity and some properties of an acid proteinase from normal and plasmodium berghei infected red cells. J. Parasitol **59**, 1004—1070.

— and A. M. ELLIOTT, 1968: Biochemical and ultrastructural changes in *Tetrahymena pyriformis* during starvation. J. Protozool. **15**, 208—222.

LIBBIN, R. M., P. PERSON, and A. S. GORDON, 1974: Renal lysosomes: role in biogenesis of erythropoietin. Science **185**, 1174—1176.

LILLIE, J. M., and S. S. HAN, 1973: Secretory protein synthesis in the stimulated rat parotid gland: temporal dissociation of the maximal response from secretion. J. Cell Biol. **59**, 708—721.

LIN, S., and I. ZABIN, 1972: β-galactosidase: rates of synthesis and degradation of incomplete chains. J. biol. Chem. **247**, 2205—2219.

LINDERSTROM-LANG, K., 1950: Structure and enzymatic breakdown of proteins. Cold Spring Harb. Symp. Quant. Biol. **4**, 117—126.

LINDY, S., 1974: Synthesis and degradation of rat liver lactate dehydrogenase M_4. J. biol. Chem. **249**, 4961—4968.

LIPETZ, J., and V. J. CRISTOFALO, 1972: Ultrastructural changes accompanying the aging of human diploid cells in culture. J. Ultrastruct. Res. **39**, 43—56.

LIPSCHITZ, D. A., M. O. Simon, S. R. LYNCH, J. DUGARD, T. H. Bothwell, and N. W. CHARLTON, 1971: Some factors affecting the release of iron from reticuloendothelial cells. Brit. J. Hematol. **21**, 289—303.

LLOYD, D., R. BRIGHTWELL, S. E. VENABLES, G. I. ROACH, and G. TURNER, 1971: Subcellular fractionation of *Tetrahymena pyriformis* ST by zonal centrifugation: changes in activities and distribution of enzymes during the growth cycle and on starvation. J. gen. Microbiol. **65**, 209—223.

LLOYD, J. B., 1973: Experimental support for the concept of lysosomal storage disease. In: Lysosomes and storage diseases. (HERS, H. G., and F. VAN HOOF, eds.), pp. 173—195. New York: Academic Press.

— and F. BECK, 1969: Teratogenesis. In: Lysosomes in biology and pathology. (DINGLE, J. T., and H. B. FELL, eds.), Vol. 1, pp. 433—449. Amsterdam: North-Holland Publishing.

LOCKE, M., and J. V. COLLINS, 1968: Protein uptake into multivesicular bodies and storage granules in the fat body of an insect. J. Cell Biol. **36**, 453—483.

— and J. T. McMAHON, 1971: The origin and fate of microbodies in the fat body of an insect. J. Cell Biol. **48**, 61—78.

— and J. V. COLLINS, 1973: Organelle turnover. In: Pathological changes in cell membranes. (TRUMP, B. J., ed.) New York: Academic Press (in press).

— and A. K. SYKES, 1975: The role of the Golgi complex in the isolation and digestion of organelles. Tissue and Cell. **7**, 143—158.

LOCKHARD, V. G., J. B. GROGAN, and J. G. BRUNSON, 1973: Alterations in bactericidal activity of rabbit alveolar macrophages as a result of tumbling stress. Amer. J. Path. **70**, 57—68.

LOCKSHIN, R. A., 1969 a: Programmed cell death. Activation of lysis by a mechanism involving the synthesis of protein. J. Insect. Physiol. **15**, 1505—1516.

— 1969 b: Lysosomes in insects. In: Lysosomes in biology and pathology. (DINGL, J. T., and H. B. FELL, eds.), Vol. **1**, pp. 363—391. Amsterdam: North-Holland Publishing. See also Life Sci. **15**, 1549—1565, for a more recent review.

— 1973 a: Degeneration of insect intersegmental muscles; electrophysiological studies of populations of fibers. J. Insect. Physiol. **19**, 2359—2373.

— 1973 b: Muscle degeneration and the escape of muscle proteins into blood. The Physiologist **16**, 379.

— and C. M. WILLIAMS, 1965: Programmed cell death. III. Neural control of the breakdown of the intersegmental muscle of silk moths. J. Insect. Physiol. **11**, 601—610.

— and J. BEAULATON, 1974: Programmed cell death. 1. Cytochemical evidence for lysosomes during the normal breakdown of the intersegmental muscles. 2. Cytochemical appearance of lysosomes when the death of intersegmental muscles is prevented. J. Ultrastruct. Res. **46**, 43—62, 63—78; Life Sci. **15**, 1549—1565.

LODI, W. R., and D. R. SONNEBORN, 1974: Protein degradation and protease activity during the life cycle of *Blastocladiella emersonii*. J. Bact. **117**, 1035—1042.

LODISH, H. F., 1973: Biosynthesis of reticulocyte membrane proteins by membrane free ribosomes. Proc. nat. Acad. Sci. (U.S.) **70**, 1526—1530.

LODISH, H., and B. SMALL, 1975: Membrane protein synthesized by rabbit reticulocytes. J. Cell Biol. **65**, 51—64.

LOEB, J. N., R. R. HOWELL, and G. M. TOMPKINS, 1965: Turnover of ribosomal RNA in rat liver. Science **149**, 1093—1095.

LONDON, M., R. McHUGH, and P. B. HUDSON, 1955: Thermal stabilization of prostatic acid phosphatase by fluoride. Arch. Biochem. Biophys. **55**, 121—129.

LORD, J. M., T. KAGAWA, T. S. MOORE, and H. BEEVERS, 1973: Endoplasmic reticulum as the site of lecithin formation in castor bean endosperm. J. Cell Biol. **57**, 659—667.

LOSPALLUTO, J., K. FEHR, and M. ZIFF, 1971: Intracellular proteases and digestion of immunoglobulins. In: Tissue proteinases. (BARRETT, A. J., and J. T. DINGLE, eds.), pp. 263—289. Amsterdam: North-Holland Publishing.

LOUIS, C., and G. GIANNOTTI, 1974: Formations lysosomales et membranaires associees a des symbiotes et a des mycoplasmes endocellulaires d'insectes, in: Concept du systeme lysosomal chez les vegetaux et chez les animaux (COULOMB, P., and N. VICENTE, eds.), pp. 43—55. Le-Brusc-Var, France: Fond. Scient. Ricard, Observatoire de la mer.

LOVAAS, E., 1974: Evidence for a proteolytic system in rat liver mitochondria. FEBS Letters **45**, 244—247.

LOW, R. B., and A. L. GOLDBERG, 1973: Non-uniform rates of turnover of myofibrillar proteins in rat diaphragm. J. Cell Biol. **56**, 590—595.

LOWRIE, D. B., P. S. JACKETT, and N. A. RATCLIFFE, 1975: Mycobacterium microti may protect itself from intracellular destruction by releasing cyclic AMP into phagosomes. Nature **254**, 600—602.

LUCHT, U., 1972 a: Absorption of peroxidase by osteoclasts as studied by electron microscope histochemistry. Histochemie **29**, 274—286.

— 1972 b: Osteoclasts and their relationship to bone as studied by electron microscopy. Z. Zellforsch. **135**, 211—228.

— 1972 c: Cytoplasmic vacuoles and bodies of the osteoclast. Z. Zellforsch. **135**, 229—244.

LUCY, J. A., 1969: Lysosomal membranes. In: Lysosomes in biology and pathology. (DINGLE, J. T., and H. B. FELL, eds.), Vol. **2**, pp. 313—341. Amsterdam: North-Holland Publishing.

LUTZNER, M. A., and C. T. LOWRIE, 1969: Ultrastructure of the development of the normal black and giant beige melanin granules in the mouse. In: Pigmentation, its genesis and biological controls. (RILEY, V., ed.); Int. Pigment Cell Conf. VII. Seattle, 1969, pp. 89—105. New York: Appleton-Century-Crofts.

Lutzner, M. A., J. H. Tierney, and E. P. Benditt, 1966: Giant granules and widespread cellular inclusions in a genetic syndrome of aleutian mink; and electron microscopic study. Lab. Invest. **14**, 2063—2079.

Luzikov, V. N., 1973: Stabilization of the enzymatic systems of the inner mitochondrial membrane and related problems: A possible approach to the problem of the regulation of mitochondrial formation. Sub. Cell Biochem. **2**, 1—31.

Ma, M. H., and L. Biempica, 1971: The normal human liver cell: Cytochemical and ultrastructural studies. Amer. J. Path. **62**, 353—390.

— W. A. Laird, and H. Scott, 1974: Cytopempsis of horseradish peroxidase in the hepatocyte. J. Histochem. Cytochem. **22**, 160—169.

Maack, T., 1968: Changes in the activity of acid hydrolases during renal reabsorption of lysozyme. J. Cell Biol. **35**, 268—272.

Mackaness, G. B., 1970: The monocyte in delayed-type hypersensitivity. In: Mononuclear phagocytes. (van Furth, R., ed.), pp. 478—484. Philadelphia: F. A. Davis.

Macomber, P. B., H. Spring, and A. J. Tousimis, 1967: Morphological effects of chloroquine on Plasmodium berghei in mice. Nature **214**, 937—939.

MacRae, E. K., and J. K. Spitznagel, 1975: Ultrastructural localization of cationic proteins in cytoplasmic granules of chicken and rabbit polymorphonuclear leukocytes. J. Cell Sci. **17**, 79—94.

Mahadevan, S., C. J. Dillard, and A. L. Tappel, 1969: Degradation of polysaccharides, mucopolysaccharides and glycoproteins by lysosomal glycosidases. Arch. Biochem. Biophys. a **29**, 525—533.

Mahlberg, P., 1972: Further observations on the phenomenon of secondary vacuolation in living cells. Amer. J. Bot. **59**, 172—179.

Majerus, P. W., and E. Kilburn, 1969: Acetyl enzyme A carboxylase, the roles of synthesis and degradation in regulation of enzyme levels in rat liver. J. biol. Chem. **744**, 6254—6262.

Malawista, S. E., J. B. L. Gee, and K. G. Bensch, 1971: Cytochalasin B reversibly inhibits phagocytosis: functional, metabolic, and ultrastructural effects in human blood leukocytes and rabbit alveolar macrophages. Yale J. Biol. Med. **44**, 286—300.

Malkoff, D. B., and D. E. Buetow, 1964: Ultrastructural changes during carbon starvation in *Euglena*. Exp. Cell Res. **35**, 58—68.

Mandell, G. L., 1970: Intraphagosomal pH of human polymorphonuclear leukocytes. Proc. Soc. Exp. Biol. Med. **134**, 447—449.

— 1973: Interaction of intraleukocytic bacteria and antibiotics. J. Clin. Invest. **52**, 1673—1679.

Mandelstam, J., 1960: The intracellular turnover of protein and nucleic acid and its role in biochemical differentiation. Bact. Rev. **24**, 289—308.

— 1969: Regulation of bacterial spore formation; Symp. Soc. General Microbiol. **19**, 377—402.

— and W. M. Waites, 1968: Sporulation in Bacillus subtilis: the role of exoprotease. Biochem. J. **109**, 793—801.

Mann, T., 1963: Biochemistry of the prostate gland and its secretion. Nat. Cancer Inst. Monogr. **12**, 235—251.

Manning, J. E., and O. C. Richards, 1972: Synthesis and turnover of *Euglena gracilis* nuclear and chloroplast DNA. Biochemistry **11**, 2036—2043.

Mant, M. J., and B. G. Firkin, 1972: Uptake of latex and thorotrast by human platelets *in vitro*: effect of various chemicals demonstrating differing mechanisms and metabolic requirements. Brit. J. Hematol. **22**, 385—391.

Margulis, L., 1970: Origin of eucaryotic cells. New Haven: Yale Univ.

Marikovsky, Y., and D. Danon, 1969: Electron microscope analysis of young and old red blood cells stained with colloidal iron for surface charge evaluation. J. Cell Biol. **43**, 1—7.

Marks, N., and A. Lajtha, 1970: Developmental changes in peptide bond hydrolases. In: Protein metabolism of the nervous system. (Lajtha, A., ed.), pp. 39—75. New York: Plenum Press.

MARKS, N., A. GALOYAN, A. GRYNBAUM, and A. LAJTHA, 1974: Protein and peptide hydrolases of the rat hypothalamus and pituitary. J. Neurochem. **22**, 735—739.

MARKS, P. A., A. B. JOHNSON, and E. HIRSCHBERG, 1958: Effects of age on enzyme activity in erythrocytes. Proc. nat. Acad. Sci. (U.S.) **44**, 529—535.

MARKUS, G., 1965: Protein-substrate conformation and proteolysis. Proc. nat. Acad. Sci. (U.S.) **54**, 253—258.

MARSH, C. A., C.-W. LIN, and W. H. FISHMAN, 1974: Golgi β-glucuronidase of androgen stimulated mouse kidney. Biochem. J. **142**, 491—497.

MARSHALL, J. M., U. N. SCHUMAKER, and P. W. BRANDT, 1959: Pinocytosis in amoebae. Annals (N.Y.) Acad. Sci. **78**, 515—523.

MARTY, F., 1970: Rôle du système membranaire vacuolaire dans la différentiation des laticiferes d'*Euphorbia charcias* L. C. R. Acad. Sci. (Paris) D **271**, 2301—2304.

— 1971 a: Vesicles autophagiques différencies d'*Euphorbia characias.* L. C. R. Acad. Sci. (Paris) D **272**, 399—402.

— 1971 b: Peroxysomes et compartment lysosomal dans les cellules du méristeme radiculaire d'*Euphorbia characias* L.: une étude cytochemique. C. R. Acad. Sci. (Paris) D **273**, 2504—2507.

— 1972: Distributions des activites phosphatasiques acides au cours du procès d'autophagie cellulaire dans les cellules du méristeme radiculaire d'*Euphorbia characias.* C. R. Acad. Sci. (Paris) D **274**, 206—209.

MARUTA, H., A. M. A. EL-ASFAHANI, and D. MIZUNO, 1973: Selective recognition in erythrophagocytosis by mouse peritoneal macrophages: isolation and purification of a possible self-marker from mouse erythrocytes. Exp. Cell Res. **80**, 143—151.

MARVALDI, J., J. PICHON, M. DELAAGE, and G. MARCHIS-MOUREN, 1974: Individual ribosomal protein pool size and turnover rate in *Escherichia coli.* J. molec. Biol. **84**, 83—96.

MARVER, H. S., A. COLLINS, T. TSHUDY, and M. RECHCIGL, 1966: Deltaaminolevulinic acid synthetase. II. Induction in rat liver. J. biol. Chem. **241**, 4323—4329.

MARVIN, D. A., and E. J. WACHTEL, 1975: Structure and assembly of filamentous bacterial viruses. Nature **253**, 19—23.

MASUR, S. K., and E. HOLTZMAN, 1969: Lysosomes and secretory granules in the normal and autotransplanted salamander pars distalis. Gen. Comp. Endocrinology **12**, 33—39.

— — I. L. SCHWARTZ, and R. WALTER, 1971: Correlation between pinocytosis and hydro-osmosis induced by neurohypophyseal hormone, and mediated by adenosine 3′, 5′ cyclic monophosphate. J. Cell Biol. **49**, 582—594.

— — and R. WALTER, 1972: Hormone-stimulated exocytosis in the toad urinary bladder; possible implications for turnover of surface membranes. J. Cell Biol. **52**, 211—219.

MATTHEWS, M. B., 1973: Mammalian messenger RNA. Essays in Biochem. **9**, 59—102.

MATTHEWS, M. B., 1973: An ultrastructural study of axonal changes following constriction of postganglionic branches of the superior cervical gangion in the rat. Phil. Trans. Roy. Soc. (Lond.) B **264**, 479—508.

— and G. RAISMAN, 1972: A light and electron microscopic study of the cellular response to axonal injury in the superior cervical ganglion of the rat. Proc. Roy. Soc. (Lond.) B **181**, 43—79.

MATILE, P., 1969: Plant lysosomes. In: Lysosomes in biology and pathology. (DINGLE, J. T., and H. B. FELL, eds.), Vol. **1**, pp. 406—430. Amsterdam: North-Holland Publishing.

— 1970: Recent progress in the study of yeast cytology. Acta. Facultatis Medicae Univ. Brunesis **37**, 71—83 (Proc. 2nd Int. Symp. on Yeast protoplasts; Brno, 1968).

— 1971: Vacuoles, lysosomes of *Neurospora.* Cytobiologie **3**, 324—330.

— (in press) 1974: Dynamics of plant ultrastructure. (ROBARDS, ed.), Chapter 5.

— and F. WINKENBACH, 1971: Functions of lysosomal enzymes in the senescing corolla of the morning glory (*Ipomoea purpurea*). J. exp. Bot. **22**, 759—770.

— A. WIEMKEN, and W. GUYER, 1971: A lysosomal amino-peptidase in differentiating yeast cells and protoplasts. Planta **96**, 43—53.

MAUEL, J., R. BEHIN, and BIROUN-VOERJASIN, 1973: Quantitative release of live micro-organisms from infected macrophages by sodium dodecyl sulfate. Nature New Biol. **244**, 93—94.

MAUNSBACH, A. B., 1969: Functions of lysosomes in kidney cells. In: Lysosomes in biology and pathology. (DINGLE, J. T., and H. B. FELL, eds.), Vol. 1, pp. 115—154. Amsterdam: North-Holland Publishing.

MAYER, M. M., 1973: The complement system. Sci. Amer. 229 (#5), 54—66.

MAZUR, A., S. GREEN, and A. CARLTON, 1960: Mechanism of plasma iron incorporation into hepatic ferritin. J. biol. Chem. 235, 595—603.

McDONALD, J. K., F. X. CALLAHAN, S. ELLIS, and R. E. SMITH, 1971: Poly-peptide degradation by dipeptidyl aminopeptidase I (Cathepsin C) and related peptidases. In: Tissue proteinases. (BARRETT, A. J., and J. T. DINGLE, eds.), pp. 69—107. Amsterdam: North-Holland Publishing. See also SMITH, in: Lysosomes in biology and pathology, Vol. 4 (in press, 1975).

— B. B. ZERTMAN, F. X. CALLAHAN, and S. ELLIS, 1974: Angiotensinase activity of dipeptidyl aminopeptidase I (Cathepsin C) of rat liver. J. biol. Chem. 249, 234—240.

McILHINNEY, A., and B. L. M. HOGAN, 1974: Rapid degradation of puromycyl peptides in hepatoma cells and reticulocytes. FEBS Letters 40, 297—301.

McKANNA, J. A., 1973 a: Cyclic membrane flow in the ingestive-digestive system of the peritrich protozoans. I. Vesicular fusion at the cytopharynx. II. Cup-shaped coated vesicles. J. Cell Sci. 13, 663—686.

— 1973 b: Membrane recycling: vesiculation of the ameba contractile vacuole at systole. Science 179, 88—90.

McLEAN, N., and N. D. HOLLAND, 1973: Absorption of ferritin by cells of the digestive gland in Nassarius tegula (Gastropoda; prosobranchia). Tissue and Cell 5, 585—590.

McNAMERA, D. J., and T. E. WEBB, 1974: Glucagon-mediated changes in the concentration of rat hepatic tyrosine-amino-transferase: an immunochemical analysis. Arch. Biochem. Biophys. 163, 777—783.

McNAMERA, P. D., O. KOLDOVSKY, and S. SEGAL, 1974: Synthesis and degradation of microvillar protein of kidney in adult rats. FEBS Letters 41, 139—142.

McQUILLAN, M. T., and V. M. TRIKOJUS, 1971: Thyroid acid proteinase. In: Tissue proteinases. (BARRETT, A. J., and J. T. DINGLE, eds.), pp. 57—166. Amsterdam: North-Holland Publishing.

MEBAN, C., 1972: Localization of phosphatidic acid phosphatase activity in granular pneumocytes. J. Cell Biol. 53, 249.

— 1973: An electron microscope study of acid hydrolase in pneumocytes of Xenopus laevis. Histochem. J. 5, 557—566.

MEGO, J. L., R. M. FARB, and J. BARNES, 1972: An adenosine triphosphate dependent stabilization of proteolytic activity in heterolysosomes: evidence for a proton pump. Biochem. J. 128, 763—769.

— 1973 a: Protein digestion in isolated heterolysosomes. In: Lysosomes in biology and pathology. (DINGLE, J. T., ed.), Vol. 3, pp. 138—168. Amsterdam: North-Holland Publishing.

— 1973 b: A biochemical method for the evaluation of heterolysosome formation and function. In: Lysosomes in biology and pathology. (DINGLE, J. T., ed.), Vol. 3, pp. 527—537. Amsterdam: North-Holland Publishing.

MEISLER, M., and K. PAIGEN, 1972: Coordinated development of β-glucuronidase and β-galactosidase in mouse liver. Science 177, 894—896.

MELDOESI, J., 1974: Dynamics of cytoplasmic membranes in guinea pig pancreatic acinar cells. I. Synthesis and turnover of membrane proteins. J. Cell Biol. 61, 1—13.

MELMED, R. N., C. J. BENITEZ, and S. J. HOLT, 1973: Intermediate cells of the pancreas. III. Selective autophagy and destruction of granules in intermediate cells of the rat pancreas induced by alloxan and streptozotocin. J. Cell Sci. 130, 297—315.

MERKOW, L., M. PARDO, S. M. EPSTEIN, E. VERNEY, and H. SIDRANSKY, 1968: Lysosomal stability during phagocytosis of Aspergillis flavis spores by alveolar macrophages of cortisone treated mice. Science 160, 79—81.

MESQUITA-GUIMARAES, A., and A. COIMBRA, 1974: Acid hydrolases of the perinuclear secretion granules of the rat preputial gland: a light and electron microscopic study. Histochem. J. 6, 685—692.

MICKENBERG, I. D., R. K. ROOT, and S. M. WOLFF, 1972: Bactericidal and metabolic properties of eosinophils. Blood **39**, 67—80.

MILLER, A. T., D. M. HALE, and K. D. ALEXANDER, 1965: Histochemical studies on the uptake of horseradish peroxidase by rat kidney slices. J. Cell Biol. **27**, 305—312.

MILLER, F., und V. HERZOG, 1969: Die Lokalisation von Peroxydase und saurer Phosphatase in eosinophilen Leukocyten während der Reifung: Elektronenmikroskopisch-cytochemische Untersuchungen am Knochenmark von Ratte und Kaninchen. Z. Zellforsch. **97**, 84—110.

— and G. E. PALADE, 1964: Lytic activity in renal protein absorption droplets: an electron microscopical cytochemical study. J. Cell Biol. **23**, 519—552.

MILLER, J. P. G., and D. J. PERKINS, 1969: Model experiments for the study of iron transfer from transferrin to ferritin. Eur. J. Biochem. **10**, 146—151.

MILSOM, J. P., and C. H. WYNN, 1973: The heterogeneous distribution of acid hydrolases within a homogeneous population of cultured mammalian cells. Biochem. J. **132**, 493—500.

MILSON, D. W., and C. H. WYNN, 1973: Protein and carbohydrate composition of lysosomal membranes. Biochem. Soc. Transact. **1**, 426—428.

MILSTEIN, C., G. C. BROWNLAE, T. M. HARRISON, and M. B. MATHEWS, 1972: A possible precursor of immunoglobulin light chains. Nature New Biol. **239**, 117—120.

MINDICH, L., 1971: Induction of *Staphlococus aureus* lactose permease in the absence of glycerolipid synthesis. Proc. nat. Acad. Sci. (U.S.) **68**, 420—424.

— 1974: Synthesis and assembly of bacterial membranes. In: Bacterial surfaces. (LEIVE, L., ed.), pp. 1—36.

— and S. DALES, 1972: Membrane synthesis in *Bacillus subtilis*. III. The morphological localization of the sites of membrane synthesis. J. Cell Biol. **55**, 32—41.

MISHIMA, Y., 1967: Lysosomes in melanin phagocytosis and synthesis. Nature **216**, 67—68.

MITCHISON, N. A., 1969: The immunogenic capacity of antigen taken up by peritoneal exudate cells. Immunology **16**, 1—14.

MOE, H., J. ROSTGAARD, and O. BEHNBE, 1965: On the morphology of virgin lysosomes in the intestinal epithelium of the rat. J. Ultrastruct. Res. **12**, 396—403.

MORELL, A. G., G. GREGORIADIS, I. H. SCHEINBERG, J. HICKMAN, and G. ASHWELL, 1971: The role of sialic acid in determining the survival of glycoproteins in the circulation. J. biol. Chem. **246**, 1461—1467.

MORGAN, C., and H. M. ROSE, 1968: Structure and development of viruses as observed in the electron microscope. VIII. Entry of influenza virus. J. Virol **2**, 925—936.

— H. S. ROSENKRANZ, and B. MEDNIS, 1969: Structure and development of viruses as observed in the electron microscope. X. Entry and uncoating of adenovirus. J. Virol **4**, 777—796.

MORGAN, E. M., and T. C. APPLETON, 1969: Autoradiographic localization of ^{125}I labelled transferrin in rabbit reticulocytes. Nature **223**, 1371—1372.

MORGAN, W. S., 1968: Ductal excretion of neutral red lysosomes in the mouse pancreas. J. Cell Biol. **36**, 261—265.

MORRÉ, D. J., W. W. FRANKE, B. DEUMLING, S. E. NYQUIST, and C. OVTRACHT, 1971: Golgi apparatus function in membrane flow and differentiation: origin of plasma membrane from endoplasmic reticulum. Biomembranes. (MANSON, L. A., ed.), Vol. **II**, pp. 95—104. New York: Plenum Press.

— W. D. MERRITT, and C. A. LEMBI, 1971: Connections between mitochondria and endoplasmic reticulum in rat liver and onion stem. Protoplasma **73**, 43—49.

MORRIS, S. J., H. J. RALSTON, and E. M. SHOOTER, 1971: Studies on the turnover of mouse brain synaptosomal protein. J. Neurochem. **18**, 2279—2290.

MORRISON, M., A. W. MICHAELS, D. R. PHILLIPS, and S. J. CHOI, 1974: Life span of erythrocyte membrane protein. Nature **248**, 763—764.

MOVAT, H. Z., W. J. WEISER, M. F. GLYNN, and J. F. MUSTARD, 1965: Platelet phagocytosis and aggregation. J. Cell Biol. **27**, 531—543.

— S. G. STEINBERG, F. M. HABAL, and N. S. RANADIVE, 1973: Demonstration of a kinin generating enzyme in the lysosomes of human polymorphonuclear leukocytes. Lab. Invest. **29**, 669—677.

Muir, H., 1973: Structure and enzymic degradation of mucopolysaccharides. In: Lysosomes and storage diseases. (Hers, H. G., and F. van Hoof, eds.), pp. 79—104. New York: Academic Press.

Müller, M., 1971: Lysosomes in *Tetrahymena pyriformis*. II. Intracellular distribution of several acid hydrolases. Acta. Biol. Acad. Sci. Hung. **22**, 179—186.

— 1972: Secretion of acid hydrolases and its intracellular source in *Tetrahymena pyriformis*. J. Cell Biol. **52**, 478—487.

— 1973: Biochemical cytology of trichomonad flagellates. J. Cell Biol. **57** 453—474.

— P. Rohlich, J. Toth, and I. Toro, 1963: Fine structure and enzymic activity of protozoan food vacuoles. In: Ciba foundation symposium on lysosomes. (de Reuck, A. V. S., and M. P. Cameron, eds.), pp. 201—225. Boston: Little Brown.

— P. Baudhuin, and C. de Duve, 1966: Lysosomes in *Tetrahymena pyriformis*. I. Some properties and lysosomal localization of acid hydrolases. J. Cell Physiol. **68**, 165—175.

Mullins, J. T., and E. A. Ellis, 1974: Sexual morphogenesis in Achlya: ultrastructural basis for the hormonal induction of antheridial hyphae. Proc. nat. Acad. Sci. (U.S.) **71**, 1347—1350.

Munnell, J. F., and R. Getty, 1968: Rate of accumulation of cardiac lipofuscin in the aging canine. J. Gerontol. **23**, 154—158.

Murata, F., and S. S. Spicer, 1973: Morphologic and cytochemical studies of rabbit heterophilic leukocytes: evidence for tertiary granules. Lab. Invest. **29**, 65—72.

— J. M. Hardin, and S. S. Spicer, 1973: Coexistence of acid phosphatase and acid mucosubstance in the nucleoid of human blood platelet granules. Histochemie **35**, 319—330.

Murphy, J. V., H. J. Wolfe, E. A. Balazs, and H. W. Moser, 1971: A patient with deficiency of arylsulfatases A, B, C, and steroid sulfatase, associated with storage of sulfatide, cholesterol sulfate and glycosaminoglycans. In: Lipid storage diseases: Enzymatic defects and clinical implications. (Bernsohn, J., and H. J. Grossman, eds.). New York: Academic Press.

Muscatini, L., 1973: Chloroplasts and algae as symbionts in molluscs. Int. Rev. Cytol. **36**, 137—169.

Nachman, R. L., B. Ferris, and J. G. Hirsch, 1971: Macrophage plasma membranes. I. Isolation and study of protein components. J. exp. Med. **133**, 785—805. II. Studies on synthesis and turnover of protein constituents. J. exp. Med. **133**, 807—820.

— J. G. Hirsch, and M. Baggiolini, 1972: Studies on isolated membrane of azurophil and specific granules from rabbit polymorphonuclear leukocytes. J. Cell Biol. **54**, 133—140.

— and B. Weksler, 1972: The platelet as an inflammatory cell. Ann. (N. Y.) Acad. Sci. **201**, 131—137.

Nadler, H. L., N. M. Dowben, and D. Y. Y. Hsia, 1965: Enzyme changes and poly-ribosome profiles in phytohemagglutinin-stimulated lymphocytes. Blood **34**, 52—62.

Nadler, S., and S. Goldfischer, 1970: The intracellular release of lysosomal contents in macrophages that have ingested silica. J. Histochem. Cytochem. **18**, 368—371.

Nagai, Y., 1973: Vertebrate collagenase: further characterization and the significance of its latent form *in vivo*. Molecular and Cellular Biochemistry **1**, 137—145.

Nagasawa, J., and W. W. Douglas, 1972: Thorum dioxide uptake into adrenal medullary cells and the problem of recapture of granule membrane following exocytosis. Brain Res. **37**, 141—145.

Nandy, K., and G. H. Bourne, 1966: Effects of centrophenoxane on the lipofuscin pigments in the neurons of senile guinea pig. Nature **210**, 313—314.

Nath, K., and A. L. Koch, 1970: Protein degradation in *Escherichia coli*. I. Measurement of rapidly and slowly decaying components. J. biol. Chem. **245**, 2889—2900.

Neely, A. N., and G. E. Mortimore, 1974: Localization of products of endogenous proteolysis in lysosomes of perfused rat liver. Biochem. biophys. Res. Commun. **59**, 680—687.

Nehemiah, J. L., and A. B. Novikoff, 1973: Unusual lysosomes in hamster hepatocytes. J. Cell Biol. **59**, 246 a. Exp. Molec. Path. **21**, 398 (1974).

Neil, M. W., and M. W. Horner, 1964: Acid p-nitrophenyl phosphate phosphohydroloses of adult guinea pig liver. Biochem. J. **92**, 217—224.

— — 1965: Studies on acid hydrolases in adult and fetal tissues. II. Acid phenyl phosphomonoesterases of adult mouse liver. Biochem. J. **93**, 220—224.

Nelson, E. M., 1972: Turnover and regulation of ciliary proteins in *Tetrahymena.* J. Cell Biol. **55**, 187 A.

Neufeld, E. F., and M. Cantz, 1973: The mucopolysaccharidoses studied in cell culture. In: Lysosomes and storage diseases. (Hers, H. G., and F. van Hoof, eds.), pp. 261—275. New York: Academic Press. See also Ann. Rev. Biochem. **44**, 357.

Newton, A. A., 1970: The requirements of a virus. Symp. Soc. Gen. Microbiol. **20**, 323—359.

Nicander, L., 1966: A study of multivesicular bodies in the epididymis. J. Ultrastruct. Res. **14**, 424—425.

Nichols, B. A., D. F. Bainton, and M. G. Farquhar, 1971: Differentiation of monocytes: origin, nature, and fate of their azurophil granules. J. Cell Biol. **50**, 498—515.

Nicol, D. M., D. Lagunoff, and P. Pritzl, 1974: Differential uptake of human β-glucuronidase isoenzymes from spleen by deficient fibroblasts. Biochem. biophys. Res. Commun. **59**, 941—946.

Nicol, T., and D. L. J. Bilbey, 1958: Elimination of macrophage cells of the reticuloendothelial system by way of the bronchial tree. Nature **182**, 192—193.

Nicholson, G. L., 1974: Ultrastructural analysis of toxin binding and entry into mammalian cells. Nature **251**, 628—630.

Niden, A. H., 1967: Bronchiolar and large alveolar cells in pulmonary phospholipid metabolism. Science **158**, 1323—1324.

Nielson, M. H., 1974: Fine structural localization of nucleotide triphosphatase and acid phosphatase activity in *Trichomonas vaginalis* Donne. Cell Tissue Res. **151**, 269—280.

Nilsson, J. R., 1970: Cytolysosomes in *Tetrahymena pyriformis.* C. R. Lab. Carlsberg **38**, 87—106, 107—121.

Nishida-Fukuda, M., and F. Egami, 1970: Enzymatic degradation of keratosulfates. Biochem. J. **119**, 39—47.

Nishioka, N., N. Takahata, and R. Iizuka, 1968: Histochemical studies on the lipopigments, in the nerve cells: a comparison with lipofuscin and ceroid pigment. Acta Neuropath. **11**, 174—181.

Nolan, J. C., and K. A. Hartman, 1973: RNase catalyzed hydrolysis of ribosomes in several functional states. Biochem. biophys. Res. Commun. **54**, 1216—1223.

Nopajaroonsri, C., and Simon, 1971: Phagocytosis of colloidal carbon in a lymph node. Amer. J. Path. **65**, 25—42.

Noseworthy, J., G. H. Smith, R. Himmelhoch, and W. H. Evans, 1975: Protein and glycoprotein patterns of enriched fractions of primary and secondary granules from guinea pig polymorphonuclear leukocytes. J. Cell Biol. **65**, 577—586.

Nossal, G. J. V., A. Abbot, and J. Mitchell, 1968 a: Antigens in immunity. XIV. Electron microscopic radioautographic studies of antigen capture in the lymph node medulla. J. exp. Med. **127**, 263—276.

— — — and Z. Lummus, 1968 b: Antigens in immunity. XV. Ultrastructural features of antigen capture in primary and secondary lymphoid follicles. J. exp. Med. **127**, 277—290.

Novikoff, A. B., 1961: Lysosomes and related particles. In: The cell. (Brachet, J., and A. E. Mirsky, eds.), Vol. 2, pp. 423—488. New York: Academic Press.

— 1963: Lysosomes in the physiology and pathology of cells: Contributions of staining methods. In: Ciba foundation symposium on lysosomes. (de Reuck, A. V. S., and M. P. Cameron, eds.), pp. 36—71. Boston: Little Brown.

— 1967 a: Lysosomes in nerve cells. In: The neuron. (Hyden, H., ed.), pp. 319—377. Amsterdam: Elsevier.

— 1967 b: Enzyme localization and ultrastructure of neurons. In: The neuron. (Hyden, H., ed.), pp. 255—318. Amsterdam: Elsevier.

— 1973: Lysosomes: a personal account. In: Lysosomes and storage diseases. (Hers, H. G., and F. van Hoof, eds.), pp. 1—41. New York: Academic Press.

— and E. Essner, 1962: Pathological changes in cytoplasmic organelles. Fed. Proc. **21**, 1130—1142.

NOVIKOFF, A. B., and E. HOLTZMAN, 1970: Cells and organelles. New York: Holt Reinhart Winston.
— and W.-Y. SHIN, 1964: The endoplasmic reticulum in the Golgi zone and its relations to microbodies, Golgi apparatus, and autophagic vacuoles in rat liver cells. J. Microscopie 3, 187—206.
— M. E. BEARD, A. ALBALA, B. SHEID, N. QUINTANA, and L. BIEMPICA, 1971: Localization of endogenous peroxidases in animal tissues. J. Microscopie 12, 381—404.
— P. M. NOVIKOFF, N. QUINTANA, and C. DAVIS, 1972: Diffusion artifacts in 3-3'-diaminobenzidine cytochemistry. J. Histochem. Cytochem. 20, 745—749.
— — — — 1973: Studies on microperoxisomes. IV. Interrelations of microperoxisomes, endoplasmic reticulum and lipofuscin granules. J. Histochem. Cytochem. 21, 1010—1020.
— — M. MA, W.-Y. SHIN, and N. QUINTANA, 1974: Cytochemical studies of secretory and other granules associated with the endoplasmic reticulum in rat thyroid epithelial cells. Adv. Cytopharmacology (CECCARELLI, B., F. CLEMENTI, and J. MELDOLESI, eds.), Vol. 2, pp. 349—367.
NOVIKOFF, P. M., and A. B. NOVIKOFF, 1972: Peroxisomes in absorptive cells of mammalian small intestine. J. Cell Biol. 53, 532—560.
— — N. QUINTANA, and J.-J. HAUW, 1971: Golgi apparatus, GERL, and lysosomes of neurons in rat dorsal root ganglia studied by thick section and thin section cytochemistry. J. Cell Biol. 50, 859—886.
— P. S. ROHEIM, A. B. NOVIKOFF, and D. EDELSTEIN, 1974: Production and prevention of fatty liver in rats fed clofibrate and orotic acid diets containing sucrose. Lab. Invest. 30, 732—750.
NOZAWA, Y., and G. A. THOMPSON, 1971: Studies of membrane formation in *Tetrahymena pyriformis*. III. Lipid incorporation into various cellular membranes of logarithmic phase culture. J. Cell Biol. 49, 722—730.
NUNN, W. D., and J. E. CRONAN, 1974: Unsaturated fatty acid synthesis is not required for induction of lactose transport in *Escherichia coli*. J. biol. Chem. 249, 724—733.
NYQUIST, S. E., and H. H. MOLLENHAUER, 1973: A Golgi apparatus acid phosphatase. Biochim. biophys. Acta. 315, 103—112.

O'BRIEN, J. S., 1973: Tay-Sachs disease: from enzyme to prevention. Fed. Proc. 32, 191—199.
— A. L. MILLER, A. W. LOVERDE, and M. L. VEATH, 1973: Sanfilippo disease Type B: enzyme replacement and metabolic correction in cultured hemoblasts. Science 181, 753—755.
O'CONNOR, T. M., and C. R. WYTTENBACH, 1974: Cell death in the embryonic chick spinal cord. J. Cell Biol. 60, 448—459.
O'DAY, D. H., 1974: Intracellular and extracellular enzyme patterns during microcyst germination in the cellular slime mold *Polyphondylium pallidum*. Develop. Biol. 36, 400—410.
OGAWA, K., 1972: Histiocytic uptake of exogenous peroxidase. Acta Histochem. Cytochem. 5, 272—275.
— K. MASUTANI, and Y. SHINONAGA, 1962: Electron histochemical demonstration of acid phosphatase in the normal rat jejunum. J. Histochem. Cytochem. 10, 228—229.
OLIVER, C., and E. ESSNER, 1975: Formation of anomalous lysosomes in monocytes, neutrophils and eosinophils from bone marrow of mice with Chediak-Higashi syndrome. Lab. Invest. 32, 17—27.
OLIVER, J. M., T. E. UKENA, and R. D. BERLIN, 1974: Effects of phagocytosis and colchicine on the distribution of lectin binding sites on cell surfaces. Proc. nat. Acad. Sci. (U.S.) 71, 394—398.
OLIVER, J., R. B. ZURIER, and R. D. BERLIN, 1975: Concanavalin A cap formation polymorphonuclear leukocytes of normal and beige (Chediak-Higashi) mice. Nature 253, 471—473.
OMURA, T., P. SIEKEVITZ, and G. E. PALADE, 1967: Turnover of constitutents of the endoplasmic reticulum of rat hepatocytes. J. biol. Chem. 242, 2389—2396.
ORCI, L., A. JUNOD, R. PICTET, A. E. RENOLD, and C. ROUILLER, 1968: Granulolysis in A-cells of endocrine pancreas in spontaneous and experimental diabetes in animals. J. Cell Biol. 38, 462—466.

ORCI, L., W. STAUFFACHER, W. E. DULIN, A. E. RENOLD, and C. ROUILLER, 1970: Ultrastructural changes in A-cells exposed to diabetic hyperglycemia. Observations made on pancreas of Chinese hamsters. Diabetalogia **6**, 199—206.
—F. MALAISSE-LAGAE, M. RAVAZZOLA, M. AMHERDT, and A. E. RENOLD, 1973: Exocytosis-endocytosis coupling in the pancreatic beta cell. Science **181**, 561—562.
ORGEL, L. E., 1973: Ageing of clones of mammalian cells. Nature **243**, 441—445.
ORLIC, D., and R. LEV, 1973: Fetal rat intestinal absorption of horseradish peroxidase from swallowed amniotic fluid. J. Cell Biol. **56**, 106—119.
ORONSKY, A., L. IGNARRO, and R. PEIPER, 1973: Release of cartilage mucopolysaccharide-degrading neutral protease from human leukocytes. J. exp. Med. **138**, 461—472.
ORREGO, F., 1971: Protein degradation in squid giant axon. J. Neurochem. **18**, 2249—2254.
ORRENIUS, S., and J. L. E. ERICSSON, 1966: Enzyme-membrane relationship in phenobarbitol induction of synthesis of drug metabolizing enzyme system and proliferation of endoplasmic membranes. J. Cell Biol. **28**, 181—198.
ORY, R. L., and K. W. HENNINGSEN, 1969: Enzymes associated with protein bodies isolated from ungerminated barley seeds. Plant Physiol. **44**, 1488—1498.
OSINCHAK, J., 1966: Electron microscopic localization of acid phosphatase and thiamine pyrophosphatase activity in hypothalamic neurosecretory cells of the rat. J. Cell Biol. **21**, 35—48.
OSSERMAN, E. F., M. KLOCKARS, J. HALPER, and R. E. FISCHEL, 1973: Effects of lysozyme on normal and transformed cells. Nature **243**, 331—335.
O'STEEN, W. K., and Z. A. KARCIOGLU, 1974: Phagocytosis in the light damaged albino rat eye: light and electron microscopic study. Amer. J. Anat. **139**, 503—518.
OVERATH, P., F. F. HILL, and I. LAMREK-HIRSCH, 1971: Biogenesis of *E. coli* membrane: evidence for randomization of lipid phase. Nature New Biol. **234**, 264—267.
OVERTON, J., 1968: The fate of desmosomes in trypsinized tissue. J. exp. Zool. **168**, 203—214.
OWENS, J. W., K. L. GAMMON, and P. D. STAHL, 1975: Multiple forms of β-glucuronidase in rat liver lysosomes and microsomes. Arch. Biochem. Biophys. **166**, 258—272.

PACKER, L., D. W. DEAMER, and R. L. HEATH, 1967: Regulation and deterioration of structure in membrane. Adv. Gerontol. **2**, 77—120.
PAGANO, J. S., 1970: Biological activity of isolated viral nucleic acids. Prog. Med. Virol. **12**, 1—48.
PAGE-THOMAS, D. P., 1969: Lysosomal enzymes in experimental and rheumatoid arthritis. In: Lysosomes in biology and pathology. (DINGLE, J. T., and H. B. FELL, eds.), Vol. 2, pp. 87—110. Amsterdam: North-Holland Publishing.
PALADE, G. E., 1972 a: Synthesis, transport, and discharge of secretory proteins. Abstracts, International Symposium on Lysosomes, Hakone, Japan (Aug.-Sept. 1972), p. 36.
— 1972 b: Biosynthesis of chloroplast membranes. Acta Histochem. Cytochem. **5**, 249—251.
PALAY, S. L., 1960: The fine structure of secretory neurons in the preoptic nucleus of the goldfish. Anat. Rec. **139**, 262—279.
PAPPENHEIMER, A. M., and D. M. GILL, 1973: Diphtheria. Science **182**, 353—358.
PARAKKAL, D. F., 1972: Macrophages: the time course and sequence of their distribution in the post partum uterus. J. Ultrastruct. Res. **40**, 284—291.
PARK, D. C., M. E. PARSONS, and R. J. PENNINGTON, 1973: Evidence for mast cell origin of proteinase in skeletal muscle homogenates. Biochem. Soc. Trans. **1**, 730—732.
PARKER, J. W., H. WAKASA, and R. J. LUKES, 1965: The morphological and cytochemical demonstration of lysosomes in lymphocytes incubated with phytohemagglutinin by electron microscopy. Lab. Invest. **14**, 1736—1748.
PARKES, J. G., and W. THOMPSON, 1973: Synthesis *in vivo* of phospholipid of liver mitochondria and endoplasmic reticulum from glycerol and fatty acids. Biochim. biophys. Acta **306**, 403—411.
PARMLEY, R. T., and S. S. SPICER, 1974: Cytochemical and ultrastructural identification of a small type granule in human late eosinophils. Lab. Invest. **30**, 557—570.
PARR, E. L., R. W. SCHAEDLER, and J. G. HIRSCH, 1967: The relationship of polymorphonuclear leukocytes to infertility in uteri containing foreign bodies. J. exp. Med. **126**, 523—537.

Parsons, D. S., and C. A. R. Boyd, 1972: Transport across the intestinal mucosal cell: hierarchies of function. Int. Rev. Cytol 32, 209—255.

Pasquini, F., A. Petris, S. Sbaraglia, R. Scopelliti, G. Cenci, and L. Frati, 1974: Biological activities in the granules isolated from the mouse submaxillary gland. Exp. Cell Res. 86, 233—236.

Pasteels, J. J., 1969: L'activite phosphatasique acide étudiée au microscope electronique dans les oeufs de *Barnea candida* (Mollusque, bivalve). Arch. Biol. (Liège) 801—807.

— 1973: Yolk and lysosomes. In: Lysosomes in biology and pathology. (Dingle, J. T., ed.), Vol. 3, pp. 216—234. Amsterdam: North-Holland Publishing.

Patrick, A. D., and B. D. Lake, 1973: Wolman's disease. In: Lysosomes and storage diseases. (Hers, H. G., and F. van Hoof, eds.), pp. 453—473. New York: Academic Press.

Paul, B. B., R. R. Straus, R. J. Selvaraj, and A. J. Sbarra, 1973: Peroxidase mediated antimicrobial activities of alveolar macrophage granules. Science 181, 849—850.

Paulson, J., and M. D. Rosenberg, 1972: The function and transportation of lining bodies in developing avian oocytes. J. Ultrastruct. Res. 40, 25—43.

Payne, D. N., and G. A. Ackerman, 1974: Tertiary granule formation in human developmental neutrophils revealed by S-35 ultrastructural autoradiography. J. Cell Biol. 63, 261 a.

Pearsall, N. N., and R. S. Weiser, 1970: The macrophage. Philadelphia: Lea and Febiger.

Pearse, A. D., 1974: The histochemical demonstration of hydrolytic enzymes in adjuvant-induced arthritis in rats. Histochem. J. 6, 431—446.

Pelletier, G., 1973: Secretion and uptake of peroxidase by rat adenohypophyseal cells. J. Ultrastruct. Res. 43, 445—459.

— and R. Puvani, 1974: Permeability of capillaries to different tracers and uptake of horseradish peroxidase by the secretory cells of the anterior pituitary gland. Z. Zellforsch. 147, 361—372.

Perkins, E. H., 1970: Digestion of antigen by peritoneal macrophages. In: Mononuclear phagocytes. (van Furth, R., ed.), pp. 562—578. Philadelphia: F. A. Davis.

Perrelet, A., L. Orci, and F. Baumann, 1971: Evidence for granulolysis in the retinula cells of a *Stomatopod crustacean, Squilla mantis.* J. Cell Biol. 48, 684—688.

Perry, R. P., and D. E. Kelley, 1973: Messenger RNA turnover in L-cells. J. molec. Biol. 79, 681—696.

Pesanti, E. L., and S. L. Axline, 1975: Colchicine effects on lysosomal enzyme induction and intracellular degradation in the cultivated macrophage. J. exp. Med. 141, 1030—1046.

Peters, T. J., 1970: The subcellular localization of di- and tri-peptidase hydrolase activity in guinea pig small intestine. Biochem. J. 120, 195—302.

— M. Muller, and C. de Duve, 1972: Lysosomes of the arterial wall. I. Isolation and subcellular fractionation of cells from normal rabbit aorta. J. exp. Med. 136, 1117—1139.

— T. Takano, and C. De Duve, 1972: In Atherogenesis: Initiating factors (Ciba Fndn. Sympos.), pp. 198—222. Amsterdam: Associated Scientific Publ.

— and C. De Duve, 1974: Lysosomes of the arterial wall. II. Subcellular fractionation of aortic cells from rabbits with experimental atheroma. Exp. Molec. Pathol. 20, 228—256.

Peterson, J. A., and H. Rubin, 1970: The exchange of phospholipids between cultured chick embryo fibroblasts as observed by autoradiography. Exp. Cell Res. 60, 383—392.

Pfeifer, U., 1969: Zur Frage der Beteiligung lysosomaler Enzyme an der fokalen Cytoplasmadegradation. Verh. dtsch. Ges. Path. 53, 344—350.

— and H. Scheller, 1975: A morphometric study of cellular autophagy including diurnal variations in kidney tubules of normal rats. J. Cell Biol. 64, 608—621.

Philippovich, I. I., I. N. Bezsmertnayer, and A. I. Oparin, 1973: On the localization of polyribosomes in the system of chloroplast lamellae. Exp. Cell Res. 79, 159—168.

Pickett-Heaps, J. D., 1967: Further observations on the Golgi apparatus and its functions in cells of the wheat seedling. J. Ultrastruct. Res. 18, 287—303.

Pickup, J. C., and D. B. Hope, 1972: The limited proteolysis of bovine neurophysin by cathepsin D. J. Neurochem. 19, 1049—1063.

Pine, M. J., 1967: Intracellular protein breakdown in the L 1210 ascites leukemia. Cancer Res. 27, 522—525.

Pine, M. J., 1970: Steady-state measurement of the turnover of amino acid in the cellular proteins of growing *Escherichia coli:* existence of two kinetically distinct reactions. J. Bact. **103**, 207—215.
— 1972: Turnover of intracellular proteins. Ann. Rev. Microbiol. **26**, 103—125.
Pitt, D., 1968: Histochemical demonstration of certain hydrolytic enzymes within cytoplasmic particles of *Botrytis cinera* Fr. J. gen. Microbiol. **52**, 67—75.
— and C. Coombes, 1968: The disruption of lysosome-like particles of *Solanum tuberosum* cells during infection by *Phytophthora erythroseptica* Pethyor. J. gen. Microbiol. **53**, 197—204.
— and M. Gilpin, 1973: Isolation and properties of lysosomes from dark grown potato shoots. Planta **109**, 233—258.
Pitt-Rivers, R., and R. R. Cavalieri, 1963: The free iodotyrosines of the rat thyroid gland. Biochem. J. **86**, 86—92.
Platt, T., J. H. Miller, and K. Weber, 1970: *In vivo* degradation of mutant lac repressor. Nature **228**, 1154—1156.
Pohl, S. L., M. J. Krans, L. Birnbaumer, and M. Radbell, 1972: Inactivation of glucagon by plasma membranes of rat liver. J. biol. Chem. **247**, 2295—2301.
Polakoski, K. C., and R. A. McRarie, 1973: Boar acrosin. I. Purification and preliminary characterization of a proteinase from boar sperm acrosomes. II. Classification, inhibition and specificity studies of a proteinase from sperm acrosomes. J. biol. Chem. **248**, 8178—8182, 8183—8188.
Polansky, J. R., B. P. Toole, and J. Gross, 1974: Brain hyaluronidase: changes in activity during chick development. Science **183**, 862—864.
Pollack, M. S., and J. W. C. Baird, 1968: Distribution and particle properties of acid hydrolases in denervated muscle. Amer. J. Physiol. **215**, 716—722.
Pontremole, S., E. Melloni, F. Sabmino, A. T. Franzi, A. de Flora, and B. L. Horecker, 1973: Change in rat liver lysosomes and fructose-1,6-bisphosphatase induced by cold and fasting. Proc. nat. Acad. Sci. (U.S.) **70**, 3674—3678.
Pontz, B. F., P. K. Muller, and W. N. Mergel, 1973: A study on the conversion of procollagen; release and recovery of procollagen peptide in the culture medium. J. biol. Chem. **248**, 7558—7564.
Poole, A. R., 1973: Tumor lysosomal enzymes and invasive growth. In: Lysosomes in biology and pathology. (Dingle, J. T., ed.), Vol. **3**, pp. 303—337. Amsterdam: North-Holland Publishing.
— J. T. Dingle, and A. J. Barrett, 1972: The immunocytochemical demonstration of cathepsin D. J. Histochem. Cytochem. **20**, 261—265.
— R. M. Hembry, and J. T. Dingle, 1974: Cathepsin D in cartilage: the immunohistochemical demonstration of extracellular enzyme in normal and pathological conditions. J. Cell Sci. **14**, 139—161.
Poole, B., 1971 a: Synthesis and degradation of proteins in relation to cellular structure. In: Enzyme synthesis and degradation in mammalian systems. (Rechcigl, M., ed.), pp. 375—402. Baltimore: University Park Press.
— 1971 b: The kinetics of disappearance of labelled leucine from the free amino acid pool of rat liver and its effect on the apparent turnover of catalase and other hepatic proteins. J. biol. Chem. **246**, 6587—6591.
— and C. de Duve, 1973: Lysosomes and protein turnover. Proc. Symposium on Intracellular Protein Catabolism, Friedrichroda (Mai, 1973) (in press).
— and M. Wibo, 1973 a: Protein degradation in cultured cells: the effect of fresh medium, fluoride and iodoacetate on the digestion of cellular protein of rat fibroblasts. J. biol. Chem. **248**, 6221—6226.
— — 1973 b: Studies on the mechanism of degradation of cellular proteins. Proc. Symposium on Intracellular Protein Catabolism, Friedrichroda (May, 1973) (in press).
Porter, K. R., K. Kenyon, and S. Badenhausen, 1967: Specializations of the unit membrane. Protoplasma **63**, 262—274.
Posalaki, Z., D. Szabo, E. Bacsi, and I. Okros, 1968: Hydrolytic enzymes during spermatogenesis in rat: an electron microscopic and histochemical study. J. Histochem. Cytochem. **16**, 249—262.

Poste, G., 1971 a: Tissue dissociation with proteolytic enzymes. Adsorption and activity of enzymes at the cell surface. Exp. Cell. Res. **65**, 359—367.

— 1971 b: Sub-lethal autolysis: modifications of cell periphery by lysosomal enzymes. Exp. Cell Res. **67**, 11—16.

— and A. C. Allison, 1971: Membrane fusion reaction: A theory. J. Theor. Biol. **32**, 165—184.

— — 1973: Membrane fusion. Biochim. biophys. Acta. **300**, 421—465.

Poux, N., 1963 a: Localisation de la phosphatase acide dans les cellules méristematique de blé (*Triticum vulgare*). J. Microscopie **2**, 485—489.

— 1963 b: Localisation des phosphates et de la phosphatase acide dans les cellules des embryons de blé (*Triticum vulgare* Villi) lors de la germination. J. Microscopie **2**, 557—568.

— 1963 c: Sur la présence d'enclaves cytoplasmiques en voie de dégénérescence dans les vacuoles des cellules végétales. C. R. Acad. Sci. (Paris) **D 257**, 736—738.

— 1965: Localisation de l'activite phosphatique acide et des phosphates dans les grains d'aleurone. I. Grains d'aleurone renfermant a la fois globiodes et cristalloides. J. Microscopie **4**, 771—782.

— 1969: Localisation d'activites enzymatiques dans les cellules du méristeme radiculaire de *Cucumis sativus*. II. Activite peroxydasique. J. Microscopie **8**, 855—866.

Poznansky, M. J., and W. B. Weglicki, 1974: Lysophospholipid induced volume changes in lysosomes and lysosomal lipid dispersions. Biochem. biophys. Res. Commun. **58**, 1016—1021.

Pressey, R., and J. K. Avants, 1973: Two forms of polygalacturonase in tomatoes. Biochim. biophys. Acta **309**, 363—369.

Price, G. M., 1973 a: Protein and nucleic acid metabolism in insect fat body. Biol. Reviews **48**, 333—375.

— 1973 b: The internal secretion of protease by the salivary gland of the larva of the blowfly, *Calliphora erythrocephala*. Biochem. Soc. Trans. **1**, 1152—1155.

Pricer, W. E., and G. Ashwell, 1971: The binding of disalylated glycoproteins by plasma membranes of rat liver. J. biol. Chem. **246**, 4825—4833.

Prieur, D. J., W. C. Davis, and G. A. Pladgett, 1972: Defective function of renal lysosomes in mice with the Chediak-Higahsi syndrome. Amer. J. Path. **67**, 227—240.

— H. M. Olson, and D. M. Young, 1974: Lysozyme deficiency—an inherited disorder of rabbits. Amer. J. Path. **77**, 283—298.

Prouty, W. F., and A. L. Goldberg, 1972 a: Effects of protease inhibitors on protein breakdown in *Escherichia coli*. J. biol. Chem. **247**, 3341—3352.

— — 1972 b: Fate of abnormal proteins in *E. coli*. Accumulation in intracellular granules before catabolism. Nature New Biol. **240**, 147—150.

— M. J. Karnovsky, and A. L. Goldberg, 1975: Fate of abnormal proteins in *E. coli*. III. Formation of protein inclusions in cells exposed to amino acid analogs. J. biol. Chem. **250**, 1112—1122.

Quarles, R. M., and R. M. C. Dawson, 1969: The hydrolysis of monolayers of phosphatidyl (Me-^{14}C) choline by phospholipase D. Biochem. J. **113**, 697—705.

Quincey, R. V., and S. U. Wilson, 1969: The utilization of genes for ribosomal RNA, 5 S RNA and transfer RNA in liver cells of adult rats. Proc. nat. Acad. Sci. (U.S.) **64**, 981—988.

Quirk, S. J., J. Byrne, and G. B. Robinson, 1973: Studies on the turnover of protein and glycoprotein components in rat kidney brush borders. Biochem. J. **132**, 501—508.

Rabinovitch, M., 1970: Phagocytic recognition. In: Mononuclear phagocytes. (van Furth, R., ed.), pp. 299—313. Philadelphia: F. A. Davis.

Rabinowitz, M., 1972: Protein synthesis and turnover in normal and hypertrophied heart. Amer. J. Cardiol. **31**, 202—210.

— and J. M. Fisher, 1964: Characteristics of the inhibition of hemoglobin synthesis in rabbit reticulocytes by threo-α-amino-β-chlorobutyric acid. Biochim. biophys. Acta **91**, 313—322.

RADFORD, S. V., and D. W. MISCH, 1971: The cytological effect of ecdysterone on the midgut cells of the flesh-fly *Sarcophaga bullata*. J. Cell Biol. **49**, 702—711.

RADIN, N. S., 1971: Brain cerebroside metabolism and possible implications for clinical problems. In: Lipid storage diseases. (BERNSOHN, J., and H. J. GROSSMAN, eds.), pp. 137—163. New York: Academic Press.

RAFF, R. A., G. GREENHOUSE, K. W. GROSS, and P. R. GROSS, 1971: Synthesis and storage of microtubule proteins by sea urchin embryos. J. Cell Biol. **50**, 516—527.

RAGHAVAN, S. S., R. A. MUMFORD, and J. N. KANFER, 1974: Synthesis of glucosyl ceramide by purified calf spleen β-glucosidase. Biochem. biophys. Res. Commun. **58**, 99—106.

RAHMAN, Y.-E., and B. J. WRIGHT, 1975: Liposomes containing chelating agents: cellular penetration and a possible mechanism of metal removal. J. Cell Biol. **65**, 112—122.

RANNELS, D. E., R. KAO, and H. E. MORGAN, 1975: Effects of insulin on protein turnover in heart muscle. J. Biol. Chem. **250**, 1694—1701.

RASMUSSAN, L., 1973: On the role of food vacuole formation in the uptake of dissolved mutants by *Tetrahymena*. Exp. Cell Res. **82**, 192—196.

RAVIOLA, E., and M. J. KARNOVSKY, 1972: Evidence for a blood-thymus barrier using electron opaque tracers. J. exp. Med. **136**, 466—498.

RAY, T. K., I. LIEBERMAN, and A. I. LANSING, 1968: Synthesis of the plasma membrane of the liver cell. Biochem. biophys. Res. Commun. **31**, 54—58.

RAZ, A., and R. GOLDMAN, 1974: Spontaneous fusion of rat liver lysosomes *in vitro*. Nature **247**, 206—207.

REAVAN, E. P., and S. G. AXLINE, 1972: Subplasmalemmal microfilaments and microtubules in resting and phagocytosing cultivated macrophages. J. Cell Biol. **59**, 12—27.

REBHUN, L. I., 1972: Polarized intracellular particle transport: Saltatory movements and cytoplasmic streaming. Int. Rev. Cytol. **32**, 93—137.

RECHCIGL, M., ed., 1971 a: Enzyme synthesis and degradation in mammalian systems. University Park Press, Baltimore (Originally published by S. Karger and A. G., Basel, Switzerland).

— 1971 b: Intracellular protein turnover and the role of synthesis and degradation in regulation of enzyme levels. In: Enzyme synthesis and degradation in mammalian systems. (RECHCIGL, M., ed.), pp. 237—310. Baltimore: University Park Press.

REDDY, J., and D. SVOBODA, 1973: Further evidence to suggest that microbodies do not exist as individual entities. Amer. J. Path. **70**, 421—438.

REED, B. L., and D. G. WENZEL, 1975: The lysosomal permeability test modified for toxicity testing with cultured heart endothelioid cells. Histochem. J. **7**, 115—126.

REEL, J. R., and F. T. KENNEY, 1968: Superinduction of tyrosineaminotransferase in hepatoma cells: differential inhibition of synthesis and turnover by actinomycin D. Proc. nat. Acad. Sci. (U.S.) **61**, 200—206.

REEVES, P., 1972: The bacteriocins (Molecular biology, biochemistry and biophysics; Vol. **11**). Berlin-Heidelberg-New York: Springer-Verlag.

REFSNES, K., S. OLSNES, and A. PHIL, 1974: On the toxic proteins abrin and ricin: studies of their binding to and entry into Ehrlich ascites cells. J. biol. Chem. **249**, 3557—3562.

REGNIER, P. H., and M. N. THANG, 1973: Membrane associated proteases in *E. coli*. FEBS Letters **36**, 31—33.

REGNIER, P. M., and M. N. THANG, 1972: Subcellular distribution and characterization of endo and exocellular protease in *E. coli*. Biochemie **54**, 1227—1236.

REICHEL, W., 1968: Lipofuscin pigment accumulation and distribution in five rat organs as a function of age. J. Gerontol. **23**, 145—153.

REIGNGOUD, D. J., and J. M. TAGER, 1973: Measurement of intralysosomal pH. Biochim. biophys. Acta. **297**, 174—178.

REIJNGOUD, D. J., and J. M. TAGER, 1975: Effects of ionophores and temperature on intra-lysosomal pH. FEBS letters **54**, 76—79.

REISS, J., 1974: Cytochemical detection of hydrolases in fungus cells. III. Aryl sulfatase. J. Histochem. Cytochem. **22**, 183—188.

REMACLE, J., S. FOWLER, H. BEAUFAY, and J. BERTHET, 1974: Ultrastructural localization of cytochromes on rat liver microsomes by means of hybrid antibodies labeled with ferritin. J. Cell. Biol. **61**, 237—240.

REYNOLDS, R. D., and S. D. THOMPSON, 1974: Irreversible inactivation of rat liver tyrosine amino transferase. Arch. Biochem. Biophys. **164**, 43—51.

RICHARDS, O. C., and R. S. RYAN, 1974: Synthesis and turnover of *Euglena gracilis* mitochondrial DNA. J. molec. Biol. **82**, 57—75.

RICKETTS, T. R., 1970: Effects of endocytosis upon acid phosphatase activity of *Tetrahymena pyriformis*. Protoplasma **71**, 127—137.

— 1971: Endocytosis in *Tetrahymena pyriformis*. The selectivity of uptake of particles and the adaptive increase in cellular acid phosphatase activity. Exp. Cell Res. **66**, 49—58.

RIECKEN, E. D., and A. G. E. PEARSE, 1966: Histochemical studies on the Paneth cells in the rat. Gut **7**, 86—93.

RIETRA, P. J. G. M., J. L. MOLENAAR, M. N. HAMERS, J. M. TAGER, and P. BORST, 1974: Investigation of the galactosidase deficiency in Fabry's disease using antibodies against the purified enzyme. Eur. J. Biochem. **46**, 89—101.

— F. A. J. T. M. VAN DEN BERGH, and J. M. TAGER, 1974: Recent developments in enzyme replacement therapy of lysosomal storage diseases. In: Enzyme Therapy in Lysosomal Storage Diseases. (J. M. TAGER, G. J. M. HOOGHWINKEL, and W. TH. DAEMS, eds.), pp. 53—79. Amsterdam: North Holland Co.

RIFKIND, R. A., 1966: Destruction of injured red cells *in vivo*. Amer. J. Med. **41**, 711—723.

RIFKIN, D. B., J. N. LOEB, G. MOORE, and E. REICH, 1974: Properties of plasminogen activators formed by neoplastic human cell cultures. J. exp. Med. **139**, 1317—1328.

RIORDAN, J. R., L. MITCHELL, and M. SLAVIK, 1974: The binding of asialoglycoprotein to isolated Golgi apparatus. Biochem. biophys. Res. Commun. **59**, 1373—1379.

ROBBINS, E., and N. K. GONATAS, 1964 a: The ultrastructure of a mammalian cell during the mitotic cycle. J. Cell Biol. **21**, 429—464.

— — 1964 b: Histochemical and ultrastructural studies on HeLa cell cultures exposed to spindle inhibitors with special reference to the interphase cell. J. Histochem. Cytochem. **12**, 704—711.

— P. I. MARCUS, and N. K. GONATAS, 1964: Dynamics of acridine-orange-cell interactions. II. Dye-induced ultrastructural changes in multivesicular bodies (acridine orange particles). J. Cell Biol. **21**, 49—61.

— E. M. LEVINE, and H. EAGLE, 1970: Morphologic changes accompanying senescence of cultured human diploid cells. J. exp. Med. **131**, 1211—1222.

ROBERTSON, B., and G. ENHORNING, 1974: The alveolar lining of the premature newborn rabbit after pharyngeal deposition of surfactant. Lab. Invest. **31**, 54—59.

ROBERTSON, P. B., R. B. RYEL, R. E. TAYLOR, K. W. SHYU, H. M. FULLMER, 1972: Collagenase: localization in polymorphonuclear leukocyte granules in rabbit. Science **177**, 64—65.

ROBINEAUX, R., and J. PIRET, 1959: An *in vitro* study of some mechanisms of antigen uptake by cells. In: Ciba Found. Symp. on Cellular Aspects of Immunity. (WOLSTENHOLME, G. E. W., and M. O'CONNOR, eds.), pp. 5—43. Boston: Little Brown and Co.

ROBINSON, A. B., J. H. McKERROW, and P. CARY, 1970: Controlled deamination of peptides and proteins: an experimental hazard and a possible biological timer. Proc. nat. Acad. Sci. (U.S.) **66**, 753—757.

RODEWALD, R., 1972: Intestinal transport of antibodies in the newborn rat. J. Cell Biol. **58**, 189—211.

— 1974: Transport of immunoglobulin fragments by the small intestine of the neonatal rat. J. Cell Biol. **63**, 287 a.

ROELENTS, G. E., and J. W. GOODMAN, 1969: The chemical nature of macrophage RNA-antigen complexes and their relevance to immune induction. J. exp. Med. **130**, 557—574.

ROELS, O. A., 1969: The influence of vitamins A and E on lysosomes. In: Lysosomes in biology and pathology. (DINGLE, J. T., and H. B. FELL, eds.), Vol. **1**, pp. 254—275. Amsterdam: North-Holland Publishing.

ROJAS-ESPINOSA, O., A. M. DANNENBERG, L. A. STERNBERGER, and T. TSUDA, 1974: The role of cathepsin D in the pathogenesis of tuberculosis. Amer. J. Path. **74**, 1—18.

ROMEO, D., N. STAGNI, and M. C. PUGLIARELLO, 1967: Protection effects of polyhydroxyl compounds on heart lysosomal structures. Experientia **23**, 247—248.

ROOT, R. K., D. S. ROSENTHAL, and D. J. BALESTRA, 1972: Abnormal bactericidal, metabolic and lysosomal functions of Chediak-Higashi syndrome lysosomes. J. Clin. Invest. **51**, 649—666.

ROSENBAUM, R. M., and M. WITTNER, 1962: The activity of intracytoplasmic enzymes associated with feeding and digestion in *Paramecium caudatum*. The possible relationship to neutral red granules. Arch. Protistenk. **106**, 223—240.

ROSENBLUTH, J., and S. L. WISSIG, 1963: The distribution of exogenous ferritin in toad spinal ganglia and the mechanisms of its uptake by neurons. J. Cell Biol. **23**, 307—325.

ROSER, B., 1970: The migration of macrophages *in vivo*. In: Mononuclear phagocytes. (VAN FURTH, R., ed.), pp. 166—174. Philadelphia: F. A. Davis.

ROSS, R., and J. A. GLOMSET, 1973: Atherosclerosis and the arterial smooth muscle cell. Science **180**, 1332—1339.

— and G. ODLAND, 1968: Human wound repair. II. Inflammatory cells, epithelial mesenchymal interrelations and fibrogenesis. J. Cell Biol. **39**, 152—170.

ROSSMAN, T., C. NORRIS, and W. TROLL, 1974: Inhibition of macromolecule synthesis in *Escherichia coli* by protease inhibitors: specific reversal by glutathione of the effects of chloromethyl ketones. J. biol. Chem. **249**, 3412—3417.

ROTH, L. E., and Y. SHIGENAKA, 1964: The resorption of cilia. Z. Zellforsch. **64**, 19—24.

ROTH, T. F., and K. R. PORTER, 1964: Yolk protein uptake in the oocyte of the mosquito, *Aedes aegypti*. J. Cell Biol. **20**, 313—332.

ROTHSCHILD, J., 1966: Ameba acid phosphatase. I. An intracellular inhibitor for particle associated acid phosphomonoesterase. II. Distribution in cell fractions from *Chaos chaos*. C. R. Lab. Carlsberg **35**, 319—418, 457—500.

ROTHSTEIN, T. L., and J. J. BLUM, 1973: Lysosomal physiology in *Tetrahymena*. I. Effects of glucose, acetate, pyruvate, and carmine on the intracellular content and extracellular release of three acid hydrolases. J. Cell Biol. **57**, 630—641.

— — 1974 a: Lysosomal physiology in *Tetrahymena*. IV. Effect of dichloroisoproteranol on the intracellular sources of released acid hydrolases. Exp. Cell Res. **87**, 168—174.

— — 1974 b: Lysosomal physiology in *Tetrahymena*. III. Pharmacological studies on acid hydrolase release and ingestion and egestion of dimethylbenzanthracene particles. J. Cell Biol. **62**, 844—859.

RUBIN, C. S., M. E. BALIS, S. PIOMELLI, P. M. BEHRMAN, and J. DANCIS, 1969: Elevated AMP pyrophosphorylase activity in congenital IMP pyrophosphorylase deficiency (Lesch-Nyhan disease). J. Lab. clin. Med. **74**, 732—741.

RUBIN, M. S., and A. TZAGOLOFF, 1973: Assembly of the mitochondrial membrane system. X. Mitochondrial synthesis of three of the subunit proteins of yeast cytochrome oxidase. J. biol. Chem. **248**, 4275—4279.

RUBY, J. R., R. F. DYER, and S. KALKO, 1970: Continuities between mitochondria and endoplasmic reticulum in the mammalian ovary. Z. Zellforsch. **97**, 30—37.

RUDZINSKA, M. A., 1961: The use of a protozoan for studies in aging. I. Differences between young and old organisms of *Tokophyra infusionum* as revealed by light and electron microscopy. J. Gerontol. **16**, 213—224.

— 1962: The use of a protozoan for studies in aging. III. Similarities between young overfed and old normally fed *Tokophyra infusionum:* a light and electron microscope study. Gerontologica **6**, 206—226.

— 1969: An electron microscopic study of acid phosphatases in *Tokophyra infusionum*. J. Cell Biol. **43**, 120 A—121 A.

— 1970: The mechanism of food intake in *Tokophyta infusionum* and ultrastructural changes in food vacuoles during digestion. J. Protozool. **17**, 626—641.

RUENWONGSA, P., and M. CHULVATANOL, 1974: A new acid protease in human seminal plasma. Biochem. biophys. Res. Commun. **59**, 44—50.

RYAN, G. B., J. Z. BORYSENKO, and M. J. KARNOVSKY, 1974: Factors affecting the redistribution of surface-bound Concanavalin A on human polymorphonuclear leukocytes. J. Cell Biol. **62**, 351—365.

RYSER, H. J.-P., 1967: Studies on protein uptake by isolated tumor cells. III. Apparent stimulations due to pH, hypertonicity, polycations or dehydration and their relations to the enhanced penetration of infectious nucleic acids. J. Cell Biol. **32**, 737—750.

Riser, H. J.-P., 1968: Uptake of protein by mammalian cells: an underdeveloped area. Science **159**, 390—396.

— M. P. Gabathuler, and A. B. Roberts, 1971: Uptake of macromolecules at the cell surface. In: Biomembranes. (Manson, L. A., ed.), Vol. **II**, pp. 197—209. New York: Plenum Press.

Ryter, A., Y. Hirota, and U. Schwarz, 1973: Process of cellular division in *Escherichia coli*. Growth pattern of *E. coli* murein. J. molec. Biol. **78**, 185—196.

Sabatini, D. D., Y. Tashiro, and G. E. Palade, 1966: On the attachment of ribosomes to microsomal membranes. J. Mol. Biol. **19**, 503—524.

Sabatini-Smith, S., 1971: Synthesis localization and renewal of lipids in mammalian tissues. In: Cellular and molecular renewal in the mammalian body. (Cameron, I. C., and J. D. Thrasher, eds.), pp. 277—329. New York: Academic Press.

Sabri, M. I., A. H. Bone, and A. N. Davidson, 1974: Turnover of myelin and other structural proteins in the developing rat brain. Biochem. J. **142**, 499—507.

Saito, T., and K. Ogawa, 1974: Lysosomal changes in rat hepatic parenchymal cells after glucagon administration. Acta Histochem. Cytochem. **7**, 1—8.

Saito, M., T. Yoshizawa, T. Aoyagi, and Y. Nagai, 1973: Involvement of proteolytic activity in early events in lymphocyte transformation by Phytohemagglutinin. Biochem. biophys. Res. Commun. **52**, 569—575.

Salthouse, T. N., and D. Pfeffer, 1965: Acid phosphatase activity in the canine gall bladder and bile duct epithelium. J. Histochem. Cytochem. **13**, 240—242.

Salzberger, B., et R. Weber, 1966: La regression du mesonephos chez l'embryon de poulet; étude d'activites de la phosphatase acide et des cathepsines. Analyse biochemique, histochemique et observations au microscope électronique. J. Embryol. exp. Morph. **15**, 397—419.

Samorajski, T., J. R. Keefe, and J. M. Ordy, 1964: Intracellular localization of lipofuscin age pigments in the nervous system. J. Gerontol. **19**, 262—276.

Samuels, S., N. K. Gonatas, and M. Weiss, 1964: Chemical and ultrastructural comparison of synthetic and pathologic membrane systems. J. Cell Biol. **21**, 148—152.

Sandhoff, K., and K. Harzer, 1973: Total hexosaminidase deficiency in Tay-Sachs' disease (variant O). In: Lysosomes and storage diseases. (Hers, H. G., and F. van Hoof, eds.), pp. 345—356. New York: Academic Press.

Sanyal, S., 1972: Changes of lysosomal enzymes during hereditary degeneration and histiogenesis of retina in mice. Histochemie **29**, 28—36.

Sapolsky, A. I., R. D. Altman, J. F. Woessner, and D. S. Howell, 1973: The action of cathepsin D in human articular cartilage on proteoglycans. J. clin. Invest. **52**, 624—633.

Sarov, I., and W. K. Jocklick, 1972: Characterization of intermediates in the uncoating of vaccinia virus DNA. Virology **50**, 593—602.

Sasazuki, T., H. Tsunov, and H. Nakajima, 1974: Interaction of human hemoglobin with haptoglobin and antihemoglobin antibody. J. biol. Chem. **249**, 2441—2446.

Sastry, P. S., and L. E. Hokin, 1966: Studies on the role of phospholipids in phagocytosis. J. biol. Chem. **241**, 3354—3361.

Satir, B., C. Schooley, and P. Satir, 1973: Membrane fusion in a model system. J. Cell Biol. **56**, 153—176.

Saunders, J. W., 1966: Death in embryonic systems. Science **154**, 604—612.

Sawant, P. L., S. Shibko, U. S. Kumita, and A. L. Tappel, 1964 a: Isolation of rat liver lysosomes and their general properties. Biochim. biophys. Acta **85**, 82—92.

— I. D. Desai, and A. L. Tappel, 1964 b: Degradative aspects of purified lysosomes. Biochim. biophys. Acta **85**, 93—102.

Schacher, S. M., E. Holtzman, and D. C. Hood, 1974: Uptake of horse-radish peroxidase by frog photoreceptor synapses in the dark and the light. Nature **249**, 261—263.

Scharrer, B., 1966: Ultrastructural study of the regressing prothoracic glands of *Blattarian* insects. Z. Zellforsch. **69**, 1—21.

Schebli, H. D., and M. M. Burger, 1972: Selective inhibition of growth of transformed cells by protease inhibitors. Proc. nat. Acad. Sci. (U.S.) **12**, 3825—3827.

SCHEIB, D., 1963: Properties and role of acid hydrolases of the mullerian ducts during sexual differentiation in the male chick embryo. In: Ciba foundation symposium on lysosomes. (DE REUCK, A. V. S., and M. P. CAMERON, eds.), pp. 264—281. Boston: Little Brown.

SCHELLENS, J. P. M., 1974: Ageing of mouse liver lysosomes; an experimental study using indigestible marker substances. Cell Tiss. Res. **155**, 455—473.

SCHENGRUND, C.-L., D. S. JENSEN, and A. ROSENBERG, 1972: Localization of sialidase in the plasma membrane of rat liver cells. J. biol. Chem. **247**, 2742—2746.

— R. N. LAUSCH, and A. ROSENBERG, 1973: Sialidase activity in transformed cells. J. biol. Chem. **248**, 4424—4428.

SCHER, B., and D. DUBNAU, 1973: A manganese stimulated endonuclease from *Bacillus subtilis*. Biochem. biophys. Res. Commun. **55**, 595—602.

SCHIAFFINO, S., and V. HANZLIKOVA, 1972 a: Autophagic degradation of glycogen in skeletal muscle of the newbron rat. J. Cell Biol. **52**, 41—51.

— — 1972 b: Studies on the effect of denervation in developing muscle. II. The lysosomal system. J. Ultrastruct. Res. **39**, 1—14.

SCHIMKE, R. T., 1973: Control of enzyme levels in mammalian tissues. Adv. Enzymol. **37**, 135—187.

— E. W. SWEENY, and C. M. BERLIN, 1965: Studies of the stability *in vivo* and *in vitro* of rat tryptophan pyrrolase. J. biol. Chem. **240**, 4609—4620.

SCHIN, K. S., and U. CLEVER, 1965: Lysosomal and free acid phosphatase in salivary glands of *Chironomus tentans*. Science **150**, 1053—1055.

— — 1968: Ultrastructural and cytochemical studies of salivary gland regression in *Chironomus tentans*. Z. Zellforsch. **86**, 262—279.

— and H. LAUFER, 1973: Studies of programmed salivary gland regression during larval pupal transformation in *Chironomus thummi*. I. Acid hydrolase activity. Exp. Cell Res. **82**, 335—340.

SCHLESSINGER, D., and F. BEN-HAMIDA, 1966: Turnover of protein in *Escherichia coli* starved for nitrogen. Biochem. biophys. Acta **119**, 171—182.

SCHMIDT-ULLRICH, R., D. F. H. WALLACH, and E. FERBER, 1974: Concanavalin A augments the turnover of electrophoretically defined thymocyte plasma membrane proteins. Biochim. biophys. Acta **356**, 288—299.

SCHNEIDER, D. L., 1970: A membraneous ATP'ase unique to lysosomes. Biochem. Biophys. Res. Comm. **61**, 882—888.

SCHNEIDER, F. H., 1970: Secretion from the bovine adrenal gland: release of lysosomal enzymes. Biochem. Pharmacol. **19**, 833—847.

— A. D. SMITH, and H. WINKLER, 1967: Secretion from the adrenal medulla: histochemical evidence for exocytosis. Brit. J. Pharmacol. **31**, 94—104.

SCHNITZER, B., T. SODERMAN, M. L. MEAD, and P. G. CONTACOS, 1972: Pitting function of the spleen in malaria: ultrastructural observations. Science **177**, 175—177.

— D. L. RUCKNAGEL, H. H. SPENCER, and M. AIKAWA, 1971: Erythrocytes: pits and vacuoles as seen with transmission and scanning electron microscopy. Science **173**, 251—252.

SCHNITZER, R. J., G. BUNESAU, and V. BADEN, 1971: Interaction of mineral fiber surfaces with cells *in vitro*. Ann. N.Y. Acad. Sci. **172**, 757—772.

SCHUEL, H., W. L. WILSON, R. S. BRESSLER, J. W. KELLY, and J. R. WILSON, 1972: Purification of cortical granules from unfertilized sea urchin egg homogenates by zonal centrifugation. Develop. Biol. **29**, 307—320.

SCHULMAN, J. D., and K. M. BRADLEY, 1970: The metabolism of amino acids, peptides, and disulfides in lysosomes of fibroblasts cultured from normal individuals and those with cystinosis. J. exp. Med. **132**, 1090—1104.

SCHWARZ, U., A. ASMUS, and H. FRANK, 1969: Autolytic enzymes and cell division of *Escherichia coli*. J. molec. Biol. **41**, 419—429.

SCHWARZ, W., and F. H. GUILDNER, 1967: Elektronenmikroskopische Untersuchungen des Kollagenabbaus im Uterus der Ratte nach der Schwangerschaft. Z. Zellforsch. **83**, 416—426.

SCORNIK, O. A., 1972: Decreased *in vivo* disappearance of labelled liver protein after partial hepatectomy. Biochem. biophys. Res. Commun. **47**, 1063—1066.

Scorza, J. V., C. de Scorza, and M. C. C. Monteiro, 1972: Cytochemical observations of three acid hydrolases in blood stages of malarial parasites. Ann. trop. Med. Parasit. **66**, 167—172.

Scott, A. P., J. G. Ratcliffe, L. H. Rees, J. Landon, M. P. J. Bennett, P. J. Lowry, and P. McMartin, 1973: Pituitary peptide. Nature New Biol. **244**, 65—67.

Seegmiller, J. E., 1973: Cystinosis. In: Lysosomes and storage diseases. (Hers, H. G., and F. van Hoof, eds.), pp. 485—518. New York: Academic Press.

Seeman, P. M., and G. E. Palade, 1967: Acid phosphatase localization in rabbit eosinophils. J. Cell Biol. **34**, 745—756.

Segal, H. L., 1969: What determines the half-life of proteins *in vivo?* Some experiments with alanine amino transferase of rat tissues. Biochem. biophys. Res. Commun. **36**, 764—770.

— J. A. Brown, and J. R. Winkler, 1974 a: Lysosomes and intracellular protein turnover. J. Cell Biol. **63**, 308 a.

— J. R. Winkler, and M. P. Miyagi, 1974 b: Relationship between degradation rates of proteins *in vivo* and their susceptibility to lysosomal proteases. J. biol. Chem. **249**, 6364—6365.

Seljelid, R., 1966: An electron microscopic study of the formation of cytosomes in rat kidney adenoma. J. Ultrastruct. Res. **16**, 569—583.

— 1967: Endocytosis in thyroid follicle cells. IV. On the acid phosphatase activity in thyroid follicle cells with special reference to the quantitative aspects. J. Ultrastruct. Res. **18**, 237—256.

— S. C. Silverstein, and. Z. A. Cohn, 1973: The effect of poly-L-lysine on the uptake of reovirus double-stranded RNA in macrophages *in vitro.* J. Cell Biol. **57**, 484—498.

Sellinger, O. Z., J. C. Santiago, M. A. Sands, B. Furinsloat, 1973: N-acetyl β-D-glucosaminidases of nerve cells: a developmental study of two molecular components. Biochim. biophys. Acta **315**, 128—146.

Sexton, R., and J. F. Sutcliffe, 1969: The distribution of β-glycerophosphatase in young roots of *Pisum sativium.* Ann. Bot. **33**, 407—419.

Seybart, H. W., H. Schmidt-Gayk, F. H. Jakobs, and E. Hackenthal, 1973: Cyclic adenosine monophosphate in phagocytosing granulocytes and alveolar macrophages. J. Cell Biol. **57**, 567—571.

Shatkin, A. J., 1971: Viruses with segmented RNA genomes: Multiplication of influenza vs reovirus. Bact. Rev. **35**, 250—266.

Shea, S. M., M. J. Karnovsky, and W. M. Bossert, 1969: Vesicular transport across endothelium: simulation of a diffusion model. J. theor. Biol. **24**, 30—42.

Shecher, U., Rafaeli-Eshbol, and A. Hershko, 1973: Influence of protease inhibitors and energy metabolism on intra-cellular protein breakdowns in starving *Escherichia coli.* Biochem. biophys. Res. Commun. **54**, 1518—1524.

Shelburne, J. D., A. M. Arista, and B. F. Trump, 1973 a: Studies in cellular autophagocytosis: cyclic AMP and dibutyryl cyclic AMP-stimulated autophagy in rat liver. Amer. J. Path. **72**, 521—540.

— A. U. Arstila, and B. F. Trump, 1973 b: Studies on cellular autophagocytosis: the relationships of autophagocytosis to protein synthesis and to energy metabolism in rat liver and flounder kidney. Amer. J. Path. **73**, 641—690.

Sheldrake, A. R., 1974: The aging, growth and death of cells. Nature **250**, 381—385.

Shepp, M., H. Toff, H. Yamada, and T. G. Gabuzda, 1972: Heterogeneous metabolism of marrow ferritins during erythroid cell maturation. Brit. J. Hematol. **22**, 377—382.

Shibko, S., Panghorn, and A. L. Tappel, 1965: Studies on the release of lysosomal enzymes from kidney lysosomes. J. Cell Biol. **25**, 479—483.

Shio, H., M. G. Farquhar, and C. de Duve, 1974: Lysosomes of the arterial wall. IV. Cytochemical localization of acid phosphatase and catalase in smooth muscle cells and foam cells from rabbit atheromatous aorta. Amer. J. Path. **76**, 1—16.

Shockman, G. D., L. Faneo-Moore, and M. C. Higgins, 1973: Problems of cell wall and membrane growth, enlargement, and division. Annals N.Y. Acad. Sci. **235**, 161—197.

Shokeir, M. H. K., and D. C. Schreiffer, 1969: Cytochrome oxidase deficiency in Wilson's disease: a suspected ceruloplasmin function. Proc. nat. Acad. Sci. (U.S.) **62**, 867—872.

SHORTMAN, K., E. DRENER, P. RUSSELL, and W. D. ARMSTRONG, 1970: The role of non-lymphoid accessory cells in the immune response to different antigens. J. exp. Med. **131**, 461—482.

SIEKEVITZ, P., 1972 a: Biological membranes: the dynamics of their organization. Ann. Rev. Physiol. **34**, 117—140.

— 1972 b: The turnover of proteins and the usage of information. J. theor. Biol. **37**, 321—334.

SIEVERS, A., 1966: Lysosomenähnliche Kompartimente in Pflanzenzellen. Naturwissenschaften **53**, 334—335.

SILVERBLATT, F. J., G. E. TYSON, and R. E. BULGER, 1974: Effects of vinblastine on the phagolysosome system of proximal tubule cells of rat kidney. Administration of horseradish peroxidase. Lab. Invest. **31**, 170—180.

SILVERSTEIN, S., 1970: Macrophages and viral immunity. Seminars in Hematology **7**, 185—214.

— and S. DALES, 1968: The penetration of reovirus RNA and initiation of its genetic function in L-strain fibroblasts. J. Cell Biol. **36**, 197—230.

— C. ASTELL, D. LEVIN, M. SCHONBERG, and G. ACS, 1973: The role of lysosomes in the uncoating and activation of the Reovirus genome. Adv. Biosciences **11**, 1—26.

— F. M. GRIFFIN, J. E. LEIDER, C. BIANCO, and J. GRIFFIN, 1975: The mechanism of phagocytosis (in press).

SIMIONESCU, N., M. SIMIONESCU, and G. E. PALADE, 1975: Permeability of muscle capillaries to small heme-peptides. Evidence for the existence of patent transendothelial channels. J. Cell Biol. **64**, 586—607.

SIMMONS, S. R., and M. L. KARNOVSKY, 1973: Iodinating ability of various leukocytes and their bactericidal activity. J. exp. Med. **138**, 44—63.

SIMON, F. R., O. O. BLUMENFELD, I. M. ARIAS, 1970: Two protein fractions obtained from hepatic plasma membranes: studies of their composition and differential turnover. Biochim. biophys. Acta **219**, 349—360.

SIMPSON, C. F., and J. M. KLING, 1968: The mechanism of mitochondrial extrusion from phenylhydrazine induced reticulocytes in the circulating blood. J. Cell Biol. **36**, 103—109.

SINGER, R. H., and S. PENMAN, 1973: Messenger RNA in HeLa cells: Kinetics of formation and decay. J. molec. Biol. **78**, 321—334.

SINGER, S. J., and L. I. ROTHFIELD, 1973: Synthesis and turnover of cell membranes. Neurosciences Research Program Bulletin **11**, #1.

SINHA, A. K., and S. P. R. ROSE, 1973: β-N-acetyl-D-galactosaminidase in bulk separated neurons and neurophil from rat recebral cortex. J. Neurochem. **20**, 39—44.

SJOBERG, I., L.-A. FRANSSON, R. MATALON, and A. DORFMAN, 1973: Biochem. biophys. Res. Commun. **54**, 1125—1132.

SKOSEY, J. L., E. DAAMGARD, D. CHOW, and L. B. SORENSON, 1974: Modification of zymosan induced release of lysosomal enzymes from human polymorphonuclear leukocytes by cytochlasin B. J. Cell Biol. **62**, 625—634.

SKUTELSKY, E., and D. DANON, 1967: An electron microscopic study of nuclear elimination from the late erythroblast. J. Cell Biol. **33**, 625—635.

— — 1969: Reduction in surface charge as an explanation of the recognition by macrophages of nuclei expelled from normoblasts. J. Cell Biol. **43**, 8—15.

SLAPIKOFF, S., J. L. SPITZER, and D. VACCARO, 1971: Sporulation in *Bacillus brevis;* studies on protease and protein turnover. J. Bact. **196**, 739—744.

SLATER, T. F., 1969: Lysosomes and experimentally induced tissue injury. In: Lysosomes in biology and pathology. (DINGLE, J. T., and H. B. FELL, eds.), Vol. **1**, pp. 469—492. Amsterdam: North-Holland Publishing.

— and P. A. RILEY, 1966: Photosensitization and lysosomal damage. Nature **209**, 151—154.

SLAUGHTER, D., and E. TRIPLETT, 1975: Amphibian embryo proteases inhibitor. Cell Differentiation **4**, 23—33.

SLOAN, H. R., J. L. BRESLOW, and D. S. FREDRICKSON, 1971: Purification and properties of two sphingolipid hydrolases. In: Sphingolipids, sphingolipidoses and allied disorders. (VOLK, B. W., and S. M. ARONSON, eds.). New York: Plenum Press.

SLOAT, B. F., and J. M. ALLEN, 1969: Soluble and membrane associated forms of acid phosphatase associated with the lysosomal fraction of rat liver. Ann. N.Y. Acad. Sci. **166**, 574—601.

SLOR, M., M. BUSTAN, and T. LEV, 1973: Deoxyribonuclease. II. Activity in relation to cell cycle in a synchronized HeLa S_3 cells. Biochem. biophys. Res. Commun. **52**, 556—561.

SMITH, A. D., 1972: The storage and secretion of hormones. In: The scientific basis of medicine-annual review. 74—102.

— and H. WINKLER, 1969: Lysosomes and chromaffin granules in the adrenal medulla. In: Lysosomes in biology and pathology. (DINGLE, J. T., and H. B. FELL, eds.), Vol. **1**, pp. 155—166. Amsterdam: North-Holland Publishing.

SMITH, H., 1972: Mechanisms of virus pathogenicity. Bact. Rev. **36**, 291—310.

SMITH, M. E., 1968: The turnover of myelin in the adult rat. Biochim. biophys. Acta **64**, 285—293.

SMITH, R. E., 1969: Phosphohydrolases in cell organelles: electron microscopy. Ann. N.Y. Acad. Sci. **166**, 525—564.

— and M. G. FARQUHAR, 1965: Preparation of non-frozen sections for electron microscope cytochemistry. RCA Sci. Instr. News **10**, 13—17.

— — 1966: Lysosome function in the regulation of the secretory process in cells of the anterior pituitary gland. J. Cell Biol. **31**, 319—336.

— and W. H. FISHMAN, 1969: p-(Acetoxymercuric)aniline diazotate, a reagent for visualizing the naphthol AS-BI-product of acid hydrolase activity at the level of the light and electron microscope. J. Histochem. Cytochem. **17**, 1—22.

SMITH, T. J., and R. R. WAGNER, 1967: Rabbit macrophage interferons. I. Conditions for biosynthesis by virus infected and uninfected cells. II. Some physiochemical properties and estimations of molecular weights. J. exp. Med. **125**, 559—577, 579—593.

SMITH, U., and J. W. RYAN, 1973: Electron microscopy of endothelial and epithelial components of the lungs: correlations of structure and function. Fed. Proc. **32**, 1957—1966.

SMITH-SONNEBORN, J., and M. KLASS, 1974: Change of DNA synthesis pattern of *Paramecium* with increased clonal age and interfission time. J. Cell Biol. **61**, 591—598.

SMOLEN, J. E., and S. B. SHOHET, 1974: Remodeling of granulocyte membrane fatty acids during phagocytosis. J. clin. Invest. **53**, 724—734.

SOFFER, R. L., R. REZA, and P. R. B. CALDWELL, 1974: Angiotensin-converting enzyme from rabbit pulmonary particles. Proc. nat. Acad. Sci. (U.S.) **71**, 720—724.

SOMMER, J. R., and J. J. BLUM, 1965: Cytochemical localization of acid phosphatases in *Euglena gracilis*. J. Cell Biol. **24**, 235—248.

SOROKIN, E., J. F. BOREL, and V. J. STECHER, 1970: Chemtaxis of mononuclear and polymorphonuclear phagocytes. In: Mononuclear phagocytes. (VAN FURTH, R., ed.), pp. 397—418. Philadelphia: F. A. Davis.

SOROKIN, H. P., and S. SOROKIN, 1968: Fluctuations in the acid phosphatase activity of spherosomes in guard cells of *Campanula persicifolia*. J. Histochem. Cytochem. **16**, 791—802.

SOROKIN, S. P., 1973: The respiratory system. In: Histology. (GREEP, R., and L. WEISS, eds.), pp. 661—711. McGraw-Hill.

SPECTOR, I. M., 1974: Animal longevity and protein turnover rate. Nature **249**, 66.

SPECTOR, W. G., and G. B. RYAN, 1970: The mononuclear phagocyte in inflammation. In: Mononuclear phagocytes. (VAN FURTH, R., ed.), pp. 219—231. Philadelphia: F. A. Davis.

SPICER, S. S., A. J. GARVIN, H. J. WOHLTMANN, and J. A. V. SIMPSON, 1974: The ultrastructure of the skin in patients with mucopolysaccharidoses. Lab. Invest. **31**, 488—502.

SPICHIGER, J. U., 1969: Isolation und Charakterisierung von Sphärosomen und Glyoxisomen aus Tabakendosperm. Planta **89**, 56—75.

SPITZNAGEL, J., 1972: Sorting out lysosomes and other cytoplasmic granules from polymorphs of rabbits and humans: a search for antibacterial factors. In: Phagocytic mechanisms in health and disease. (WILLIAMS, R. C., and H. H. FUDENBERG, eds.), pp. 83—106. New York: Intercont. Med. Books.

Spitznagel, J. K., F. G. Dalldorf, M. S. Leffell, J. D. Folds, I. R. M. Welsh, M. H. Connet, and B. S. Martin, 1974: Character of azurophil and specific granules purified from human polymorphonuclear leukocytes. Lab. Invest. **30**, 774—785.

Sprick, M., 1956: Phagocytosis of *M. tuberculosis* and *M. smegmatis* stained with indicator dyes. Amer. Rev. Tuberc. Pulm. Dis. **74**, 552—589.

Sridharar, S., and L. Levenbook, 1974: The contribution of the fat body to RNA and ribosome changes during development of the blow fly *Calliphora erythrocephala* (Meig). Dev. Biol. **38**, 64—72.

Srivastava, S. K., and E. Bentler, 1974: Studies on human β-D-N-Acetylhexosaminidases. III. Biochemical genetics of Tay Sachs and Sandhoff's diseases. J. biol. Chem. **249**, 2054—2057.

— A. Yoshida, T. C. Awasthi, and E. Beutler, 1974: Studies on human β-D-N-Acetyl-hexosaminidases. J. biol. Chem. **249**, 2043—2048, 2049—2053.

Stackpole, C. W., J. B. Jacobsen, and M. P. Lardis, 1974: Two distinct types of capping of surface receptors on mouse lymphoid cells. Nature **248**, 232—234.

Staehelin, L. A., 1974: Structure and function of intercellular junctions. Int. Rev. Cytol. **39**, 191—283.

Stahn, R., K. P. Maier, and K. Hannig, 1970: A new method for the preparation of rat liver lysosomes: separation of cell organelles of rat liver by carrier-free continuous electrophoresis. J. Cell Biol. **46**, 576—591.

Stambaugh, R., and J. Buckley, 1969: Identification and subcellular localizations of the enzyme effecting penetration of the zona pellucida by rabbit spermatozoa. J. Reprod. Fert. **19**, 423—432.

Stang-Voss, C., 1972: Ultrastrukturen der zellularen Autophagie. Elektronenmikroskopische Beobachtungen an Spermatiden von *Eisenia foetida* (Annelidae) während der cyto-plasmatischen Reduktionsphase. Z. Zellforsch. **127**, 580—590.

Stein, Y., C. Widnell, and O. Stein, 1968: Acylation of lysosphosphatides by plasma membrane fractions of rat liver. J. Cell Biol. **39**, 185—192.

Steinberg, R. A., B. B. Levinson, and G. M. Tompkins, 1975: Superinduction of tyrosine aminotransferase by Actinomycin D: a reevaluation. Cell **5**, 29—35.

Steiner, D. F., W. Kemmler, H. S. Tager, and J. D. Peterson, 1974: Proteolytic processing in the biosynthesis of insulin and other proteins. Fed. Proc. **33**, 2105—2115.

Steinman, R. M., and Z. A. Cohn, 1972: The interaction of soluble horse-radish peroxidase with mouse peritoneal macrophages *in vitro*. J. Cell Biol. **55**, 186—204, 616—634.

— J. M. Silver, and Z. A. Cohn, 1974: Pinocytosis in fibroblasts: quantitative studies *in vitro*. J. Cell Biol. **63**, 949—969.

Stern, J., 1972: The induction of ganglioside storage in nervous system cultures. Lab. Invest. **26**, 509—514.

— A. B. Novikoff, and R. D. Terry, 1971: The induction of sulfatide, ganglioside and cerebroside storage in organized nervous system cultures. In: Sphingolipids, sphingo-lipidoses and allied disorders (Adv. exp. Med. Biol. **19**). (Volk, B. W., and S. M. Aronson, eds.), pp. 651—660. New York: Plenum Press.

Sternberger, L. A., E. F. Osserman, and A. M. Seligman, 1970: Lysozyme and fibrinogen in normal and leukemic blood cells: a quantitative electron immunocytochemical study. Johns Hopkins Med. J. **126**, 188—209.

Stewart, J. R., and R. A. Weisman, 1972: Exocytosis of latex beads during the encystment of *Acanthamoeba*. J. Cell Biol. **52**, 117—130.

Stieglitz, H., and H. Stern, 1973: Regulation of β-1,3-glucanase activity in developing anthers of *Lilium*. Dev. Biol. **34**, 169—173.

Stiffel, C., D. Monton, and G. Biozzi, 1970: Kinetics of the phagocytic function of reticuloendothelial macrophages *in vivo*. In: Mononuclear phagocytes. (van Furth, R., ed.), pp. 335—381. Philadelphia: F. A. Davis.

Stoltze, M. J., N. S. T. Lui, O. R. Anderson, and O. A. Roels, 1969: The influence of the mode of nutrition on the digestive system of *Ochromonas malhamensis*. J. Cell Biol. **43**, 396—409.

Stossel, T. P., 1973: Quantitative studies of phagocytosis: kinetic effects of cations and heat labile opsonins. J. Cell Biol. **58**, 346—356.

STOSSEL, T. P., and T. D. POLLARD, 1973: Myosin in polymorphonuclear leukocytes. J. biol. Chem. **248**, 8288—8294.

— — R. J. MASON, and M. VAUGHAN, 1971: Isolation and properties of phagocytic vesicles from polymorphonuclear leukocytes. J. clin. Invest. **50**, 1745—1757.

STRAUS, W., 1967: Lysosomes, phagosomes, and related particles. In: Enzyme cytology. (ROODYN, D. P., ed.), pp. 239—319. New York: Academic Press.

STROBER, W., R. P. MOGIELNICKI, and T. A. WALDMAN, 1973: The role of the kidney in the metabolism of serum proteins. In: Protein turnover (CIBA Foundation Symposium 9, New Series). (WOLSTENHOLME, G. E. W., and M. O'CONNOR, eds.) Amsterdam: Elsevier.

STUART, A. E., 1970: Phylogeny of mononuclear phagocytes. In: Mononuclear phagocytes. (VAN FURTH, R., ed.), pp. 316—334. Philadelphia. F. A. Davis.

SUBBIAH, P. V., and G. A. THOMPSON, 1974: Studies of membrane formation in *Tetrahymena pyriformis*. The biosynthesis of proteins and their assembly into membrane of growing cells. J. biol. Chem. **249**, 1302—1310.

SUKENO, T., A. L. TARENTINO, T. H. PLUMMER, and F. MALEY, 1972: Purification and properties of α-D and β-D mannosidases from hen oviduct. Biochemistry **11**, 1493—1501.

SUNG, J. H., J. P. MEYERS, E. M. STADLAN, D. COWEN, and A. WOLF, 1969: Neuropathological changes in Chediak-Higashi disease. J. Neuropath. exp. Neurol. **28**, 86—118.

SUSSMAN, A. J., and C. GILVARG, 1971: Peptide transport and metabolism in bacteria. Ann. Rev. Biochem. **40**, 397—408.

SUSSMAN, M., and R. SUSSMAN, 1969: Patterns of RNA synthesis and of enzyme accumulation and disappearance during cellular slime mold differentiation. Symp. Soc. General Microbiol. **19**, 403—435.

SUZUKI, K., and K. SUZUKI, 1973: Globoid Cell Leukodystrophy (Krabbe's disease). In: Lysosomes and storage diseases. (HERS, H. G., and F. VAN HOOF, eds.), pp. 395—410. New York: Academic Press.

SVOBODA, D., and J. REDDY, 1972: Microbodies in experimentally altered cells. IX. The fate of microbodies. Amer. J. Path. **67**, 541—554.

SWANK, R. T., and K. PAIGEN, 1973: Genetic evidence for a macromolecular β-glucuronidase complex in microsomal membranes. J. molec. Biol. **77**, 371—390.

SWEELEY, C. C., C. A. MAPES, R. L. ANDERSON, R. J. DESNICK, and W. KRIVIT, 1971: Fabry's disease: Chemical and enzymatic abnormalities and a pilot study of enzyme replacement. In: Lipid storage diseases. (BERSOHN, J., and H. J. GROSSMAN, eds.), pp. 165—182. New York: Academic Press.

SWICK, R. W., 1958: Measurement of protein turnover in rat liver. J. biol. Chem. **231**, 751—764.

— A. K. REXROTH, and J. L. STANGE, 1968: The metabolism of mitochondrial proteins. III. The dynamic state of rat liver mitochondria. J. biol. Chem. **243**, 3581—3587.

SWIFT, H., and Z. HRUBAN, 1964: Focal degradation as a biological process. Fed. Proc. **23**, 1026—1037.

SYLVEN, B., C. A. TOBIAS, H. MALMGREN, R. OTTOSON, and B. THORELL, 1959: Cyclic variations in the peptidase and catheptic activity of yeast cultures synchronized with respect to cell multiplication. Exp. Cell Res. **16**, 75—87.

SZEGO, C. M., and B. J. SELLER, 1973: Hormone induced activation of target specific lysosomes: acute translocation to the nucleus after administration of gonadal hormone *in vivo*.

TABER, R., R. WERTHEIMER, and J. GOLRICH, 1973: Effect of an inhibitor of proteolysis on the size of newly synthesized protein in HeLa cells. J. molec. Biol. **80**, 367—372.

TAGER, H., and D. F. STEINER, 1974: Peptide hormones. Ann. Rev. Biochem. **43**, 509—538.

TAI, R.-C., and B. D. DAVIS, 1974: Activity of colicin E 3-treated ribosomes in initiation and in chain elongation. Proc. Nat. Acad. Sci. **71**, 1021—1025.

TALLMAN, J. F., R. O. BRADY, and K. SUZUKI, 1971: J. Neurochem. **18**, 1775—1777.

TANAKA, Y., and G. BRECHER, 1971: Micropinocytotic ferritin in erythroid cells: species dependency and relation to serum iron levels. Blood **38**, 431—444.

TAPPEL, A. L., 1968 a: Lysosomes. In: Comprehensive biochemistry. (FLORKIN, M., and E. H. STOTZ, ed.), **23**, pp. 77—98.

— 1968 b: Will antioxidant nutrients slow aging process. Geriatrics **23** (10), 97—105.

— 1969: Lysosonal enzymes and other components. In: Lysosomes in biology and pathology. (DINGLE, J. T., and H. B. FELL, eds.), Vol. **2**, pp. 207—244. Amsterdam: North-Holland Publishing.

— 1973: Lipid peroxidation damage to cell components. Fed. Proc. **32**, 1870—1874.

— P. L. SAWANT, and S. SHIBKO, 1963: Lysosomes: Distribution in animals, hydrolytic capacity and other properties. In: Ciba foundation symposium on lysosomes. (DE REUCK, A. V. S., and M. P. CAMERON, eds.), pp. 78—107. Boston: Little Brown.

TATA, J. R., M. J. HAMILTON, and R. D. COLE, 1972: Membrane phospholipids associated with nuclei and chromatin: melting profile, template activity and stability of chromatin. J. molec. Biol. **67**, 231—246.

TAVASSOLI, M., and W. M. CROSBY, 1973: Fate of the nucleus of the mature erythroblast. Science **179**, 912—913.

TAYLOR, G. B., F. BAILEY, and W. BARTLEY, 1967: Studies on the biosynthesis of protein and lipid components of rat liver mitochondria. Biochem. J. **105**, 605—609.

TAYLOR, J. M., P. J. DEHLINGER, J. F. DICE, and R. T. SCHIMKE, 1973: The synthesis and degradation of membrane proteins. Drug metabolism and disposition **1**, 84—91.

TEICHBERG, S., and E. HOLTZMAN, 1973: Axonal agranular reticulum and synaptic vesicles in cultured embryonic chick sympathetic neurons. J. Cell Biol. **57**, 88.

— — S. M. CRAIN, and E. R. PETERSON, 1974: Circulation of synaptic vesicle membrane in neurons of spinal cord explants. Abstract 4th Ann. Meet. Soc. for Neurosciences, 448.

— — — — 1975: Circulation and turnover of synaptic vesicle membrane in cultured spinal cord neurons. J. Cell Biol. **67**, 215—230.

TENG, M.-H., and A. KAPLAN, 1974: Purification and properties of rat liver lysosomal lipase. J. biol. Chem. **249**, 1064—1070.

TERJUNG, R. L., W. W. WINDER, K. M. BALDWIN, and J. O. HOLLOSZY, 1973: Effect of exercise on the turnover of cytochrome c in skeletal muscle. J. biol. Chem. **248**, 7404—7406.

TERRY, R. D., 1971: Some morphologic aspects of the lipidoses. In: Lipid storage diseases. (BERNSOHN, J., and H. J. GROSSMAN, eds.), pp. 3—25. New York: Academic Press.

TESSENOW, W., 1969: Lytic processes in development of fresh-water sponges. In: Lysosomes in biology and pathology. (DINGLE, J. T., and H. B. FELL, eds.), Vol. **1**, pp. 392—405. Amsterdam: North-Holland Publishing.

THINES-SEMPOUX, D., 1973: A comparison between the lysosomal and the plasma membrane. In: Lysosomes in biology and pathology. (DINGLE, J. T., ed.), Vol. **3**, pp. 278—299. Amsterdam: North-Holland Publishing.

THOMASSEN, W. A., and H. K. MITCHELL, 1972: Hormonal control of protein granule accumulation in fat bodies of *Drosophila melanogaster* larvae. J. Insect Physiol. **18**, 1886—1899.

THOMPSON, J., and R. VAN FURTH, 1970: The effects of glucocorticosteroids on the kinetics of monocytes and peritoneal macrophages. In: Mononuclear phagocytes. (VAN FURTH, ed.), pp. 255—264. Philadelphia: F. A. Davis.

THOMPSON, J. N., A. L. STOOLMILLER, R. MATALON, and A. DORFMAN, 1973: N-acetyl-β-hexosaminidase role in degradation of glycosaminoglycans. Science **181**, 866—867.

THORNTON, N. M., 1968: The fine structure of *Phycomyces*. I. Autophagic vacuoles. J. Ultrastruct. Res. **21**, 269—280.

THRASER, J. D., 1971: Turnover of intracellular proteins. In: Cellular and molecular renewal in the mammalian body. (CAMERON, I. C., and J. D. THRASHER, eds.), pp. 153—219. New York: Academic Press.

THYBERG, J., S. MOKALEWSKI, and U. FRIBERG, 1975: Electron microscopic studies on the uptake of colloidal thorium dioxide particles by isolated fetal guinea pig chondrocytes and the distribution of labelled lysosomes in cartilage formated by transplanted chondrocytes. Cell Tiss. Res. **156**, 317—344.

Tiffon, Y., 1971: Detection cytochemique d'hydrolases lysosomiales dans les clooisons septates steriles de *Cerianthus lloydi* (Coeletere, Anthozoaire), C. R. Acad. Sci. (Paris) D 273, 1953—1956.

— R. Rasmont, L. de Vos, and J. Bouillon, 1973: Digestion in lower metazoa. In: Lysosomes in biology and pathology. (Dingle, J. T., ed.), Vol. 3, pp. 49—68. Amsterdam: North-Holland Publishing.

Tizard, I. R., 1971: Macrophage cytophilic antibodies and the functions of macrophage bound immunoglobulins. Bact. Rev. 35, 365—378.

Tomino, S., and K. Paigen, 1975: Egasyn: a protein complexed with microsomal β-glucuronidase. J. Biol. Chem. 250, 1146—1148.

Tooze, J., and H. G. Davies, 1965: Cytolysomes in amphibian erythrocytes. J. Cell Biol. 24, 146—150.

Topping, T. M., and D. F. Travis, 1974: An electron cytochemical study of mechanisms of lysosomal activity in the rat left ventricular mural myocardium. J. Ultrastruct. Res. 46, 1—22.

Toro, I., and S. Z. Vinagh, 1966: The fine structure of the liver cell in the bat (*Myotis myotis*) during hibernation, arousal and forced feeding. Z. Zellforsch. 69, 403—417.

Toth, S. E., 1968: The origin of lipofuscin age pigment. Exp. Gerontol. 3, 19—30.

Touster, O., 1973: Some aspects of the cellular biochemistry of lysosomal and related glycosidases. Molecular and Cellular Biochemistry 2, 169—177.

Trager, W., 1974: Some aspects of intracellular parasitism. Science 183, 269—273.

Trouet, A., 1969: Caracteristiques et proprietes antigeniques des lysosomes du foie. Vander, Louvain. See also Berzins, K., F. Blomberg, and P. Perlmann, 1975: Soluble and membrane bound enzyme active antigens of rat liver lysosomes. Eur. J. Biochem. 51, 181—191.

— D. Deprez-de Campeneere, and C. de Duve, 1972: Chemotherapy through lysosomes with a DNA-daunorubicon complex. Nature New Biol. 239, 110—112.

Truchet, G., et P. Coulomb, 1971: Localisation de la phosphatase acide et de la peroxydase dans les cellules de nodules radiculaires de pois (*Pisum sativum*). Relations entre la cellule hôte et la bacterie. C. R. Acad. Sci. (Paris) D 272, 1499—1502.

— — 1973: Mise en evidence et evolution des differentes zones de nodules radiculaires de Pois (*Pisum sativum* L.); notion d'héterophagie. J. Ultrastruct. Res. 43, 36—57.

Trump, B. F., 1961: An electron microscope study of the uptake, transport, and storage of colloidal materials by cells of the vertebrate nephron. J. Ultrastruct. Res. 5, 291—310.

— and R. E. Bulger, 1965: Effect of cyanide on ultrastructure of isolated nephrons *in vitro*. Fed. Proc. 24, 616.

— J. M. Valigorsky, A. U. Arstila, W. J. Mergener, and T. D. Kinney, 1973: The relationship of intracellular pathways of iron metabolism to cellular iron overload and the iron storage diseases. Amer. J. Path. 72, 295—336.

Tsan, M., and R. D. Berlin, 1971: Effects of phagocytosis on membrane transfer of non-electrolytes. J. exp. Med. 134, 1016—1035.

Tsurugi, K., T. Morita, and K. Ogata, 1974: Mode of degradation of ribosomes in regenerating rat liver *in vivo*. Eur. J. Biochem. 45, 119—126.

Tulkens, P., A. Trouet, F. van Hoof, 1970: Immunological inhibition of lysosome function. Nature 228, 1282—1285.

Turner, W. A., J. D. Taylor, and T. T. Tchen, 1975: Melanosome formation in the goldfish: the role of multivesicular bodies. J. Ultrastruct. Res. 51, 16—31.

Tweto, J., M. Liberati, and A. R. Larrabee, 1971: Protein turnover and 4'-phosphopantetheine exchange in rat liver fatty acid synthetase. J. biol. Chem. 246, 2468—2471.

— P. Dehlinger, and A. R. Larrabee, 1972: Relative turnover rates of subunits of rat liver fatty acid synthetase. Biochem. biophys. Res. Commun. 48, 1371—1377.

Uchida, T., A. M. Pappenheimer, and A. A. Harper, 1973: Diphtheria toxin and related proteins. II. Kinetic studies on intoxication of HeLa cells by diphtheria toxin and related proteins. J. biol. Chem. 248, 3845—3850.

Uhlig, G., H. Komnick, and K.-E. Wohlfarth-Bottermann, 1965: Intrazellulare Zellzotten in Nahrungsvakuolen von Ciliaten. Helgol. wiss. Meeresunters. 12, 61—77.

Ullyot, J. L., and D. R. Bainton, 1974: Azurophil and specific granules of blood neutrophils in chronic myelogenous leukemia: an ultrastructural and cytochemical analysis. Blood **44**, 469—482.

Ulsamer, A. G., F. R. Smith, and E. D. Korn, 1969: Lipids of *Acanthameba castellanii*: composition and effects of phagocytosis on incorporation of radioactive precursors. J. Cell Biol. **43**, 105—114.

Unanue, E. R., and J. C. Cerottini, 1970: The immunogenicity of antigen bound to the plasma membrane of macrophages. J. exp. Med. **131**, 711—725.

Unkeless, J., K. Dan, G. M. Kellerman, and E. Reich, 1974: Fibrinolysis associated with oncogeneic transformation. Partial purification and characterization of the cell factor, a plasminogen activator. J. biol. Chem. **249**, 4295—4304.

Unkeless, J. C., S. Gordon, and E. Reich, 1974: Secretion of plasminogen activator by stimulated macrophages. J. exp. Med. **139**, 834—850.

Usuku, G., and P. Gross, 1965: Morphologic studies of connective tissue resorption in the tail fin of metamorphosing bull frog tadpole. Dev. Biol. **11**, 352—369.

Vacquier, V. D., M. J. Tegner, and D. Epel, 1972: Protease activity establishes the block against polyspermy in sea urchin eggs. Nature **240**, 352—353.

— — — 1973: Protease released from sea urchin eggs at fertilization alters the vitelline layer and aids in preventing polyspermy. Exp. Cell Res. **80**, 111—119.

Vaes, G., 1969: Lysosomes and the cellular physiology of bone resorption. In: Lysosomes in biology and pathology. (Dingle, J. T., and H. B. Fell, eds.), Vol. **1**, pp. 217—253. Amsterdam: North-Holland Publishing.

— 1972: The release of collagenase as an inactive proenzyme by bone explants in culture. Biochem. J. **126**, 275—289.

— 1973: Digestive capacity of lysosomes. In: Lysosomes and storage diseases. (Hers, H. G., and F. van Hoof, eds.), pp. 43—77. New York: Academic Press.

Van den Hamer, C. J. A., A. G. Morell, I. H. Scheinberg, J. Hickman, and G. Ashwell, 1970: Physical and chemical studies on ceruloplasmin. IX. The role of galactosyl residues in the clearance of ceruloplasmin from the blood. J. Biol. Chem. **245**, 4397—4402.

Van der Woude, W. J., D. J. Morré, and C. C. Bracker, 1971: Isolation and characterization of secretory vesicles in germinated pollen of *Lilium lingiflorum*. J. Cell Sci. **8**, 331—351.

Van Furth, R. (ed.), 1970: Mononuclear phagocytes. Philadelphia: F. A. Davis.

— J. G. Hirsch, and M. E. Fedorko, 1970: Morphology and peroxidase cytochemistry of mouse promonocytes, monocytes, and macrophages. J. exp. Med. **142**, 794—805.

— M. M. C. Diesselhoff-den Dulk, and H. Mattie, 1973: Quantitative study on the production and kinetics of mononuclear phagocytes during an acute inflammatory reaction. J. exp. Med. **138**, 1314—1330.

Vanha-Perttula, T., V. K. Hopsu, V. Nonninen, and G. G. Glenner, 1965: Cathepsin C activity as related to some histochemical substrates. Histochemie **5**, 107—181.

Van Hoof, F., 1973: Mucopolysaccharidoses. In: Lysosomes and storage diseases (Hers, H. G., and F. van Hoof, eds.), pp. 217—259. New York: Academic Press.

— P. Evard, and H. G. Hers, 1971: An unusual case of G_{M_2}-gangliosidosis with deficiency of hexosaminidases A and B. In: Sphingolipids, sphingolipidoses and allied disorders (Adv. Exp. med. Biol. **19**). (Volk, B. W., and S. M. Aronson, eds.) New York: Plenum Press.

Van Kleef, F. S. M., M. J. C. M. Nijzink-Maas, and H. J. Hoenders, 1974: Intracellular degradation of α-crystallin. Fractionation and characterization of degraded α-A chains. Eur. J. Biochem. **48**, 563—570.

Van Lancker, J. L., and P. L. Lentz, 1970: Study on the rate of biosynthesis of β-glucuronidase and its appearance in lysosomes in normal and hypoxic rats. J. Histochem. Cytochem. **18**, 529—541.

Van Snick, J. L., P. L. Masson, and J. F. Heremans, 1974: The involvement of lactoferrin in the hyposideremia of acute inflammation. J. exp. Med. **140**, 1068—1084.

VARANDANI, P. T., 1973: Insulin degradation. X. Identification of insulin degrading activity of rat liver plasma membrane as glutathione-insulin transhydrogenase. Biochem. biophys. Res. Commun. **55**, 689—696.

— 1974: Insulin degradation in insulinoma: evidence for the occurrence of an inactive form of glutathione-insulin transhydrogenase and for the absence of insulin A and B chain degrading proteases. Biochem. biophys. Res. Commun. **60**, 1119—1126.

VASSAF, A. A., J. B. FLORA, J. G. WEEKS, B. S. BIBBS, and R. R. SCHMIDT, 1973: The effects of enzyme synthesis and stability and of DNA replication on the cellular levels of aspartate transcarbamylase during the life cycle of the eucaryote, *Chlorella*. J. biol. Chem. **248**, 1976—1985.

VATTER, A. E., O. K. REISS, J. K. NEWMAN, K. LINDQUIST, and E. GROENEBOER, 1968: Enzymes of the lung. I. Detection of esterase with a new cytochemical method. J. Cell Biol. **38**, 80—98.

VAUGHAN, R. B., and S. V. BOYDEN, 1964: Interactions of macrophages and erythrocytes. Immunology **7**, 118—126.

VENGER, R., M. C. E. NJERAS, G. M. DE HAAS, 1973: Action of phospholipase A at interfaces. J. biol. Chem. **248**, 4023—4034.

VENSEL, W. H., J. KOMENDER, and E. A. BARNARD, 1971: Non-pancreatic proteases of the chymotrypsin family. II. Two proteases from a mouse mast cell tumor. Biochim. biophys. Acta **250**, 395—407.

VERITY, M. A., and W. J. BROWN, 1968: Structure linked activity of lysosomal enzymes in the developing mouse brain. J. Neurochem. **15**, 69—80.

VERNON-ROBERTS, B., 1972: The macrophage. Cambridge: Cambridge University Press.

VESELL, S., and P. J. FRITZ, 1971: Factors affecting the activity, tissue distribution and degradation of isozymes. In: Enzyme synthesis and degradation in mammalian systems. (RECHCIGL, M., ed.), pp. 339—374. Baltimore: University Park Press.

VIAN, B., et J. C. ROLAND, 1972: Différenciation des cytomembranes et renouvellment du plasmalemme dans les phénomènes de secretions végétales. J. Microscopie **13**, 119—136.

VIGIL, E. L., 1970: Cytochemical and developmental changes in microbodies (glyoxysomes) and related organelles of castor bean endosperm. J. Cell Biol. **46**, 435—454.

— 1973: Structure and function of plant microbodies. Subcellular Biochem. **2**, 237—285.

VILLIERS, T. A., 1967: Cytolysosomes in long dormant plant embryo cells. Nature **214**, 56—57.

— 1972: Cytological studies in dormancy. II. Pathological ageing changes during prolonged dormancy and recovery upon dormancy release. New Phytol. **71**, 145—152.

VITELTA, E. S., and J. W. UHR, 1974: Cell surface immunoglobulin. IX. A new method for the study of synthesis, intracellular transport and exteriorization in murine splenocytes. J. exp. Med. **139**, 1599—1620.

VLADUTICU, G. D., and N. R. ROSE, 1974: Intracellular distribution of a primate specific esterase in cultured cells and tissues. J. Cell Biol. **62**, 560—566.

VOLK, B. W., and S. M. ARONSON (eds.), 1971: Sphingolipids, sphingolipidoses and allied disorders (Adv. Exp. med. Biol. **19**). New York: Plenum Press.

VON FIGURA, K., and M. KRESSE, 1974 a: Quantitative aspects of pinocytosis and the intracellular fate of N-acetyl-D-glucosaminidase in Sanfilippo B fibroblasts. J. clin. Invest. **53**, 85—90.

— — 1974 b: Inhibition of pinocytosis by cytochalasin B. Decrease in intracellular lysosomal enzyme activities and increased storage of glycosaminoglycans. Eur. J. Biochem. **48**, 357—363.

VORBRODT, A., 1961: Histochemical studies on the intracellular localization of acid deoxyribonuclease. J. Histochem. Cytochem. **9**, 647—655.

— S. GRUCA, and S. GRUCA-KRZYZOWSKA, 1971: Cytochemical studies on the participation of endoplasmic reticulum in the formation of lysosomes (dense bodies) in rat hepatocytes. J. Microscopie **12**, 73—82.

WADDELL, W. J., and R. G. BATES, 1969: Intracellular pH. Physiol. Rev. **49**, 285—329.

WAHL, L. M., S. M. WAHL, S. E. MERGENHAGEN, and G. R. MARTIN, 1974: Collagenase production by endotoxin activated macrophages. Proc. nat. Acad. Sci. (U.S.) **71**, 3598—3601.

WAITE, M., and P. SISSON, 1973: Solubilization by heparin of the phospholipase A from the plasma membrane of rat liver. J. biol. Chem. **248**, 7201—7206.

WAKE, K., 1974: Development of Vitamin A rich lipid droplets in multivesicular bodies of rat liver stellate cells. J. Cell Biol. **63**, 683—691.

WALLINGFORD, W. R., and D. J. McCARTY, 1971: Differential membranolytic effects of microcrystalline sodium urate and calcium pyrophosphate dehydrate. J. exp. Med. **133**, 100—112.

WARD, P., 1968: Chemotoxis of mononuclear cells. J. exp. Med. **128**, 1201—1221.

WARDROP, A. B., 1968: Occurrence of structures with lysosomes-like function in plant cells. Nature **218**, 978—980.

WARHURST, D. C., 1973: Chemotherapeutic agent and malaria research. In: Chemotherapeutic agents in the study of parasites: Symposia of the British Society for Parasitology 11. (TAYLOR, A. E. R., and R. MULLER, eds.), pp. 1—28. Oxford: Blackwell Scientific Publications.

— and D. J. HOCKLEY, 1967: Mode of action of chloroquine on *Plasmodium berghei* and *P. Cynomolgi*. Nature **214**, 935—936.

WARREN, L., and M. C. GLICK, 1968: Membranes of animal cells. II. The metabolism and turnover of the surface membranes. J. Cell Biol. **37**, 729—746.

WASSERMANN, F., and T. F. McDONALD, 1960: Electron microscopic investigation of the surface membrane structures of the fat cell and of their changes during depletion of the cell. Z. Zellforsch. **52**, 778—800.

WASTESON, A., U. LINDAHL, and A. MALTEN, 1972: Mode of degradation of the chrondroitin sulfate proteoglycan in rat costal cartilage. Biochem. J. **130**, 729—738.

WATTIAUX, R., M. WIBO, and P. BAUDHUIN, 1963: Influence of the injection of Triton WR-1339 on the properties of rat-liver lysosomes. In: Ciba foundation symposium on lysosomes. (DE REUCK, A. V. S., and M. P. CAMERON, eds.), pp. 176—200. Boston: Little Brown.

WEBER, R., 1969: Tissue involution and lysosomal enzymes during anuran metamorphosis. In: Lysosomes in biology and pathology. (DINGLE, J. T., and H. B. FELL, eds.), Vol. **2**, pp. 437—461. Amsterdam: North-Holland Publishing.

WEISS, R. L., 1973: Intracellular localization of ornithine and arginine pools in *Neurospora*. J. biol. Chem. **248**, 5409—5413.

WEINSTOCK, I. M., and A. A. IODICE, 1969: Acid hydrolase activity in muscular dystrophy and denervation atrophy. In: Lysosomes in biology and pathology. (DINGLE, J. T., and H. B. FELL, eds.), Vol. **1**, pp. 450—468. Amsterdam: North-Holland Publishing.

WEISSMANN, G., 1965: Lysosomes. New Eng. J. Med. **273**, 1084—1090.

— 1967: The role of lysosomes in inflammation and disease. Ann. Rev. Med. **18**, 97—112.

— 1969: The effects of steroids and drugs on lysosomes. In: Lysosomes in biology and pathology. (DINGLE, J. T., and H. B. FELL, eds.), Vol. **1**, 276—295.

— 1972: Lysosomal mechanisms of tissue injury in arthritis. New Eng. J. Med. **286**, 141—147.

— and G. A. RITA, 1972: Molccular basis of gouty inflammation: the interaction of monosodium urate crystals with lysosomes and liposomes. Nature New Biol. **240**, 167—172.

— I. SPILBERG, and K. KRAKAUER, 1969: Arthritis induced in rabbits by lysates of granulocyte lysosomes. Arthr. Rheum. **12**, 103—116.

— R. B. ZURIER, and S. HOFFSTEIN, 1972: Leukocytic proteases and the immunological release of lysosomal enzymes. Amer. J. Path. **68**, 539—559.

WENTHOLD, R. O., H. R. MAHLER, and W. J. MOORE, 1974: The half-life of acetyl-cholinesterase in mature rat brain. J. Neurochem. **22**, 941—944.

WERB, Z., and Z. A. COHN, 1971: Cholesterol metabolism in the macrophage. I. The regulation of cholesterol exchange. II. Alteration of subcellular exchangeable cholesterol compartments and exchange in other cell types. J. exp. Med. **134**, 1545—1569, 1570—1590.

— — 1972 a: Cholesterol metabolism in the macrophage. III. Ingestion and intracellular fate of cholesterol and cholesterol esters. J. exp. Med. **135**, 21—49.

— 1972 b: Plasma membrane synthesis in the macrophage following phagocytosis of polystyrene latex particles. J. biol. Chem. **247**, 2439—2446.

— and M. C. Burleigh, 1974: A specific collagenase from rabbit fibroblasts in monolayer culture. Biochem. J. **137**, 373—385.

— and J. J. Reynolds, 1974: Stimulation by endocytosis of the secretion of collagenase and neutral proteinase from rat synovial fibroblasts. J. exp. Med. **140**, 1482—1497.

— M. C. Burleigh, Barrett, and P. M. Starkbey, 1974: The interaction of mammalian collagenase and other metal proteinases. Biochem. J. **139**, 359—368.

Weston, P. D., and A. R. Poole, 1973: Antibodies to enzymes and their uses, with specific reference to cathepsin D and other lysosomal enzymes. In: Lysosomes in biology and pathology. (Dingle, J. T., ed.), Vol. 3, pp. 425—464. Amsterdam: North-Holland Publishing.

Wettstein, R., and J. R. Sotelo, 1963: Electron microscopic study of the regenerative process of peripheral nerves of mice. Z. Zellforsch. **59**, 708—730.

Wetzel, B. K., S. S. Spicer, and S. H. Wollman, 1965: Change in fine structure and acid phosphatase localization in rat thyroid cells following thyrotropin administration. J. Cell Biol. **25**, 593—618.

— R. G. Horn, and S. S. Spicer, 1967: Fine structural studies on the development of heterophil, eosinophil and basophil granulocytes in rabbit. Lab. Invest. **16**, 349—382.

Whitaker, S., and F. S. La Bella, 1972: Ultrastructural localization of acid phosphatase in the posterior pituitary of the dehydrated rat. Z. Zellforsch. **125**, 1—15.

White, J. G., 1966: The Chediak-Higashi syndrome: a possible lysosomal disease. Blood **28**, 143—156.

— 1972 a: Uptake of latex particles by blood platelets: phagocytosis or sequestration? Amer. J. Path. **69**, 439—458.

— 1972 b: Ultrastructural defects in congenital disorders of platelet function. Ann. N.Y. Acad. Sci. **201**, 205—233.

— and R. D. Estensen, 1974 a: Selective labilization of specific granules in polymorphonuclear leukocytes by phorbol myristate acetate. Amer. J. Path. **75**, 45—60.

— — 1974 b: Cytochemical electron microscope studies of the action of phorbol myristate acetate on platelets. Amer. J. Path. **75**, 301—314.

— V. I. French, and J. M. Stark, 1970: A study of the localization of a protein antigen in the chicken spleen and its relation to the formation of germinal centers. J. med. Microbiol. **3**, 65—83.

Whur, P., N. E. Payne, and H. Koppel, 1974: Effect of protease and protease inhibitors on the adhesion of Ehrlich ascites tumour cells to plastic. Exp. Cell Res. **86**, 422—427.

Wibo, M., and B. Poole, 1974: Protein degradation in cultured cells. II. The uptake of chloroquine by rat fibroblasts and the inhibition of cellular protein degradation and cathepsin B_1. J. Cell Biol. **63**, 430—440.

Widmann, J.-J., R. S. Cotran, and H. D. Fahimi, 1972: Mononuclear phagocytes (Kupffer cells) and endothelial cells; Identification of two functional cell types in rat liver sinusoids by endogenous peroxidase activity. J. Cell Biol. **52**, 159—170.

Widnell, C. C., 1972: Localization of nucleotidease activities in the lysosome membrane. J. Cell Biol. **55**, 280 A.

Wiegers, K. J., and G. Koch, 1972: Interaction of polio virus-induced double-stranded RNA with HeLa cells. Arch. Biochem. Biophys. **148**, 89—96.

Wiesmann, U. N., J. Lightbody, F. Vassella, and N. N. Herschkowitz, 1972: Multiple lysosomal enzyme deficiency due to enzyme leakage. New Eng. J. Med. **284**, 109—110.

Wiggins, P. M., 1972: Intracellular pH and the structure of cell water. J. theor. Biol. **37**, 363—371.

Wild, A. E., 1973: Transport of immunoglobulins and other proteins from mother to young. In: Lysosomes in biology and pathology. (Dingle, J. T., ed.), Vol. 3, pp. 169—215. Amsterdam: North-Holland Publishing.

Willetts, N. S., 1967: Intracellular protein breakdown in non-growing cells of *Escherichia coli*. Biochem. J. **103**, 453—461, 462—466.

Williams, D. S., and T. F. Slater, 1973: Photosensitization of isolated lysosomes. Biochem. Soc. Transactions 1, 200—202.

Williams, K. E., J. B. Lloyd, M. E. Kidston, and F. Beek, 1973: Biphasic effects of trypan blue on pinocytosis. Biochem. Soc. Trans. 1, 203—206.

Williams, N. E., and E. M. Nelson, 1973: Regulation of microtubules in *Tetrahymena*. II. Relations between turnover of microtubule protein and microtubule dissociation and assembly during oral replacement. J. Cell Biol. 56, 458—465.

Williams, R. C., and H. H. Fudenberg (eds.), 1972: Phagocytic mechanism in health and disease. New York: Intercont., Med. Books.

Williams, R. M., and F. Beck, 1969: Demonstration of alkaline phosphatase in the lysosomes of neonate rat ileum. Histochem. J. 1, 531—538.

Wills, E. J., P. Davies, A. C. Allison, and A. D. Haswell, 1972: Cytochalasin B fails to inhibit pinocytosis by macrophages. Nature New Biol. 240, 59—60.

Wilson, C. L., 1973: A lysosomal concept for plant pathology. Ann. Rev. Phytopath. 11, 247—267.

— D. L. Stiers, and G. G. Smith, 1970: Fungal lysosomes, spherosomes. Phytopath. 60, 216—227.

Wilson, G., and C. F. Fox, 1971: Biogenesis of microbial transport systems: evidence for coupled incorporation of newly synthesized lipids and proteins into membranes. J. molec. Biol. 55, 49—60.

Winkler, H., J. A. L. Schopf, H. Hortnagl, and H. Hortnagl, 1972: Bovine adrenal medulla: subcellular distribution of newly synthesized catecholamines, nucleotides, and chromogranins. Nauyn-Schmiedeberg's Arch. Pharmacol. 273, 42—61.

Wise, G. E., and C. J. Flickinger, 1970: Relation of the Golgi apparatus to the cell coat in amoebae. Exp. Cell Res. 61, 13—23.

Wisse, E., 1974: Kupffer cell reactions in rat liver under various conditions as observed in the electron microscope. J. Ultrastruct. Res. 46, 499—520.

Wissig, S. L., 1960: Anatomy of secretion in the follicular cells of the thyroid glands. I. The fine structure of the gland in the normal rat. J. biophys. biochem. Cytol. 7, 419—432.

Wittekind, D., R. L. Snipes, and V. Kretschmer, 1974: Lysosomal formation of crystalloid structures in living cells induced by the dye azur A. Exp. Cell Res. 84, 143—152.

Woebar, K. A., G. F. Doherty, and S. M. Ingbar, 1972: Stimulation by phagocytosis of the deiodination of L-thyroxine in human leukocytes. Science 76, 1039—1041.

Woessner, J. F., 1969: The physiology of the uterus and mammary gland. In: Lysosomes in biology and pathology. (Dingle, J. T., and H. B. Fell, eds.), Vol. 1, pp. 299—329. Amsterdam: North-Holland Publishing.

— 1971: Cathepsin D: Enzymic properties and role in connective tissue breakdown. In: Tissue proteinases. (Barrett, A. J., and J. T. Dingle, eds.), pp. 291—311. Amsterdam: North-Holland Publishing.

— 1973: Cartilage cathepsin D and its action on matrix components. Fed. Proc. 32, 1485—1488.

— and J. N. Ryan, 1973: Collagenase activity in homogenates of the involuting rat uterus. Biochim. biophys. Acta 309, 397—405.

Wolfe, L. S., R. G. Senior, and N. M. K. Ng-Ying-Kin, 1974: The structure of oligosaccharides accumulating in the liver of Gm_1 gangliosidoses Type I. J. biol. Chem. 249, 1828—1838.

Wolff, K., and H. Honigsmann, 1971: Permeability of the epidermis and the phagocytic activity of keratinocytes. J. Ultrastruct. Res. 36, 176—190.

— and S. Shreiner, 1971: Melanosomal acid phosphatase. Arch. Dermatol. Forsch. 241, 255—271.

— K. Jimbow, and T. B. Fitzpatrick, 1974: Experimental pigment donation *in vivo*. J. Ultrastruct. Res. 47, 400—419.

Wolin, E. M., H. Laufer, and D. F. Albertini, 1973: Uptake of the yolk protein, lipovitellin by developing Crustacean oocytes. Develop. Biol. 35, 160—170.

Wolinsky, H., S. Goldfischer, M. M. Daly, L. E. Kasak, and B. Coltoff-Schiller, 1975: Arterial lysosomes and connective tissue in primate atherosclerosis and hypertension. Circulation Res. **36**, 553—560.

Wollman, S. H., 1969: Secretion of thyroid hormones. In: Lysosomes in biology and pathology. (Dingle, J. T., and H. B. Fell, eds.), Vol. **2**, pp. 483—512. Amsterdam: North-Holland Publishing.

Wolman, M., 1965: Study of the nature of lysosomes and of their acid phosphatase. Z. Zellforsch. **65**, 1—9.

Wolstenholme, G. E. W., and M. O'Connor (eds.), 1973: Protein turnover (Ciba Foundation Symposium 9, new series). Amsterdam: Elsevier.

Wood, K. M., F. S. Wusteman, and C. G. Curtis, 1973: The degradation of intravenously injected chondroitin 4-sulphate in the rat. Biochem. J. **134**, 1009—1013.

Wooden, A. M., and A. A. Wienke, 1964: The participation of calcium, adenosine triphosphate and adenosine tyrophosphate and adenosine tryphosphatase in the extrusion of the granule proteins from the polymorphonuclear leukocytes. Biochem. J. **90**, 498—509.

Woodin, A. M., 1973: The leucocidin-treated leucocyte. In: Lysosomes in biology and pathology. (Dingle, J. T., ed.), Vol. **3**, pp. 395—422. Amsterdam: North-Holland Publishing.

Woodside, K. H., W. F. Ward, and G. E. Mortimore, 1974: Effects of glucagon on general protein degradation and synthesis in perfused rat liver. J. biol. Chem. **249**, 5458—5463.

Wooley, D. M., and D. W. Fawcett, 1973: The degeneration and disappearance of the centrioles during the development of the rat spermatozoan. Anat. Rec. **177**, 289—302.

Workman, E. F., and G. W. Bates, 1974: Mobilization of iron from reticulocyte ghosts by cytoplasmic agents. Biochem. biophys. Res. Commun. **58**, 787—791.

Wyllie, J. C., 1973: A study of ferritin iron formation in the rabbit alveolar macrophage. Exp. Cell Res. **78**, 98—102.

Yaari, A., 1969: Mobility of human red blood cells of different age groups in an electric field. Blood **33**, 159—163.

Yang, R. S. H., J. G. Pelliccia, and C. F. Wilkinson, 1973: Age dependent aryl sulfatase activities in the southern army worm: a possible insect endocrine regulatory mechanism. Biochem. J. **136**, 817—820.

Yannarell, A., and N. N. Aronson, 1973: Localization of an RNA degrading system of enzymes on the rat liver plasma membrane. Biochim. biophys. Acta **311**, 191—204.

Yates, E. M., H. W. Reading, L. Bitensky, and J. Chayen, 1970: The state of pigment epithelium lysosomes in the retina of normal rats and those with retinitis pigmentosa. Biochem. J. **119**, 57 P—58 P.

Yatsu, L. Y., and T. J. Jacks, 1968: Association of lysosomal activity with aleurone grains in plant seeds. Arch. Biochem. Biophys. **124**, 466—471.

Yoshinaga, M., B. Waksman, S. E. Malawista, 1972: Cytochalasin B inhibits lymphotoxin production by antigen stimulated lymphocytes. Science **176**, 1147—1149.

Young, R. W., 1967: The renewal of photoreceptor cell outer segments. J. Cell Biol. **33**, 61—72.

— 1971: The renewal of rod and cone outer segments in the rhesus monkey. J. Cell Biol. **49**, 303—318.

— 1973: Biogenesis and renewal of visual cell outer segment membranes. Exp. Eye Res. (in press).

— and D. Droz, 1968: The renewal of protein in retinal rods and cones. J. Cell Biol. **39**, 169—184.

— and D. Bok, 1969: Participation of the retinal pigment epithelium in the rod outer segment renewal process. J. Cell Biol. **42**, 392—403.

Zabucchi, G., M. R. Soranzo, F. Rossi, and D. Romeo, 1975: Exocytosis in human polymorphonuclear leucocytes induced by A 23187 and calcium. FEBS letters **54**, 44—48.

Zahlten, R. N., A. A. Hochberg, F. W. Stratman, and H. A. Lardy, 1972: Glucagon-stimulated phosphorylation of mitochondrial and lysosomal membranes of rat liver *in vivo*. Proc. nat. Acad. Sci. (U.S.) **69**, 800—804.

ZANEVELD, L. J. D., B. M. DRAGOJE, and G. F. B. SCHUMACHER, 1972: Acrosome proteinase and proteinase inhibitor of human spermatozoa. Science **177**, 702—703.

ZELDIN, M. H., and W. SKEA, 1973: Proteases in *Euglena* chloroplast development. J. Cell Biol. **59**, 376 A.

— — and D. MATTESON, 1973: Organelle formation in the presence of a protease inhibitor. Biochem. biophys. Res. Commun. **52**, 544—549.

ZELICKSON, A. S., D. B. WINDHORST, J. G. WHITE, and R. A. GOOD, 1967: The Chedick-Higashi syndrome: formation of giant melanosomes and the bases of hypopigmentation. J. Invert. Dermatol. **49**, 575—581.

ZEMAN, W., and A. N. SIAKOTOS, 1973: The neuronal ceroid-lipofuscinoses. In: Lysosomes and storage diseases. (HERS, H. G., and F. VAN HOOF, eds.), pp. 519—551. New York: Academic Press.

ZEYA, H. I., and J. K. SPITZNAGEL, 1966: Cationic proteins of polymorphonuclear leukocyte lysosomes. II. Composition, properties and mechanism of anti-bacterial action. J. Bact. **91**, 755—762.

— and LAZLO, 1973: Granule assembly in precursors of human leukemia granulocytes. Amer. J. Path. **71**, 467—476.

ZIFF, M., 1973: Pathophysiology of rheumatoid arthritis. Fed. Proc. **32**, 131—133.

ZIGMOND, S. M., and J. G. HIRSCH, 1973: Leukocyte locomotion and chemotoxis: new methods for evaluation and demonstration of a cell-derived chemotactic factor. J. exp. Med. **137**, 387—410.

ZURIER, R. B., S. HOFFSTEIN, and G. WEISSMANN, 1973 a: Cytochalasin B: effect on lysosomal enzyme release from human leukoccytes. Proc. nat. Acad. Sci. (U.S.) **70**, 844—848.

— — — 1973 b: Mechanisms of lysosomal enzyme release from human leukocytes. I. Effect of cyclic nucleotides and colchicine. J. Cell Biol. **58**, 27—41.

ZWEIFACH, B. W., L. GRANT, and R. T. McCLUSKY (ed.), 1974: The inflammatory process, 2nd ed. New York: Academic Press.

Subject Index